AF352874

Entropy in Image Analysis III

Entropy in Image Analysis III

Editor

Amelia Carolina Sparavigna

MDPI • Basel • Beijing • Wuhan • Barcelona • Belgrade • Manchester • Tokyo • Cluj • Tianjin

Editor
Amelia Carolina Sparavigna
Department of Applied Science
and Technology,
Polytechnic University of Turin
Italy

Editorial Office
MDPI
St. Alban-Anlage 66
4052 Basel, Switzerland

This is a reprint of articles from the Special Issue published online in the open access journal *Entropy* (ISSN 1099-4300) (available at: https://www.mdpi.com/journal/entropy/special_issues/entropy_image_III).

For citation purposes, cite each article independently as indicated on the article page online and as indicated below:

LastName, A.A.; LastName, B.B.; LastName, C.C. Article Title. *Journal Name* **Year**, *Volume Number*, Page Range.

ISBN 978-3-0365-3515-9 (Hbk)
ISBN 978-3-0365-3516-6 (PDF)

Contents

About the Editor

Amelia Carolina Sparavigna is a physics researcher, working mainly in the field of condensed matter physics and image processing. She graduated from the University of Torino in 1982 and obtained a Ph. D. in Physics at Politecnico di Torino in 1990. Since 1993, she has carried out teaching and research activities at the Politecnico. Her scientific researches cover the fields of thermal transport and Boltzmann equation, liquid crystals, and the related image processing of polarized light microscopy. She has proposed new methods of image processing inspired by physical quantities, such as coherence lengths. Her recent works mainly concern the problem of image segmentation. She is also interested in the history of physics and science. The papers that she has published in international journals are mainly on the topics of phonon thermal transport, the elastic theory of nematic liquid crystals, and the texture transitions of liquid crystals, investigated by means of image processing.

Preface to "Entropy in Image Analysis III"

Image analysis basically refers to any extraction of information from images, including those contained in large and complex datasets, such as the collections used for biometric identification or the sets of satellite surveys employed in the monitoring of Earth's climate changes. Image analysis is also necessary when providing methods for hiding information. When images play a major role in data transmission, it is imperative to protect them. Therefore, increasingly sophisticated algorithms, supported by artificial intelligence methods, are indispensable in the encryption and decryption of images, as they are up-to-date with their secure transmission. In addition to the extraction and encryption of information, another task for image analysis is related to the computer vision, where there are many applications that can only be properly managed using computers. These assorted scenarios tell us that "image analysis" is not just what we can imagine by taking our human vision system as a model; it is all the bulk of methods that computers use at present and the body of knowledge that they will be able to manage in a totally unsupervised manner in future, thanks to their artificial intelligence. In fact, the articles published in this book evidence that encryption and decryption, neural networks and machine learning are the leitmotifs of advanced image analysis. The guest editor hopes that the readers can receive, from the published articles, fruitful hints and inspiration for future research and publications, for which the Topical Collection "Entropy in Image Analysis" could provide a proper publication place.

Amelia Carolina Sparavigna
Editor

Editorial

Entropy in Image Analysis III

Amelia Carolina Sparavigna

Department of Applied Science and Technology, Polytechnic University of Turin, 10129 Turin, Italy;
amelia.sparavigna@polito.it

Citation: Sparavigna, A.C. Entropy in Image Analysis III. *Entropy* **2021**, *23*, 1648. https://doi.org/10.3390/e23121648

Received: 2 December 2021
Accepted: 6 December 2021
Published: 8 December 2021

Image analysis basically refers to any extraction of information from images, which can be as simple as QR codes required in logistics and digital certifications or related to large and complex datasets, such as the collections of images used for biometric identification or the sets of satellite surveys employed in the monitoring of Earth's climate changes. At the same time, image analysis is necessary for providing methods for hiding information too. When images are playing a major role in data transmission, it is imperative to protect them. Therefore, increasingly sophisticated algorithms, supported by specific artificial intelligence methods as well, are indispensable in encryption and decryption of images for being up to date with secure performances in their transmission.

Regarding the image analysis involved in computer vision, we are generally used to comparing it with the human visual system and its ability of extracting high-level and relevant information. However, we currently have many applications which can be properly managed only by means of computers. Let us consider, for instance, the face recognition applied to find a profile in a huge database. This is an impossible task for humans alone, but computers are turning it into a possible one. Consequently, "image analysis" is not just what we can imagine by taking as a model our human vision system; it is all the bulk of methods that computers are using today in several multivariate applications and the body of knowledge that they will be able to manage, in future, in a totally unsupervised manner thanks to their artificial intelligence.

Together with encryption and decryption, the use of neural networks and machine learning is one of the leitmotifs contained in the articles proposed in this Special Issue, such as in the previously published Issues regarding Entropy in Image Analysis.

Let us describe the articles of the Issue shortly.

In [1], the problem of tomographic image reconstruction is addressed. In the proposed method, an extended class of power-divergence measures, which are including a large set of distances and relative entropy measures, are involved in an iterative reconstruction algorithm. The authors introduced in the method a system of nonlinear differential equations based on Lyapunov functions. Actually, the resulting iterative algorithm proposed in [1] represents a natural extension of the maximum-likelihood expectation-maximization method.

As told before, the secure transmission of digital images in the current network and big data environment is one of the main tasks of image analysis. This is the problem considered in Ref. [2]. In their article, the authors propose a security-enhanced image communication scheme, requiring cellular neural network (CNN) and cryptanalysis. The features of CNN are applied to create pseudorandom sequences which are used in the image encryption. By means of a plain image, the cipher image is obtained in the CNN-based sequence. In [2], cryptanalysis demonstrates the safety of the performance.

In [3], modified Hilbert space-filling curves, related to rectangles and cuboids, are applied in an entropy coding of images and for video compression. By means of these Hilbert curves, an efficient run-length-based entropy coding has been developed, which is suitable for a series of high-efficiency image compression algorithms. As observed by authors, the 2-D Hilbert curves are defined on squares while subband image compression requires rectangles of arbitrary sizes. In [3], the authors provide details about the construction of the required modified 2-D Hilbert curves and 3-D cuboids.

The word "retinex" comes from "retina" and "cortex", since both the eyes and brain are involved in human vision. Then, in image analysis, retinex methods are those which mimic how human beings perceive their surrounding environment. In [4], we find proposed a retinex fast algorithm to enhance low-light images in order to restore the information which is hidden by low illuminance. The experimental results proposed in [4] demonstrate that the images, enhanced by the proposed retinex method, have better performance with respect to those obtained by means of other state-of-the-art methods.

As asserted in [5], automated video surveillance systems are offering today some solutions to avoid any human intervention, which could result in inefficient tasks. In this framework, properly devised methods and models are strictly necessary. If we need a crowd surveillance, for instance, it is fundamental to analyze the human crowd behavior (HCB) by means of systems possessing robust feature extraction methods and reliable decision-making classifiers. In [5], the authors describe an approach based on a particles force model for multi-person tracking, the performance of which has been tested on publicly available benchmark datasets.

In [6], we find the methodology referred to as full waveform inversion (FWI) applied to subsurface investigations. This methodology is commonly used in the petroleum industry, mainly to characterize oil reservoirs. The authors of [6] propose the addition of a relative entropy in the formalism of FWI. In the article, the authors show some features of this entropy and propose three different ways to add information through it in the inverse problem. This prior information, conveyed by the addition of entropy relative to FWI, can provide a result with better resolution.

Diagnostic radiography designates a technical mode of acquiring medical images by means of X-rays. In the used devices, an electron beam is converted into X-rays by means of a target material. As a result of the mechanisms of emission, the field intensity towards the cathode is larger than the intensity towards the anode. This is the so-called anode heel effect, addressed in [7], intended to cause a non-uniform image quality. The purpose of the study proposed in [7] is that of evaluating the non-uniformity in digital radiographs. The author is also giving a novel method, based on circular step-wedge phantom and normalized mutual information, which outperforms the conventional visible ratio metrics.

In Ref. [8], we can find again the problem of image security. According to the proposed discussion, hyperchaotic image encryption is the method which is today generally used to secure images. In this framework, article [8] is proposing a novel encryption scheme based on multiple bit permutation and diffusion (MBPD). The method is described in detail, starting from a four-dimensional hyperchaotic system with Lyapunov exponents and ending with permutation and diffusion. After experiments, it is concluded that MBPD can effectively resist different types of attacks with better performance than popular encryption methods.

A computerized tomography (CT) scan is a medical imaging technique which combines a series of X-ray images taken from different angles around the body, processing them to create cross-sectional images (slices) of it. In [9] we can find CT scan used to evidence the spleen injuries, and an automated method based on machine learning for processing it. In fact, computer-assisted diagnosis systems exist for other conditions, but for spleen injuries the current methodology is based on detecting them by manual inspection. The results proposed by [9] suggest that a quantitative computerized analysis of spleen injuries can help in providing a faster triage (with a consequent improvement of patient outcomes).

In [10], we can find another method addressing the protection of digital information, in particular the digital visual information. The method is based on a six-dimensional hyper-chaotic encryption scheme and a three-dimensional transformed Zigzag diffusion with RNA operation. With respect to the previous literature, the specificity of the method proposed in [10] is in its focusing on the encryption of color images. The encryption starts with three pseudo-random matrices generated from a 6D hyper-chaotic system. It continues with a permutation and a transformation by means of Zigzag diffusion. The final step is RNA conversion. Experiments show that the proposed encryption has high resistance against generally used attacks.

The fractional calculus, made by means of operators of non-integer order, is mainly used in the area of nonlinear and chaotic systems. In [11], we can find it in a fractional-order hyperchaotic system applied to secure communication, in particular for the encryption of color images. Experiments reported in [11] are supporting the method as cryptographically secure in general. However, the method can be broken in some cases. Therefore, the final suggestion given by authors is that algorithm designers have to pay some specific attention in securing this kind of cryptosystems.

Article [12] explains that, to encrypt/decrypt images, most researchers are using chaotic systems, whereas others prefer non-chaotic methods. In [12], a new encryption algorithm is proposed, which combines a non-chaotic Newton-Raphson's method with a hyperchaotic two-dimensional map of a general Bischi-Naimzadah duopoly system. The multiple security experiments made for measuring the efficiency of the method (among which we find entropy analysis) show that the proposed algorithm possesses a good security efficiency.

In concluding the review of the Issue, a great interest in image encryption and decryption has been demonstrated. However, as shown by the different topics and problems addressed in the other published articles, the research field of image analysis is quite larger and variegated and not solely limited to problems of cryptanalysis. Therefore, the guest editor hopes that the readers can receive, from these published articles, fruitful hints and inspirations for future research and publications, of which the Topical Collection "Entropy in Image Analysis" (https://www.mdpi.com/journal/entropy/special_issues/entropy_image_TC, accessed on 5 December 2021) could represent the proper publication place.

Funding: This research received no external funding.

Institutional Review Board Statement: Not applicable.

Informed Consent Statement: Not applicable.

Acknowledgments: I express my thanks to the authors of the above contributions, and to the journal Entropy and MDPI for their constant and precious support during this work.

Conflicts of Interest: The author declares no conflict of interest.

References

1. Kasai, R.; Yamaguchi, Y.; Kojima, T.; Abou Al-Ola, O.M.; Yoshinaga, T. Noise-Robust Image Reconstruction Based on Minimizing Extended Class of Power-Divergence Measures. *Entropy* **2021**, *23*, 1005. [CrossRef] [PubMed]
2. Wen, H.; Xu, J.; Liao, Y.; Chen, R.; Shen, D.; Wen, L.; Shi, Y.; Lin, Q.; Liang, Z.; Zhang, S.; et al. A Security-Enhanced Image Communication Scheme Using Cellular Neural Network. *Entropy* **2021**, *23*, 1000. [CrossRef]
3. Rong, Y.; Zhang, X.; Lin, J. Modified Hilbert Curve for Rectangles and Cuboids and Its Application in Entropy Coding for Image and Video Compression. *Entropy* **2021**, *23*, 836. [CrossRef] [PubMed]
4. Liu, S.; Long, W.; He, L.; Li, Y.; Ding, W. Retinex-Based Fast Algorithm for Low-Light Image Enhancement. *Entropy* **2021**, *23*, 746. [CrossRef]
5. Abdullah, F.; Ghadi, Y.Y.; Gochoo, M.; Jalal, A.; Kim, K. Multi-Person Tracking and Crowd Behavior Detection via Particles Gradient Motion Descriptor and Improved Entropy Classifier. *Entropy* **2021**, *23*, 628. [CrossRef] [PubMed]
6. Cruz, D.S.; de Araújo, J.M.; da Costa, C.A.N.; da Silva, C.C.N. Adding Prior Information in FWI through Relative Entropy. *Entropy* **2021**, *23*, 599. [CrossRef]
7. Chou, M.-C. Evaluation of Non-Uniform Image Quality Caused by Anode Heel Effect in Digital Radiography Using Mutual Information. *Entropy* **2021**, *23*, 525. [CrossRef] [PubMed]
8. Li, T.; Zhang, D. Hyperchaotic Image Encryption Based on Multiple Bit Permutation and Diffusion. *Entropy* **2021**, *23*, 510. [CrossRef]
9. Wang, J.; Wood, A.; Gao, C.; Najarian, K.; Gryak, J. Automated Spleen Injury Detection Using 3D Active Contours and Machine Learning. *Entropy* **2021**, *23*, 382. [CrossRef]
10. Zhang, D.; Chen, L.; Li, T. Hyper-Chaotic Color Image Encryption Based on Transformed Zigzag Diffusion and RNA Operation. *Entropy* **2021**, *23*, 361. [CrossRef] [PubMed]

11. Wen, H.; Zhang, C.; Huang, L.; Ke, J.; Xiong, D. Security Analysis of a Color Image Encryption Algorithm Using a Fractional-Order Chaos. *Entropy* **2021**, *23*, 258. [CrossRef] [PubMed]
12. Karawia, A. Cryptographic Algorithm Using Newton-Raphson Method and General Bischi-Naimzadah Duopoly System. *Entropy* **2021**, *23*, 57. [CrossRef] [PubMed]

Article

Noise-Robust Image Reconstruction Based on Minimizing Extended Class of Power-Divergence Measures

Ryosuke Kasai [1], Yusaku Yamaguchi [2], Takeshi Kojima [3], Omar M. Abou Al-Ola [4] and Tetsuya Yoshinaga [3,*]

[1] Graduate School of Health Sciences, Tokushima University, 3-18-15 Kuramoto, Tokushima 770-8509, Japan; kasai-r@tokushima-u.ac.jp
[2] Shikoku Medical Center for Children and Adults, National Hospital Organization, 2-1-1 Senyu, Zentsuji 765-8507, Japan; yamaguchi.yusaku.sf@mail.hosp.go.jp
[3] Institute of Biomedical Sciences, Tokushima University, 3-18-15 Kuramoto, Tokushima 770-8509, Japan; kojima@medsci.tokushima-u.ac.jp
[4] Faculty of Science, Tanta University, El-Giesh Street, Tanta 31527, Egypt; omar26_7@yahoo.com
* Correspondence: yosinaga@medsci.tokushima-u.ac.jp

Abstract: The problem of tomographic image reconstruction can be reduced to an optimization problem of finding unknown pixel values subject to minimizing the difference between the measured and forward projections. Iterative image reconstruction algorithms provide significant improvements over transform methods in computed tomography. In this paper, we present an extended class of power-divergence measures (PDMs), which includes a large set of distance and relative entropy measures, and propose an iterative reconstruction algorithm based on the extended PDM (EPDM) as an objective function for the optimization strategy. For this purpose, we introduce a system of nonlinear differential equations whose Lyapunov function is equivalent to the EPDM. Then, we derive an iterative formula by multiplicative discretization of the continuous-time system. Since the parameterized EPDM family includes the Kullback–Leibler divergence, the resulting iterative algorithm is a natural extension of the maximum-likelihood expectation-maximization (MLEM) method. We conducted image reconstruction experiments using noisy projection data and found that the proposed algorithm outperformed MLEM and could reconstruct high-quality images that were robust to measured noise by properly selecting parameters.

Keywords: power-divergence measure; computed tomography; iterative reconstruction; maximum-likelihood expectation-maximization method; continuous-time image reconstruction

Citation: Kasai, R.; Yamaguchi, Y.; Kojima, T.; Abou Al-Ola, O.M.; Yoshinaga, T. Noise-Robust Image Reconstruction Based on Minimizing Extended Class of Power-Divergence Measures. *Entropy* **2021**, *23*, 1005. https://doi.org/10.3390/e23081005

Academic Editor: Jiayi Ma

Received: 1 June 2021
Accepted: 24 July 2021
Published: 31 July 2021

Publisher's Note: MDPI stays neutral with regard to jurisdictional claims in published maps and institutional affiliations.

1. Introduction

Image reconstruction in computed tomography (CT) is the process of estimating unknown density images from measured projections. When the system of a tomographic inverse problem is ill-posed, iterative reconstruction algorithms [1,2] based on the optimization strategy provide significant improvements over transform methods, including the filtered back-projection [3,4] (FBP) procedure. In recent years, iterative reconstruction has received much attention because of its ability to reduce radiation doses [5–9] in X-ray CT. Iterative algorithms implemented in, e.g., the algebraic reconstruction technique [1], maximum-likelihood expectation-maximization [10] (MLEM) method, and multiplicative algebraic reconstruction technique, have been used for reconstructing CT images. The MLEM algorithm, which is the most popular method used in emission CT and is derived for the maximum likelihood estimation of a Poisson distribution, reconstructs high-quality images even for noisy projection data, but it is slow to converge [11–14] under iteration. The ordered-subsets EM algorithm [11], in which the EM iteration is performed in each subset by dividing the projection into subsets or blocks, is known to be useful for accelerating MLEM [13,15,16]. However, divergence or oscillation of solutions may occur in the iterative process when the subset balance is not satisfied. Because of the high quality of image reconstruction afforded by the MLEM algorithm, improved MLEM methods have been presented for accelerating convergence. Some schemes accelerate the convergence rate by increasing a relaxation parameter or the step-size in iterative operations [14,17,18] or by introducing a parameter with a power exponent related to the projection for controlling the noise model [19,20]. However, no theory has explained the divergence and oscillation phenomena affecting solutions when the step-size parameter is large.

The convergence of iterative solutions and the quality of images are governed by the underlying objective function that has to be minimized. Hence, the base objective function is one of the most important decisions when designing an iterative algorithm. In this paper, we present an extended class of power-divergence measures [21–24] (PDMs) and derive a novel iterative algorithm based on the minimization of the extended PDM (EPDM) as an objective function for the optimization strategy. Let us define the parameterized function $\Phi_{\gamma,\alpha}(p,q)$ of vectors p and q with nonnegative elements p_i and q_i, respectively, as

$$\Phi_{\gamma,\alpha}(p,q) := \sum_i \int_{p_i}^{q_i} \frac{s^\gamma - p_i^\gamma}{s^{\gamma\alpha}} ds \tag{1}$$

where γ and α indicate positive and nonnegative parameters, respectively. The extension is performed by incorporating the parameter α in the conventional class of PDMs, which includes a large set of distance and relative entropy measures. By fixing the parameter $\alpha = 1$, $\Phi_{\gamma,1}$ gives the family of PDMs with a single parameter γ. Therefore, the measure coincides with the Kullback–Leibler (KL), or relative entropy, divergence [25] if $\gamma = 1$, Neyman's χ^2 distance if $\gamma = 2$, and the generalized Hellinger distance otherwise. Moreover, it corresponds to the squared L^2 norm when $\gamma = 1$ and $\alpha = 0$ and the reverse KL-divergence when $\gamma = 1$ and $\alpha = 2$. Thus, the parameters γ and α provide a smooth connection among the forward and reverse KL-divergences, the Hellinger distance, the χ^2 distance, and the L^2 distance and can control the trade-off between robustness and asymptotic efficiency of the estimators, in a similar way as in other families of distance measures [26–29].

By exploiting the vectors p and q in Equation (1) as the measured and forward projections, respectively, for the tomographic inverse problem, it is expected that we can create a high-performance iterative reconstruction algorithm thanks to the high degree of freedom. For constructing this novel iterative algorithm, we introduce a nonlinear differential equation whose numerical discretization is equivalent to the iterative system. The purpose of applying a dynamical method [30–35] to tomographic inverse problems [36–39] is as follows: it enables us to prove the stability of the equilibrium corresponding to the desired solution of the system of differential equations by using the Lyapunov stability theorem [40] if a proper Lyapunov function can be found; then, since the step-size used to discretize the set of differential equations corresponds to the relaxation or scaling parameter of the system of difference equations, a family of iterative algorithms incorporating a parameter for acceleration is naturally derived. Moreover, it provides a methodology for systematically designing a new iterative reconstruction algorithm based on optimization of an objective function depending on the features of the image to be reconstructed.

Since the EPDM family includes the KL-divergence, the resulting iterative algorithm with power exponents corresponding to the parameters γ and α is a natural extension of the MLEM method with $(\gamma, \alpha) = (1, 1)$. The convergence of solutions to the continuous analog of the proposed iterative algorithm is theoretically shown using the EPDM as a Lyapunov function when the tomographic inverse problem is consistent.

We conducted image reconstruction experiments using numerical and physical phantoms with noisy projections and found that the proposed algorithm outperformed the conventional MLEM method with respect to reconstructing high-quality images that are robust to measured noise when selecting a set of proper parameter values.

2. Definitions and Notations

Image reconstruction is a problem of obtaining unknown pixel values $x \in R_+^J$ satisfying

$$y = Ax + \delta \tag{2}$$

where $y \in R_+^I$, $A \in R_+^{I \times J}$, and $\delta \in R^I$ denote the measured projection, projection operator, and noise, respectively, with R_+ representing the set of nonnegative real numbers. When the system in Equation (2) without noise, i.e., $\delta = 0$, has a solution $e \in R_+^J$, it is consistent; otherwise, it is inconsistent. The tomographic inverse problem can be reduced to one of finding x, which can be accomplished using an optimization approach such as an iterative method or a continuous-time system by minimizing an objective function.

Here, we introduce the notation that will be used below. The superscript $\top$ stands for the transpose of a matrix or vector, θ_k indicates the kth element of the vector θ, Θ_i and Θ_{ij} indicate the ith row vector and the element in the ith row and jth column of the matrix Θ, respectively, $\mathrm{Log}(\theta)$ and $\mathrm{Exp}(\theta)$ are, respectively, the vector-valued functions $\mathrm{Log}(\theta) := (\log(\theta_1), \log(\theta_2), \ldots, \log(\theta_i))^\top$ and $\mathrm{Exp}(\theta) := (\exp(\theta_1), \exp(\theta_2), \ldots, \exp(\theta_i))^\top$ of each element in vector $\theta = (\theta_1, \theta_2, \ldots, \theta_i)^\top$, and $\mathrm{diag}(\theta)$ indicates the diagonal matrix in which the diagonal entries are the elements of the vector θ.

3. Proposed System

3.1. Definition

The proposed methods for obtaining a solution to the tomographic inverse problem can be described as an iterative algorithm and dynamical system.

We present an iterative reconstruction method with a relaxation or scaling parameter $h > 0$:

$$z_j(n+1) = z_j(n) \left(f_j(z(n)) \right)^h \tag{3}$$

with

$$f_j(w) := \frac{\sum\limits_{i=1}^{I} A_{ij} \left(\dfrac{y_i}{(A_i w)^\alpha} \right)^\gamma}{\sum\limits_{i=1}^{I} A_{ij} \left(\dfrac{A_i w}{(A_i w)^\alpha} \right)^\gamma} \tag{4}$$

for $j = 1, 2, \ldots, J$ and $n = 0, 1, 2, \ldots, N-1$, where $\gamma > 0$, $\alpha \geq 0$, and $z(0) = z^0 \in R_{++}^J$, with R_{++} representing the set of positive real numbers. The accompanying system derived from a continuous analog based on the dynamical method is described by a dynamical system:

$$\frac{dx_j(t)}{dt} = x_j(t) \log \left(f_j(x(t)) \right) \tag{5}$$

for $j = 1, 2, \ldots, J$ at $t \geq 0$, where the function f_j is in Equation (4) and $x(0) = z^0$. The system in Equation (5) can be equivalently written as

$$\begin{aligned}
\frac{dx(t)}{dt} = \ & X \Big(\mathrm{Log}(A^\top \mathrm{Exp}(\gamma(\mathrm{Log}(y) - \alpha\,\mathrm{Log}(Ax(t))))) \\
& - \mathrm{Log}(A^\top \mathrm{Exp}(\gamma(1 - \alpha)\,\mathrm{Log}(Ax(t)))) \Big)
\end{aligned} \tag{6}$$

where $X := \mathrm{diag}(x)$. The iterative formula in Equation (3) is obtained by discretizing the differential equation of Equation (5) by using the multiplicative Euler method [41,42] with a step-size of h. Note that the iterative formula in Equation (3) with $h = 1$ is equivalent to the algorithm presented by Zeng [19] when $\gamma = 1$, to the algorithm in Reference [20] when $\alpha = 1$, and to the MLEM algorithm when $(\gamma, \alpha) = (1, 1)$.

We apply the proposed divergence in Equation (1) to the tomographic objective function consisting of measured and forward projections. Namely, we define

$$V(x) := \Phi_{\gamma,\alpha}(y, Ax) = \sum_{i=1}^{I} \int_{y_i}^{A_i x} \frac{s^\gamma - y_i^\gamma}{s^{\gamma\alpha}} ds \tag{7}$$

which can be written as

$$
\begin{aligned}
V(x) &= \sum_{i=1}^{I} \int_{y_i}^{A_i x} \frac{s^\gamma - y_i^\gamma}{s} ds \\
&= \frac{1}{\gamma} \sum_{i=1}^{I} y_i^\gamma \left(\log\left(\frac{y_i}{A_i x}\right)^\gamma + \left(\frac{A_i x}{y_i}\right)^\gamma - 1 \right)
\end{aligned}
$$

if $\gamma\alpha = 1$;

$$
\begin{aligned}
V(x) &= \sum_{i=1}^{I} \int_{y_i}^{A_i x} \frac{s^\gamma - y_i^\gamma}{s^{1+\gamma}} ds \\
&= \frac{1}{\gamma} \sum_{i=1}^{I} \log\left(\frac{A_i x}{y_i}\right)^\gamma + \left(\frac{y_i}{A_i x}\right)^\gamma - 1
\end{aligned}
$$

if $\gamma\alpha = 1 + \gamma$; and

$$
\begin{aligned}
V(x) &= \sum_{i=1}^{I} \int_{y_i}^{A_i x} \frac{s^\gamma - y_i^\gamma}{s^{\gamma\alpha}} ds \\
&= \sum_{i=1}^{I} \frac{1}{1 - \gamma\alpha} y_i^{1+\gamma(1-\alpha)} \left(1 - \left(\frac{A_i x}{y_i}\right)^{1-\gamma\alpha} \right) \\
&\quad + \frac{1}{1 + \gamma(1 - \alpha)} y_i^{1+\gamma(1-\alpha)} \left(\left(\frac{A_i x}{y_i}\right)^{1+\gamma(1-\alpha)} - 1 \right)
\end{aligned}
$$

otherwise.

3.2. Theoretical Results

This section provides a theoretical result on the dynamical system defined in the preceding section. We show that any solution to the continuous analog converges to the desired solution of the system in Equation (2) with $\delta = 0$ when the inverse problem is consistent.

Theorem 1. *Assume there exists $e \in R_{++}^J$ satisfying $y = Ae$. Then, e is an equilibrium observed in the continuous-time system in Equation (6) and is asymptotically stable.*

Proof. We see that e is an equilibrium of the system and the solutions to the system are in R_{++}^J because the initial state value at $t = 0$ belongs to R_{++}^J and the flow cannot pass through the invariant subspace $x_j = 0$ for $j = 1, 2, \ldots, J$ in the state space according to the uniqueness of solutions for the initial value problem. The nonnegative function $V(x)$ of $x_j > 0$ in Equation (7) is well-defined as a candidate of a Lyapunov function. Then, we have the derivative of V along the solutions to Equation (6):

$$
\begin{aligned}
\frac{dV}{dt}(x)\Big|_{(6)} &= -\sum_{i=1}^{I} \left(\left(\frac{y_i}{(A_i x)^\alpha}\right)^\gamma - (A_i x)^{\gamma(1-\alpha)} \right) A_i \frac{dx}{dt} \\
&= -(\xi - \zeta)^\top X(\mathrm{Log}(\xi) - \mathrm{Log}(\zeta)) \tag{8} \\
&\leq 0
\end{aligned}
$$

where

$$\begin{aligned}
\xi &:= A^\top \mathrm{Exp}(\gamma(\mathrm{Log}(y) - \alpha\,\mathrm{Log}(Ax))) \\
\zeta &:= A^\top \mathrm{Exp}(\gamma(1-\alpha)\,\mathrm{Log}(Ax)).
\end{aligned}$$

Therefore, V is a Lyapunov function and the equilibrium e is asymptotically stable. $\square$

This theorem guarantees that the proposed difference system in Equation (3) as a first-order approximation to the differential equation in Equation (6) has a stable fixed point e when the chosen step-size h is sufficiently small to ensure numerical stability.

4. Experimental Results and Discussion

We will illustrate the effectiveness of the extended MLEM algorithm based on the EPDM family in Equation (3) with the parameter set (γ, α) (in what follows, the iterative algorithm except for MLEM with $(\gamma, \alpha) = (1,1)$ is referred to as PDEM) by using examples from numerical and physical CT experiments. The proposed systems were executed using a 6-core processor and computing tools provided by MATLAB (MathWorks, Natick, MA, USA).

We set $h = 1$ and a constant initial value z_j^0 for $j = 1, 2, \ldots, J$. Note that variation of h is out of the scope of this paper, although setting $h > 1$ would accelerate convergence. In addition, in the numerical simulation, the PDEM algorithm in Equation (3) with $h = 1$ as a simple forward Euler discretization qualitatively approximates the solutions to the differential equation in Equation (6), which were calculated by a standard MATLAB ODE solver ode113 implementing a variable step-size variable order method.

4.1. Reconstruction Using Numerical Phantom

We used a numerical phantom image consisting of $e \in [0,1]^J$ with 128×128 pixels ($J = 16,384$), as shown in Figure 1. For our experiment, a Shepp–Logan phantom [43], which is a popular test image for developing reconstruction algorithms, was modified by changing the density values for ellipses so that the resulting image had better visual perception with high contrast. The noise-free and noisy projections $y \in R_+^I$ derived from the phantom image were, respectively, simulated using Equation (2) without and with δ denoting white Gaussian noise such that the signal-to-noise ratio (SNR) was 30 dB and by setting the number of view angles and detector bins to 180 and 184 ($I = 33,120$) with 180-degree sampling.

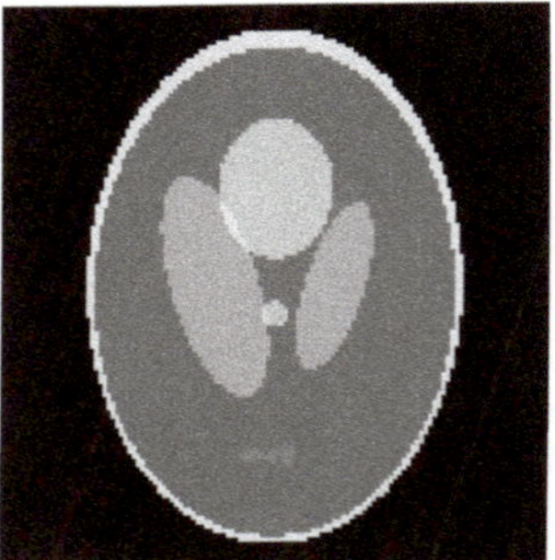

Figure 1. Image of numerical phantom.

For directly evaluating the quality of the reconstructed images, for comparing the reconstructed image compared against the true i

$$D_j(z) := |e_j - z_j|$$

for $j = 1, 2, \ldots, J$ and

$$E(z) := ||e - z||_2 = \left(\sum_{j=1}^{J} (D_j(z))^2 \right)^{\frac{1}{2}}. \tag{10}$$

First, we considered the case of a noise-free projection. Figure 2 shows the evaluation functions $E(z(n))$ of the iterative points $z(n)$ for MLEM and PDEM with the sets of parameters (γ, α) being $(0.3, 1.2)$, $(0.5, 1.2)$, $(0.8, 1.2)$, and $(1.3, 1.2)$ for $n = 0, 1, 2, \ldots, 200$. All algorithms monotonically decreased the evaluation function, as supported by the theoretical result that the solutions of the continuous analog converge to the true value. Indeed, another experiment confirmed that the monotonic decrease continued as the number of iterations exceeded 200 iterations. We can see that PDEM with the parameter set $(1.3, 1.2)$ reduces the evaluation function much more than MLEM does. To put it another way, the PDEM algorithm takes less computation time than MLEM for obtaining the same evaluation values.

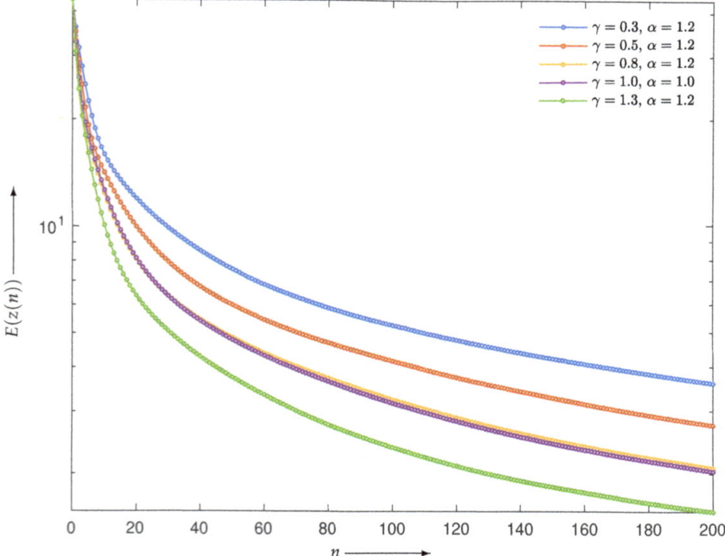

Figure 2. Evaluation functions for MLEM and PDEM algorithms at each iteration in the experiment using numerical phantom with noise-free projection. Note that because the values of PDEM with $(\gamma, \alpha) = (0.8, 1.2)$ and MLEM are very similar, the plotted points for PDEM are almost invisible.

Figure 3 shows contour plots of the evaluation values on a logarithmic scale, $\log_{10}(E(z(N)))$ for $N = 50$, 100, and 200, in the parameter plane (γ, α). The parameters γ and α were, respectively, sampled from 0.1 to 1.5 and 0 to 1.4 with a sampling interval of 0.1. We can see that, at least in the range examined, the evaluation function becomes smaller as the values of $\gamma \geq 1$ and $\alpha \geq 1$ increase.

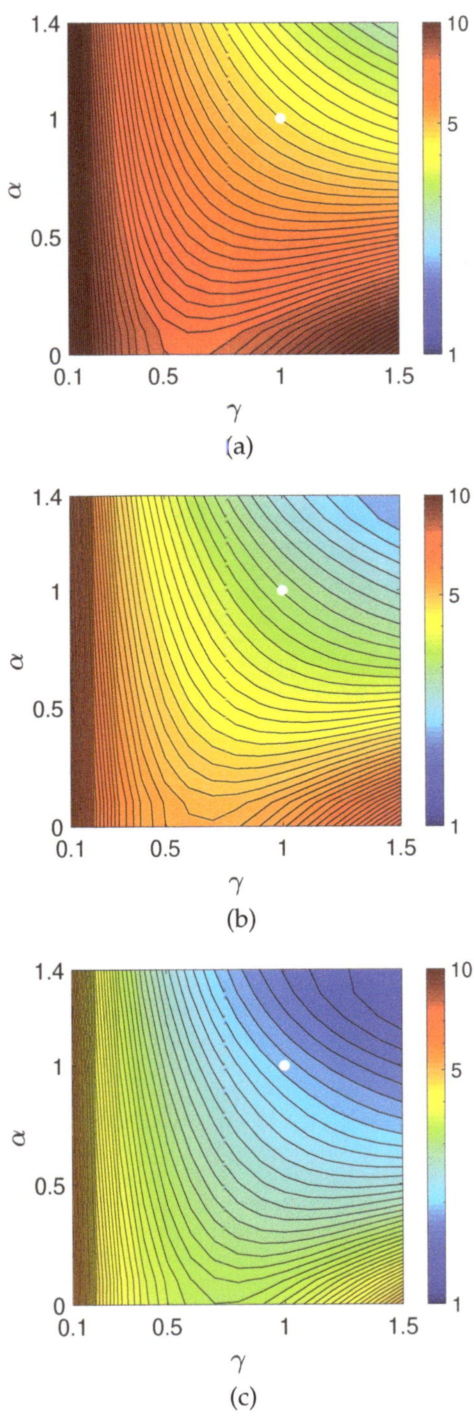

Figure 3. Contour plots of evaluation functions $\log_{10}(E(z(N)))$ with N being (**a**) 50, (**b**) 100, and (**c**) 200 using numerical phantom with noise-free projections. The white dot indicates the position of MLEM.

Figure 4 illustrates images reconstructed by MLEM and PDEM with $(\gamma, \alpha) = (1.3, 1.2)$ at the 200th iteration and the corresponding subtraction images $D_j(z(200))$ (displayed in the range from 0 to 0.2) at every jth pixel, for $j = 1, 2, \ldots, J$. By comparing the values of the subtraction between MLEM and PDEM, e.g., the edges of the high-density objects in the image, we can see that PDEM produces high-quality reconstructions, as is quantitatively indicated by its small evaluation value between the reconstructed and phantom images.

MLEM   PDEM

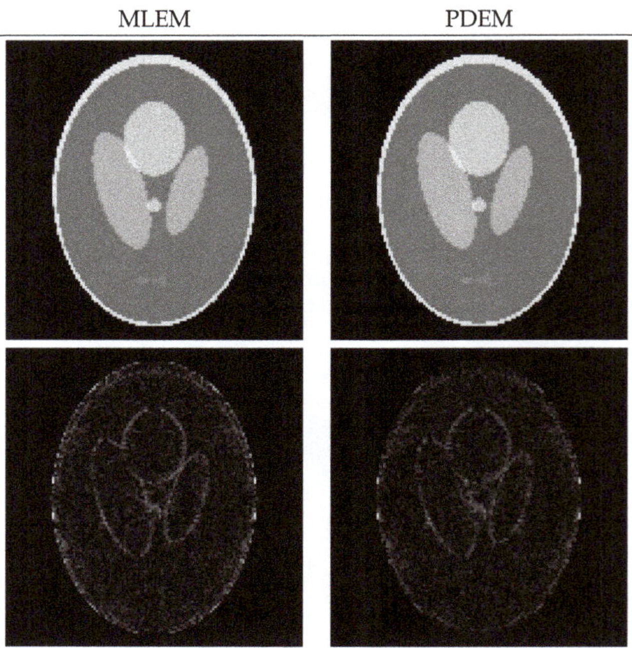

Figure 4. Reconstructed images (**upper panel**) and images of the subtraction (**lower panel**) for MLEM and PDEM with $(\gamma, \alpha) = (1.3, 1.2)$ at 200th iteration using numerical phantom with noise-free projection.

Next, let us consider the effect of the measured noise. Figure 5 is a graph of the evaluation $E(z(n))$ as a function of iteration number n with $n = 0, 1, 2, \ldots, 200$. Given noisy projection data, the algorithm with each parameter set decreases the evaluation function in the early iterations. However, the time course does not show a monotonic decrease in further iterations. Similar characteristics are known to exist and have been considered for the alternative MLEM [19] that is described as Equation (3) with $\gamma = 1$. We can see that a set of parameters (γ, α) exists at which the PDEM algorithm reduces the evaluation function more than MLEM does for any iteration number. Additionally, the smallest value of the evaluation function among the iteration numbers for a fixed set of the parameters becomes smaller with decreasing γ in the set $\{0.3, 0.5, 0.8, 1, 1.3\}$ considered for this example. The parameter dependence of the evaluation function is clearly visible in contour plots of Figure 6, showing the values of $\log_{10}(E(z(N)))$ for $N = 50, 100,$ and 200 in the parameter plane. When designing a parameterized PDEM algorithm, a relatively large value of α and a small value of γ compared with the reference values of $(\gamma, \alpha) = (1, 1)$ provide the best performance in early and sufficient iterations, respectively. The best choices of (γ, α) depending on the termination iteration number are approximately $(0.8, 1.2)$ at the 50th iteration, $(0.5, 1.2)$ at the 100th iteration, and $(0.3, 1.2)$ at the 200th iteration. The evaluation values under these conditions are indicated in Table 1, showing that PDEM with each parameter set gives a smaller value than MLEM does. The reconstructed images and subtraction images (displayed in the range from 0 to 0.3) are shown in Figure 7. The figure reveals lots of artifacts in the reconstructed image due to the presence of noise in the measured projection. In terms of a quantitative evaluation, the structural similarity index measure [44] (SSIM) between the reconstructed and the true images was calculated and summarized in Table 2. A higher value of SSIM, which is a perception-based quality metric, provides higher image quality. By comparing the images reconstructed by MLEM and PDEM at the 100th and 200th iterations (see Figure 7 and Table 2), we can see that the PDEM with a proper set of parameters is able to reconstruct high-quality images while reducing the effects of noise in the projections, which means that PDEM is more robust to noise than MLEM.

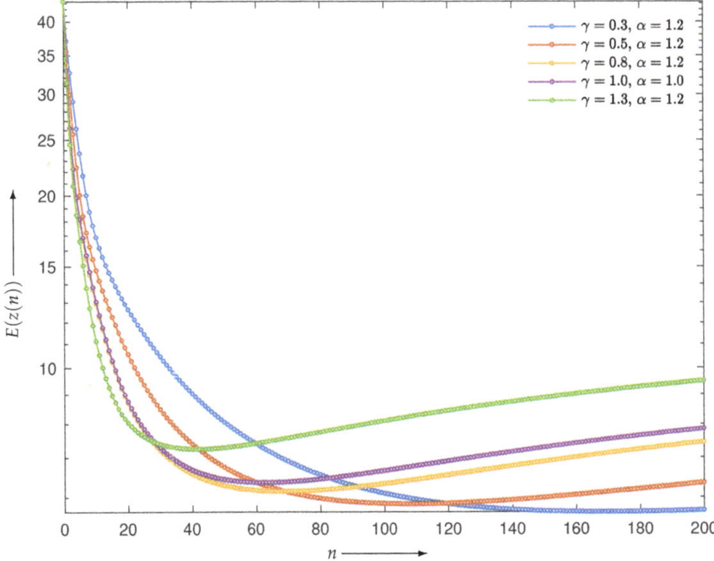

Figure 5. Evaluation functions for MLEM and PDEM algorithms at each iteration in the experiment using numerical phantom with noisy projection.

Table 1. Values of the evaluation function for MLEM and PDEM with (γ, α) equal to $(0.8, 1.2)$ at 50th iteration, $(0.5, 1.2)$ at 100th iteration, and $(0.3, 1.2)$ at 200th iteration in the experiment using numerical phantom with noisy projection.

N	$E(z(N))$		
	MLEM	**PDEM with (γ, α)**	
50	6.44	6.29	$(0.8, 1.2)$
100	6.65	5.85	$(0.5, 1.2)$
200	7.86	5.70	$(0.3, 1.2)$

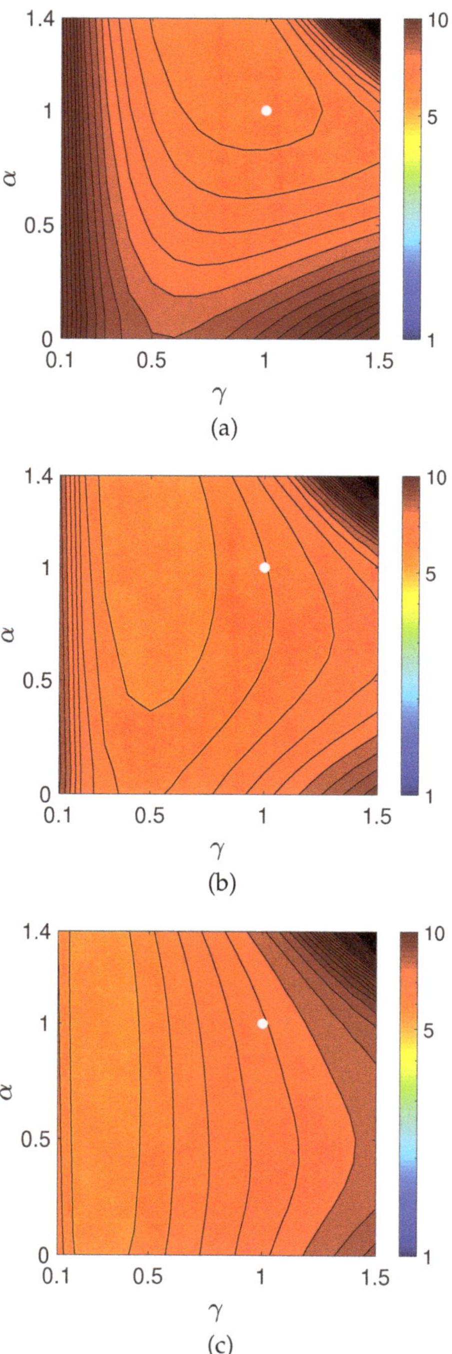

Figure 6. Contour plots of evaluation functions $\log_{10}(E(z(N)))$ with N equal to (**a**) 50, (**b**) 100, and (**c**) 200 in the experiment using numerical phantom with noisy projection. The white dot indicates the position of MLEM.

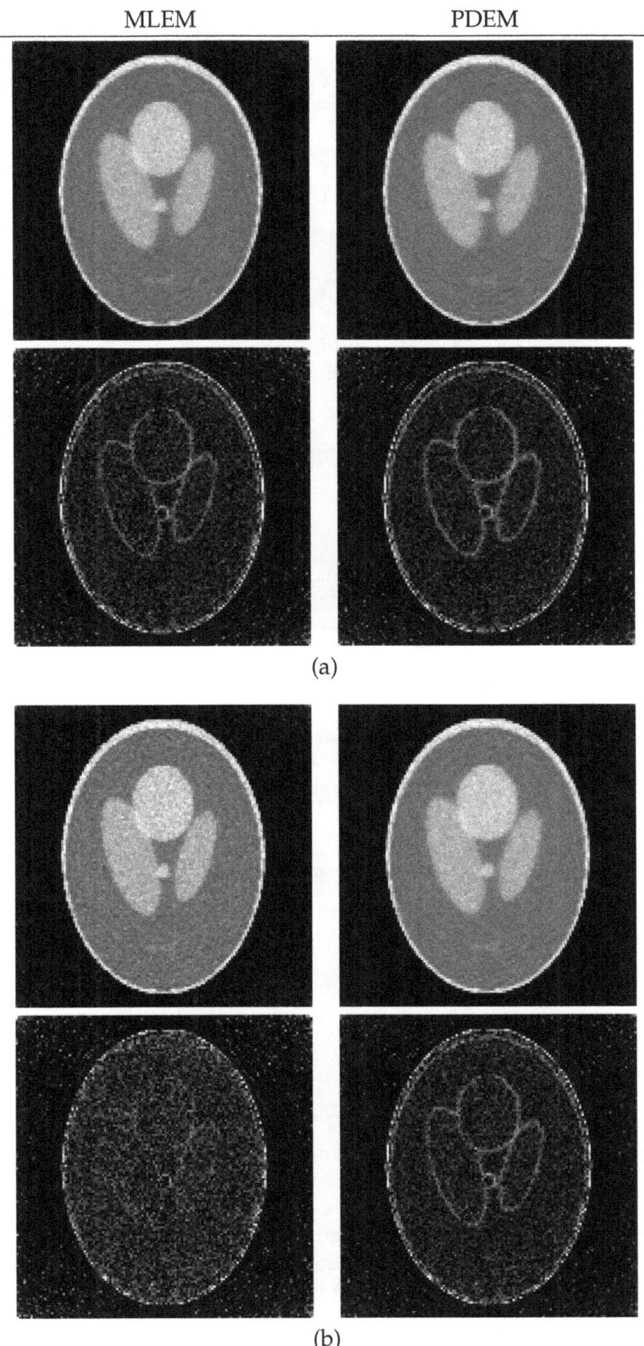

MLEM   PDEM

(a)

(b)

Figure 7. *Cont.*

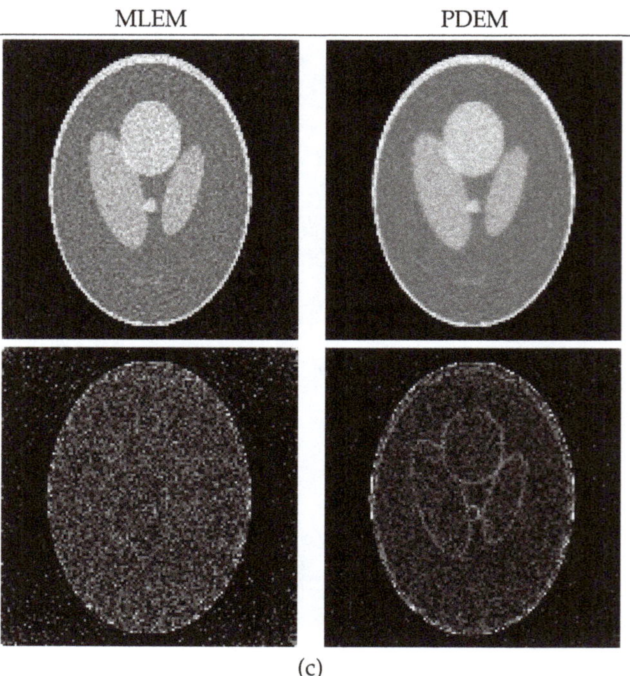

MLEM   PDEM

(c)

Figure 7. Reconstructed images (**upper panel**) and subtraction images (**lower panel**) for MLEM and PDEM with (γ, α) equal to (**a**) $(0.8, 1.2)$ at 50th iteration, (**b**) $(0.5, 1.2)$ at 100th iteration, and (**c**) $(0.3, 1.2)$ at 200th iteration in the experiment using numerical phantom with noisy projection.

Table 2. SSIM for MLEM and PDEM with the same parameters, as shown in Table 1 at Nth iteration in the experiment using numerical phantom with noisy projection.

N	SSIM		
	MLEM	**PDEM with (γ, α)**	
50	0.651	0.689	$(0.8, 1.2)$
100	0.581	0.726	$(0.5, 1.2)$
200	0.531	0.772	$(0.3, 1.2)$

4.2. Reconstruction Using Physical Phantom

A physical experiment was carried out to further validate the effectiveness of the proposed method, although the true image is not available for a quantitative evaluation. The projections were physically acquired from an X-ray CT scanner (Canon Medical Systems, Tochigi, Japan) with a body-simulated phantom [45] (Kyoto Kagaku, Kyoto, Japan) using 80 kVp tube voltage, 30 mA tube current, and an exposure time of 0.75 s per rotation. Figure 8 represents the sinogram, a two-dimensional array of data containing the projections $y \in R_+^I$, with $I = 430{,}200$ (956 acquisition bins and 450 projection directions in 180 degrees) and a reconstructed image created by FBP using a Shepp–Logan filter with $J = 454{,}276$ (674×674 pixels). Images reconstructed by MLEM and PDEM with $(\gamma, \alpha) = (0.5, 1.2)$ are shown in Figure 9. The parameter values were referred to as the results of the numerical phantom with noisy projection. Figure 10, which shows the density profiles along horizontal lines (indicated by white) in the reconstructed images of Figures 8b and 9, verifies that the PDEM has a lower density deviation on a flat distribution of the X-ray absorption in the physical phantom than either MLEM or FBP. The parameter values of the power exponents in the PDEM algorithm make it more robust to noise in spite of the higher noise level due to the low-dose X-ray exposure condition. This fact implies that the proposed method contributes to reducing patient doses during X-ray CT examinations in clinical practice by adjusting the parameter values depending on the noise levels of the projection data.

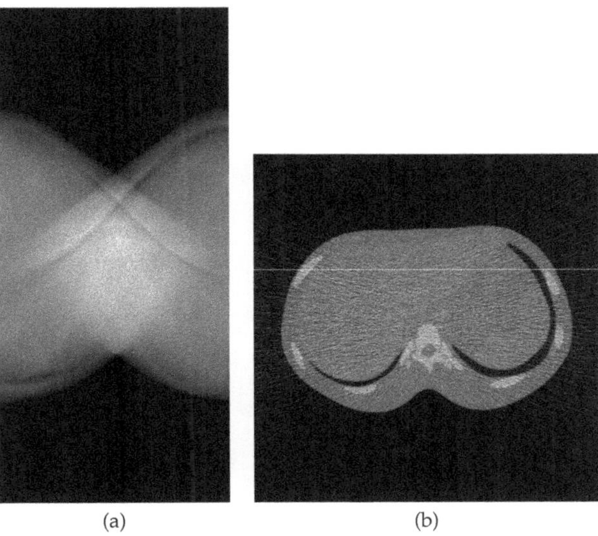

(a)   (b)

Figure 8. (**a**) Sinogram and (**b**) reconstructed image by FBP in the experiment using physical phantom.

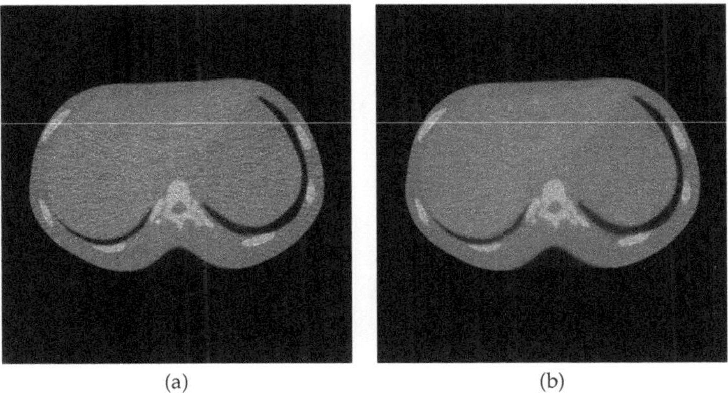

(a)   (b)

Figure 9. Reconstructed images for (**a**) MLEM and (**b**) PDEM with $(\gamma, \alpha) = (0.5, 1.2)$ at 200th iteration in the experiment using physical phantom.

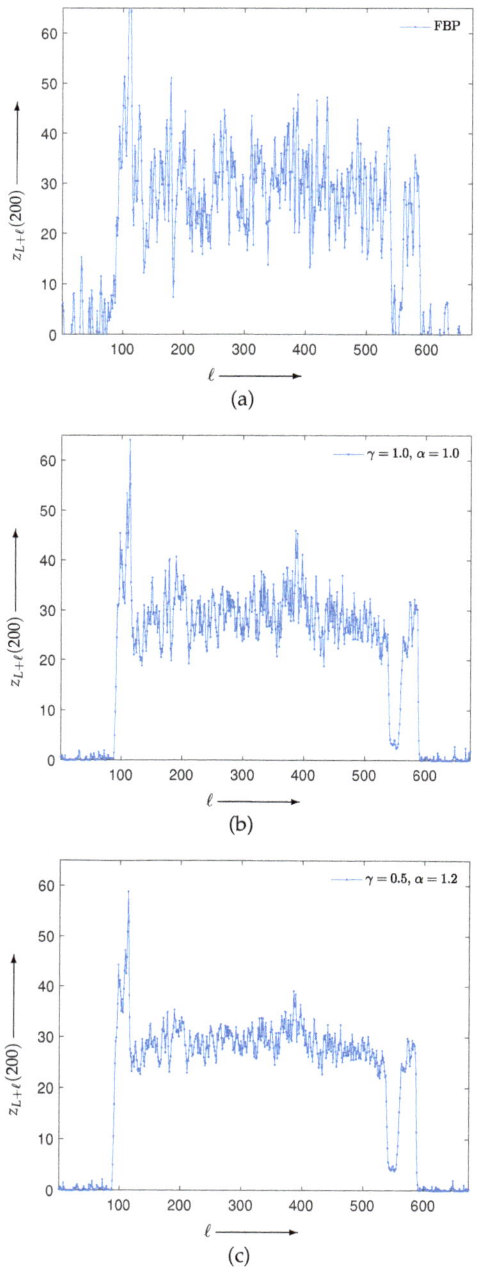

Figure 10. Density profiles for (**a**) FBP, (**b**) MLEM, and (**c**) PDEM of reconstructed images along horizontal line with $L = 674 \times 224$ and $\ell = 1, 2, \ldots, 674$.

5. Conclusions

We presented an extension of the PDM family with two parameters, γ and α, and proposed a novel iterative algorithm based on minimization of the divergence measure as an objective function of the reconstructed images. The theoretical results show the convergence of solutions to the continuous analog of the iterative algorithm owing to the objective function decreasing as the time proceeds. Numerical experiments illustrated that the proposed algorithm, which is considered to be an extended MLEM with two power exponents γ and x, has advantages over MLEM, which is the most popular and suitable iterative method of image reconstruction from noisy measured projections. The algorithm is of practical importance because its image quality is superior to that of MLEM. Our results suggest that a larger value of α accelerates convergence and a smaller value of γ improves its robustness to measured noise. An investigation of the direct relation between the parameter variation in the EPDM family and the quality of images reconstructed by the proposed algorithm based on minimization of the EPDM is a future work to be considered. Moreover, we will use techniques such as machine learning to determine the most appropriate parameter depending on the noise level of the projections, number of projections, number of pixels, etc.

Author Contributions: Conceptualization, T.Y.; data curation, Y.Y. and T.Y.; formal analysis, R.K., Y.Y., O.M.A.A.-O. and T.Y.; methodology, R.K., Y.Y., T.K., O.M.A.A.-O. and T.Y.; software, R.K., Y.Y., T.K. and T.Y.; supervision, T.Y.; validation, O.M.A.A.-O. and T.Y.; writing—original draft, R.K., Y.Y. and T.Y.; writing— review and editing, R.K., Y.Y., T.K., O.M.A.A.-O. and T.Y. All authors have read and agreed to the published version of the manuscript.

Funding: This research was partially supported by JSPS KAKENHI, grant number 21K04080.

Data Availability Statement: All data used to support the findings of this study are included within the article.

Conflicts of Interest: The authors declare no conflict of interest regarding the publication of this paper.

References

1. Gordon, R.; Bender, R.; Herman, G.T. Algebraic reconstruction techniques (ART) for three-dimensional electron microscopy and X-ray photography. *J. Theor. Biol.* **1970**, *29*, 471–481. [CrossRef]
2. Badea, C.; Gordon, R. Experiments with the nonlinear and chaotic behaviour of the multiplicative algebraic reconstruction technique (MART) algorithm for computed tomography. *Phys. Med. Biol.* **2004**, *49*, 1455–1474. [CrossRef] [PubMed]
3. Kak, A.C.; Slaney, M. *Principles of Computerized Tomographic Imaging*; IEEE Press: New York, NY, USA, 1988.
4. Stark, H. *Image Recovery: Theory and Applications*; Academic Press, New York, NY, USA, 1987.
5. Prakash, P.; Kalra, M.K.; Kambadakone, A.K.; Pien, H.; Hsieh, J.; Blake, M.A.; Sahani, D.V. Reducing abdominal CT radiation dose with adaptive statistical iterative reconstruction technique. *Investig. Radiol.* **2010**, *45*, 202–210. [PubMed]
6. Singh, S.; Kalra, M.K.; Gilman, M.D.; Hsieh, J.; Pien, H.H.; Digumarthy, S.R.; Shepard, J.O. Adaptive statistical iterative reconstruction technique for radiation dose reduction in chest CT: A pilot study. *Radiology* **2011**, *259*, 565–573. [CrossRef] [PubMed]
7. Singh, S.; Kalra, M.K.; Do, S.; Thibault, J.B.; Pien, H.; O'Connor, O.J.; Blake, M.A. Comparison of hybrid and pure iterative reconstruction techniques with conventional filtered back projection: Dose reduction potential in the abdomen. *J. Comput. Assist. Tomogr.* **2012**, *36*, 347–353. [PubMed]
8. Beister, M.; Kolditz, D.; Kalender, W.A. Iterative reconstruction methods in X-ray CT. *Phys. Med.* **2012**, *28*, 94–108. [PubMed]
9. Huang, H.M.; Hsiao, I.T. Accelerating an Ordered-Subset Low-Dose X-Ray Cone Beam Computed Tomography Image Reconstruction with a Power Factor and Total Variation Minimization. *PLoS ONE* **2016**, *11*, e0153421.
10. Shepp, L.A.; Vardi, Y. Maximum likelihood reconstruction for emission tomography. *IEEE Trans. Med. Imaging* **1982**, *1*, 113–122.
11. Hudson, H.M.; Larkin, R.S. Accelerated image reconstruction using ordered subsets of projection data. *IEEE Trans. Med. Imaging* **1994**, *13*, 601–609.
12. Fessler, J.A.; Hero, A.O. Penalized maximum-likelihood image reconstruction using space-alternating generalized EM algorithms. *IEEE Trans. Image Process.* **1995**, *4*, 1417–1429.
13. Byrne, C. Accelerating the EMML algorithm and related iterative algorithms by rescaled block-iterative methods. *IEEE Trans. Image Process.* **1998**, *7*, 100–109. [CrossRef]
14. Hwang, D.; Zeng, G.L. Convergence study of an accelerated ML-EM algorithm using bigger step size. *Phys. Med. Biol.* **2006**, *51*, 237–252. [CrossRef]
15. Byrne, C. Block-iterative methods for image reconstruction from projections. *IEEE Trans. Image Process.* **1996**, *5*, 792–794. [CrossRef]
16. Byrne, C. Block-iterative algorithms. *Int. Trans. Oper. Res.* **2009**, *16*, 427–463. [CrossRef]
17. Tanaka, E.; Nohara, N.; Tomitani, T.; Yamamoto, M. Utilization of Non-Negativity Constraints in Reconstruction of Emission Tomograms. In *Information Processing in Medical Imaging*; Stephen, L.B., Ed.; Springer: Dordrecht, The Netherlands, 1986; Volume 1, pp. 379–393.

18. Tanaka, E. A fast reconstruction algorithm for stationary positron emission tomography based on a modified EM algorithm. *IEEE Trans. Med. Imaging* **1987**, *6*, 98–105. [CrossRef]
19. Zeng, G.L. The ML-EM algorithm is not optimal for poisson noise. In Proceedings of the 2015 IEEE Nuclear Science Symposium and Medical Imaging Conference (NSS/MIC), San Diego, CA, USA, 31 October–7 November 2015; pp. 1–3.
20. Yamaguchi, Y.; Kudo, M.; Kojima, T.; Abou Al-Ola, O.M.; Yoshinaga, T. Extended Ordered-subsets Expectation-maximization Algorithm with Power Exponent for Noise-robust Image Reconstruction in Computed Tomography. *Radiat. Environ. Med.* **2019**, *8*, 105–112.
21. Read, T.R.C.; Cressie, N.A.C. *Goodness-of-Fit Statistics for Discrete Multivariate Data*; Springer: New York, NY, USA, 1988.
22. Pardo, L. *Statistical Inference Based on Divergence Measures*; Chapman and Hall/CRC: New York, NY, USA, 2005.
23. Liese, F.; Vajda, I. On Divergences and Informations in Statistics and Information Theory. *IEEE Trans. Inf. Theory* **2006**, *52*, 4394–4412. [CrossRef]
24. Pardo, L. New Developments in Statistical Information Theory Based on Entropy and Divergence Measures. *Entropy* **2019**, *21*, 391. [CrossRef]
25. Kullback, S.; Leibler, R.A. On information and sufficiency. *Ann. Math. Stat.* **1951**, *22*, 79–86. [CrossRef]
26. Beran, R. Minimum Hellinger distance estimates for parametric models. *Ann. Stat.* **1977**, *5*, 445–463. [CrossRef]
27. Basu, A.; Lindsay, B.G. Minimum disparity estimation for continuous models: Efficiency, distributions and robustness. *Ann. Inst. Stat. Math.* **1994**, *46*, 683–705. [CrossRef]
28. Basu, A.; Harris, I.R.; Hjort, N.L.; Jones, M.C. Robust and efficient estimation by minimising a density power divergence. *Biometrika* **1998**, *85*, 549–559. [CrossRef]
29. Ghosh, A.; Basu, A. Estimation of Multivariate Location and Covariance using the S-Hellinger Distance. *arXiv* **2014**, arXiv:1403.6304.
30. Schropp, J. Using dynamical systems methods to solve minimization problems. *Appl. Numer. Math.* **1995**, *18*, 321–335. [CrossRef]
31. Airapetyan, R.G.; Ramm, A.G.; Smirnova, A.B. Continuous analog of Gauss-Newton method. *Math. Models Methods Appl. Sci.* **1999**, *9*, 463–474. [CrossRef]
32. Airapetyan, R.G.; Ramm, A.G. Dynamical systems and discrete methods for solving nonlinear ill-posed problems. In *Applied Mathematics Reviews*; Anastassiou, G.A., Ed.; World Scientific Publishing: Singapore, 2000; Volume 1, pp. 491–536.
33. Airapetyan, R.G.; Ramm, A.G.; Smirnova, A.B. Continuous methods for solving nonlinear ill-posed problems. In *Operator Theory and Its Applications*; Ramm, A.G., Shivakumar, P.N., Strauss, A.V., Eds.; American Mathematical Society: Providence, RI, USA, 2000; Volume 25, pp. 111–138.
34. Ramm, A.G. Dynamical systems method for solving operator equations. *Commun. Nonlinear Sci. Numer. Simul.* **2004**, *9*, 383–402. [CrossRef]
35. Li, L.; Han, B. A dynamical system method for solving nonlinear ill-posed problems. *J. Comput. Appl. Math.* **2008**, *197*, 399–406. [CrossRef]
36. Fujimoto, K.; Abou Al-Ola, O.M.; Yoshinaga, T. Continuous-time image reconstruction using differential equations for computed tomography. *Commun. Nonlinear Sci. Numer. Simul.* **2010**, *15*, 1648–1654. [CrossRef]
37. Abou Al-Ola, O.M.; Fujimoto, K.; Yoshinaga, T. Common Lyapunov function based on Kullback–Leibler divergence for a switched nonlinear system. *Math. Probl. Eng.* **2011**, *2011*, 723509:1–723509:12. [CrossRef]
38. Yamaguchi, Y.; Fujimoto, K.; Abou Al-Ola, O.M.; Yoshinaga, T. Continuous-time image reconstruction for binary tomography. *Commun. Nonlinear Sci. Numer. Simul.* **2013**, *18*, 2081–2087. [CrossRef]
39. Tateishi, K.; Yamaguchi, Y.; Abou Al-Ola, O.M.; Yoshinaga, T. Continuous Analog of Accelerated OS-EM Algorithm for Computed Tomography. *Math. Probl. Eng.* **2017**, *2017*, 1564123:1–1564123:8. [CrossRef]
40. Lyapunov, A.M. *Stability of Motion*; Academic Press: New York, NY, USA, 1966.
41. Aniszewska, D. Multiplicative Runge–Kutta methods. *Nonlinear Dyn.* **2007**, *50*, 265–272. [CrossRef]
42. Bashirov, A.E.; Kurpinar, E.M.; Oezyapici, A. Multiplicative calculus and its applications. *J. Math. Anal. Appl.* **2008**, *337*, 36–48. [CrossRef]
43. Shepp, L.A.; Logan, B.F. The Fourier reconstruction of a head section. *IEEE Trans. Nucl. Sci.* **1974**, *21*, 21–43. [CrossRef]
44. Wang, Z.; Bovik, A.C.; Sheikh, H.R.; Simoncelli, E.P. Image quality assessment: From error visibility to structural similarity. *IEEE Trans. Image Process.* **2004**, *13*, 600–612. [CrossRef]
45. Kyoto Kagaku. CT Whole Body Phantom PBU-60. Available online: https://kyotokagaku.com/en/products_introduction/ph-2b/ (accessed on 1 June 2021).

Article

A Security-Enhanced Image Communication Scheme Using Cellular Neural Network

Heping Wen [1,2,3], Jiajun Xu [1], Yunlong Liao [1], Ruiting Chen [1], Danze Shen [1], Lifei Wen [1], Yulin Shi [1], Qin Lin [1], Zhonghao Liang [1], Sihang Zhang [1], Yuxuan Liu [1], Ailin Huo [1], Tong Li [1], Chang Cai [1] and Jiaqian Wen [1] and Chongfu Zhang [1,2,*]

[1] Zhongshan Institute, University of Electronic Science and Technology of China, Zhongshan 528402, China; wenheping@uestc.edu.cn (H.W.); xujiajun@stu.zsc.edu.cn (J.X.); teamoyan@stu.zsc.edu.cn (Y.L.); ruitingchen@stu.zsc.edu.cn (R.C.); shendanze@stu.zsc.edu.cn (D.S.); wenlifei@stu.zsc.edu.cn (L.W.); shiyulin@stu.zsc.edu.cn (Y.S.); liqin@stu.zsc.edu.cn (Q.L.); zhonghaoliang@stu.zsc.edu.cn (Z.L.); sihangzhang@stu.zsc.edu.cn (S.Z.); liuyuxuan@stu.zsc.edu.cn (Y.L.); ailinhuo@stu.zsc.edu.cn (A.H.); litong@stu.zsc.edu.cn (T.L.); caichang@stu.zsc.edu.cn (C.C.); jiaqianwen@stu.zsc.edu.cn (J.W.)
[2] School of Information and Communication Engineering, University of Electronic Science and Technology of China, Chengdu 611731, China
[3] Guangdong Provincial Key Laboratory of Information Security Technology, Guangzhou 510006, China
* Correspondence: cfzhang@uestc.edu.cn

Abstract: In the current network and big data environment, the secure transmission of digital images is facing huge challenges. The use of some methodologies in artificial intelligence to enhance its security is extremely cutting-edge and also a development trend. To this end, this paper proposes a security-enhanced image communication scheme based on cellular neural network (CNN) under cryptanalysis. First, the complex characteristics of CNN are used to create pseudorandom sequences for image encryption. Then, a plain image is sequentially confused, permuted and diffused to get the cipher image by these CNN-based sequences. Based on cryptanalysis theory, a security-enhanced algorithm structure and relevant steps are detailed. Theoretical analysis and experimental results both demonstrate its safety performance. Moreover, the structure of image cipher can effectively resist various common attacks in cryptography. Therefore, the image communication scheme based on CNN proposed in this paper is a competitive security technology method.

Keywords: secure communication; image encryption; chaos; cellular neural network

Citation: Wen, H.; Xu, J.; Liao, Y.; Chen, R.; Shen, D.; Wen, L.; Shi, Y.; Lin, Q.; Liang, Z.; Zhang, S.; et al. A Security-Enhanced Image Communication Scheme Using Cellular Neural Network. *Entropy* **2021**, 23, 1000. https://doi.org/10.3390/e23081000

Academic Editor: Amelia Carolina Sparavigna

Received: 23 June 2021
Accepted: 28 July 2021
Published: 31 July 2021

Publisher's Note: MDPI stays neutral with regard to jurisdictional claims in published maps and institutional affiliations.

1. Introduction

With the rapid development of cloud computing, big data, blockchain and other emerging technologies, the privacy and sharing of messages provides convenience for people in their work and daily lives [1–4]. However, the convenience also threatens the security of cyberspace [5–8]. In particular, as a significant transmission medium, digital images may include a lot of personal privacy, confidential information and other important data, so their privacy protection gets more attention [9–12]. Encryption technology is a common means to assure the security of digital images, and has been widely used in various fields of digital image security [13–17]. Currently, there exist many mature block encryption schemes that are widely used in text encryption and these schemes have brilliant effects [18,19]. Nevertheless, due to the uniqueness of the image, such as being two-dimensional, redundancy and a strong correlation of two adjacent pixels, traditional text encryption faces severe challenges [20–22]. Moreover, the problem of real-time transmission should be considered in image encryption to improve the communication performance [9,23,24]. Therefore, it is quite necessary to study the new technologies and methods of image encryption.

In current international studies, digital image encryption is a research hotspot [25–27]. Various mechanisms and methods are introduced to enhance the security of algorithms [28,29].

In 2015, the authors of [16] proposed a multibiometric template protection scheme based on fuzzy commitment and a chaos-based system, as well as a security analysis method of unimodal biometrics leakage. The chaos-based system is used to encrypt the dual iris feature vectors. The experimental results show that the security of BCH ECC (1,023,123,170) based on multibiometrics template is improved from 80.53 bits to 167.80 bits. In 2017, the authors of [30] designed a special image encryption scheme based on the second-order Henon mapping hyperchaos and the fifth-order CNN. Experimental results show that the scheme features high security and is suitable to spread in the network. At the same time, in [31] a new image encryption method was proposed, based on the biological DNA sequences operation and the third-order CNN. The method could effectively enhance the plaintext sensitivity and features large key space and high security. In 2019, the authors of [17] proposed a new privacy protection encryption mechanism for medical systems based on the Internet of Things. Experimental results show that the encryption mechanism is robust and effective to protect the privacy of patients. In 2020, Zhang and Zhang [32] used the Chen chaos-based system and two-dimensional logistic mapping to propose a multi-image encryption system based on bitplane and chaos. The experiment also proved its high efficiency. At the same time, in [15] a new and effective color image cryptosystem was proposed. The experimental results show that the cryptosystem has high security efficiency and can be effectively applied to the IoHT framework of secure medical image transmission. In summary, more and more theories and technological achievements have been made in digital image encryption. However, in current studies, most digital images are regarded as a two-dimensional matrix to encrypt, meaning that only the spatial domain is processed [6,33–35]. However, two defects were exposed: (1) Some encryption algorithms have security flaws and are not associated with plaintext, so it is difficult for them to resist chosen-plaintext attack (CPA); (2) The cost of attacking the encryption algorithm is relatively low because chaos-based systems are relatively simple.

Aimed at solving the existing problems, we put forward a digital image encryption algorithm based on CNN in this paper. On the one hand, a CNN chaos-based system is selected to generate a chaos-based key sequence. The CNN chaos-based system has more complex behavioral characteristics, so it has better security performance than other encryption systems. On the other hand, the scheme adopts the security mechanism of generating a chaos-based key sequence by plaintext correlation. Therefore, compared with other encryption schemes based on a CNN chaos-based system, it effectively enhances the ability to resist CPA. Theoretical analysis and experimental results show that the proposed algorithm can effectively enhance the confusion, diffusion and avalanche effect of encryption. Therefore, the image encryption algorithm based on CNN is reliable.

2. Correlation Theory

The idea of a cellular neural network (CNN) was conceived by Chua and Yang in 1988 [34]. The basic units of CNN are called cells, and each cell is a nonlinear first-order circuit which is composed of a linear resistor, a linear capacitor and a voltage-controlled current source [36,37].

In order to make the mathematical model of CNN more comprehensible, a simplified CNN cell model is adopted:

$$\frac{dx_j}{dt} = -x_j + A_j p_j + G_o + G_s + I_j \tag{1}$$

where j is used as a cell marker, x_j represents the state variable, A_j represents a constant number, I_j represents the threshold value, G_s and G_o separately represent the linear combination of the state variables of the cell and the output value of the connecting cell, and p_j represents the output of the cell.

The fourth-order fully interconnected CNN equation can be defined as follows:

$$\begin{cases} \dfrac{dx_j}{dt} = -x_j + A_j p_j + \displaystyle\sum_{k=1;k\neq j}^{4} A_{jk} p_j + \sum_{k=1}^{4} S_{jk} x_k + I_j \\[2mm] p_j = 0.5|x_j + 1| - 0.5|x_j - 1| \end{cases} \tag{2}$$

where S represents a matrix of $j \times k$, A_j and I_j both represent a matrix of $j \times 1$, $A_{jk} = 0 (j \neq k, j = 1,2,3,4; k = 1,2,3,4)$ and it can be described by the equation of state in Equation (2) [38]:

$$\begin{cases} \dfrac{dx_1}{dt} = -x_3 - \varepsilon x_4 \\[2mm] \dfrac{dx_2}{dt} = 2x_2 + x_3 \\[2mm] \dfrac{dx_3}{dt} = 14x_1 - 14x_2 \\[2mm] \dfrac{dx_4}{dt} = 200p_4 + 100x_1 - 100x_4 \end{cases} \tag{3}$$

where ε is the control parameter of the CNN model, which can control the size and quantity of Lyapunov exponents, and the range of values for ε is 0 to 2. At this moment, the system is in a chaos-based state, and four aperiodic chaos-based sequences can be generated from it, which are very sensitive to the initial conditions $x_1(0), x_2(0), x_3(0)$ and $x_4(0)$. By calculating the Lyapunov exponents of Equation (3), it can be seen that the Lyapunov exponents of the four chaos-based sequences tend to 42.8487, 2.0230, -0.0230 and -49.0391, respectively, two of which are positive. Therefore, the CNN model is a hyperchaotic system, and the Lyapunov exponents are shown in Figure 1. When the initial values of $x_1(0), x_2(0)$, $x_3(0)$ and $x_4(0)$ are 0.2, 0.2, 0.2 and 0.2, respectively, we use the fourth-order Runge–Kutta algorithm with the step size of $h = 0.005$ to get the two-dimensional chaos-based attractor, as shown in Figure 2a–d and the three-dimensional chaos-based attractor, as shown in Figure 2e–h.

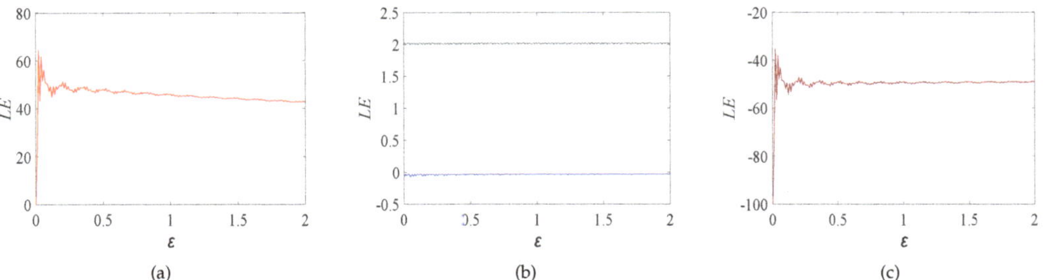

Figure 1. Lyapunov exponents spectrum. The exponents tend to 42.8487, 2.0230 and -0.0230, and -49.0391, as can be seen in (**a**–**c**), respectively.

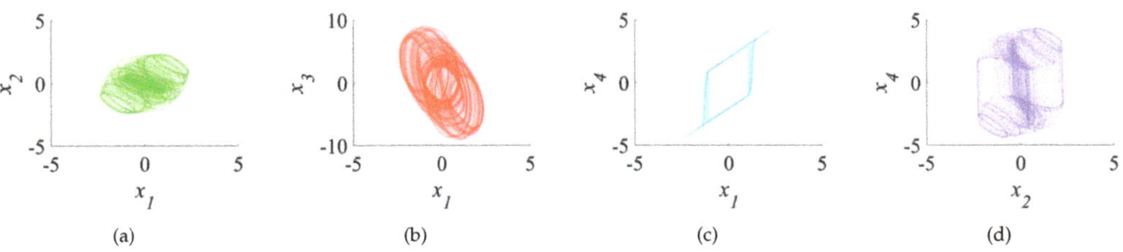

Figure 2. *Cont.*

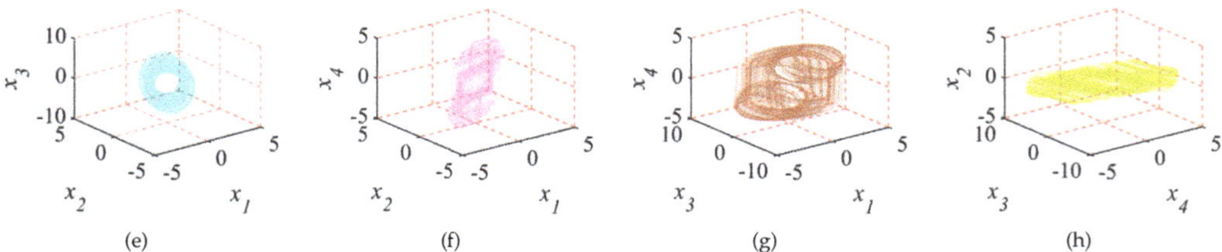

Figure 2. Chaos-based attractors generated by the fourth-order CNN: (**a**) x_1, x_2; (**b**) x_1, x_3; (**c**) x_1, x_4; (**d**) x_2, x_4; (**e**) x_1, x_2, x_3; (**f**) x_1, x_2, x_4; (**g**) x_1, x_3, x_4; (**h**) x_4, x_3, x_2.

3. The Proposed Encryption Algorithm

The encryption algorithm of chaos-based image usually adopts the classical structure "permutation–diffusion" [39,40]. However, due to the lack of security, a chaos-based image encryption algorithm based on a "confusion–permutation–diffusion" structure is proposed in this paper [35].

The encryption and decryption processes are shown in Figure 3. IEA-CNN represents the image encryption algorithm based on a cellular neural network, IDA-CNN represents the image decryption algorithm based on a cellular neural network. In order to enhance the ability to resist CPA, the image encryption system of this paper adopts the security mechanisms of chaos-based key sequences produced by plaintext association and ciphertext feedback diffusion encryption. The specific steps of the encryption algorithm are given as follows:

Step 1: *Preprocessing Sequences*

The secret key of the image encryption algorithm contains the Message-Digest Algorithm 5 (MD5) value of plain image, the initial value of the fourth-order CNN and the controlling parameters. The MD5 can be used to disturb the initial value key parameters of CNN chaos; so that the key sequence changes with different plain images, the specific treatment methods are calculated using the following formulas:

$$\begin{cases} x_1'(0) = x_1(0) + (m_1 \oplus m_2 \oplus m_3 \oplus m_4)/256 \\ x_2'(0) = x_2(0) + (m_5 \oplus m_6 \oplus m_7 \oplus m_8)/256 \\ x_3'(0) = x_3(0) + (m_9 \oplus m_{10} \oplus m_{11} \oplus m_{12})/256 \\ x_4'(0) = x_4(0) + (m_{13} \oplus m_{14} \oplus m_{15} \oplus m_{16})/256 \end{cases} \tag{4}$$

where $\oplus$ is bitwise XOR operation, $x_1(0), x_2(0), x_3(0)$ and $x_4(0)$ are the initial values of the fourth-order CNN key parameters; $x_1'(0), x_2'(0), x_3'(0)$ and $x_4'(0)$ are the initial values updated after the disturbance from MD5. Obviously, the new initial values will change with the different plain images. Then, a preprocessing operation is adopted for the chaos-based sequences. The generating methods of obfuscated sequences are shown as follows:

$$\begin{cases} real\_X = [x_1; x_2; x_3; x_4] \\ K_c' = floor(\bmod(real\_X \times 10^{10}, 256)) \\ K_c = reshape(K_c', H, W) \end{cases} \tag{5}$$

where $real\_X$ is composed of four sequences produced by the fourth-order CNN chaos-based system. The sequences diagram of four sequences generated by chaos-based mapping of the fourth-order CNN is shown in Figure 4. The size of K_c is equal to $H \times W$, H

and W are pixel rows and pixel columns of the plain images for image confusion. The generating method of permutation sequences is shown as follows:

$$\begin{cases} seq\_H = x_2(1, 1 : H) \\ seq\_W = x_3(2, 1 : 8 \times W) \\ [value_1, K_{pr}] = sort(seq\_H) \\ [value_2, K_{pc}] = sort(seq\_W) \end{cases} \tag{6}$$

where *sort* is the sorting function of array elements; x_2 represents a two-dimensional sequence of *real_X*; x_3 represents the three-dimensional sequence of *real_X*; seq_H represents the chaos-based sequence of length H extracted from x_2; *real_W* represents the chaos-based sequence of length $8 \times W$ extracted from x_3; K_{pr} means that the pixel row is generated by the sorting function and the length is H; K_{pc} means that the pixel column is generated by the sorting function and the length is $8 \times W$; $value_1$ and $value_2$ are the sorted chaos-based sequence values.

The generating method of diffusion sequences is shown as follows:

$$\begin{cases} K_d = mod(floor([x_1, x_3, x_2, x_4] \times 10^5), 256) \\ K_d{}' = mod(floor([x_3, x_4, x_1, x_2] \times 10^5), 256) \end{cases} \tag{7}$$

where the lengths of K_d and $K_d{}'$ are $H \times W$, and the key sequences of K_d and $K_d{}'$ are used for diffusion.

Step 2: *Confusion*

The key sequence K_c is used to obfuscate the plain image P. The image can be visualized and hidden to get the obfuscated image I_1, the method is shown as follows:

$$I_1(i) = K_c(i) \oplus P(i), i = (1, 2, \cdots, H \times W) \tag{8}$$

Step 3: *Permutation*

The key sequences $K_{pr}(i)$ and $K_{pc}(j)$ are used to replace the pixels in I_1 to get I_3, the method is shown as follows:

$$\begin{cases} I_2 = swap(I_1(:, K_{pc}(i)), I_1(:, i)) \\ I_3 = swap(I_2(K_{pr}(j), :), I_2(j, :)) \end{cases} \tag{9}$$

where *swap* function is used to swap the values of two pixels. The number of bit level rows is equal to the number of pixel level rows, and the number of bit level columns is equal to 8 times the number of pixel level columns, thus, $i = 1, 2, \cdots, H$ and $j = 1, 2, \cdots, 8 \times W$. I_2 and I_3 are the images after double bit column transform and row transform permutation, respectively.

Step 4: *Diffusion*

All the ciphertext pixels in I_3 are diffused dynamically. K_d and K_d' are used for the image diffusion operation to generate the final ciphertext image C.

The first ciphertext pixel $C(1)$ is generated, and the diffusion encryption equation is shown as follows:

$$\begin{cases} C(1) = I_3(1) \oplus K_d(1) \oplus (sum(1) \dotplus K_d{}'(1)) \\ sum(1) = \sum\limits_{i=1}^{L} I_3(i) \end{cases} \tag{10}$$

where the operator $\dotplus$ can be defined as $a \dotplus b \triangleq mod(a + b, 256)$, $I_3(1)$ is the first pixel of the permutation image I_3, $K_d(1)$ and $K_d{}'(1)$ are the first element of the diffusion encryption sequences, and $sum(1)$ represents the sum of all pixels of the permutation image I_3.

Then ciphertext pixel $C(i)$ is produced and its diffusion formula is shown as follows:

$$\begin{cases} C(i) = I_3(i) \oplus (C(i-1)+K_d(i)) \oplus (sum(i)+K_d{'}(i)) \\ sum(i) = sum(i-1) - I_3(i) \end{cases} \tag{11}$$

where $i = 2,3,\ldots,L$ and the i represents the ith pixel of the permutation image I_3. $C(i-1)$ is the $(i-1)$th ciphertext pixel. $sum(i)$ is the sum of the $(L-i+1)$ pixels of the permutation image I_3. According to Equation (11), starting from the second ciphertext pixel $C(2)$, the cipher image C is generated by computing iteratively $C(i)$, i in $\{1,2,\cdots,L\}$, until the Lth ciphertext $C(L)$ is generated.

Decryption is the inverse process of encryption, whose process is first confusion, then permutation, and finally diffusion. While the decryption process is to first reverse diffuse the encrypted image, then reverse permutate the reverse diffuse image, and finally reverse confuse the reverse permutation image to get the decrypted image. When the decryption key and the encryption key are matched, the image can be restored correctly. However, when the decryption key is not equal to the encryption key, even if there is a small error, the correct image cannot be decrypted.

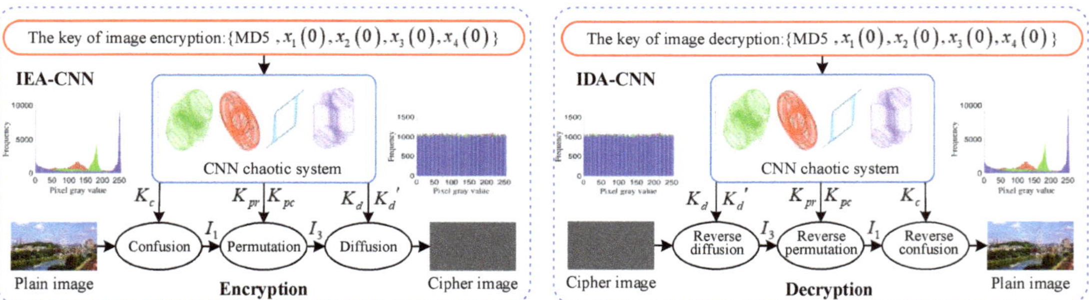

Figure 3. Principle and mechanism of image encryption and decryption.

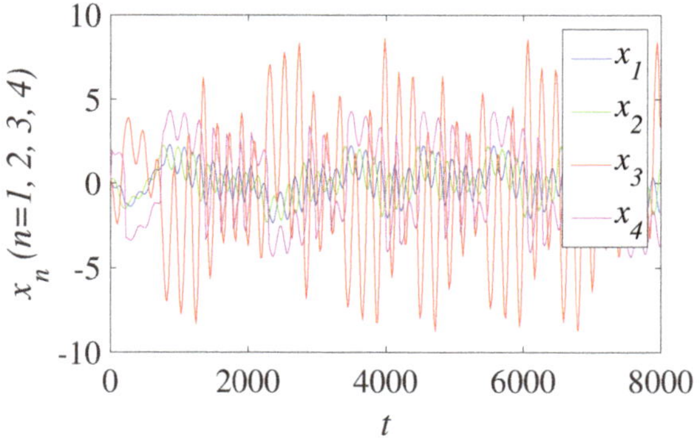

Figure 4. Sequence diagram of the fourth-order CNN.

4. Experimental Verification and Discussion

In the analysis of the experimental results, we use MATLAB 2020b to simulate and validate the proposed image encryption system which is executed on a PC with Windows 10 64 bit operating system, Intel (R) Core (TM) i7-8250 CPU @ 1.60 GHz 1.80 GHz processor and 8 GB memory. In order to prove the effectiveness and practicability of the proposed image encryption scheme, we selected the images from "USC-SIPI Image Database" and "Ground Truth Database" as the test images [41,42].

4.1. Key Space Analysis

In the encryption system, the range of valid value of key can be expressed by key space. The image encryption algorithm designed in this paper uses a fourth-order CNN system and the secret key parameters involved are the initial values of the fourth-order CNN chaos-based system $x_1(0), x_2(0), x_3(0), x_4(0)$. Because the computer precision used in experimental simulation is 10^{-15}, the size of this part of encryption system key space is $(10^{15})^4 = 10^{60} \approx 2^{199}$. Considering that MD5 of 128 bits can also be used as part of the secret key, the total secret key space 2^{327} and the encryption system can resist the exhaustive attack effectively [43,44].

4.2. Nist 800-22 Test

The NIST 800-22 test is an internationally recognized random number test. It consists of 16 different tests. As long as the 16 test results are greater than or equal to 0.001, the random array can be considered to be qualified. In this test, we divide the generated 3,000,000 bits of byte stream data into 10 segments of 300,000 bits. The K_c, K_{pr}, K_{pc}, K_d and K_d' sequences needed in encryption passed the test successfully, and the test results of the K_d' sequence are shown in Table 1. The experimental results show that the random numbers generated by our algorithm fully conform to the international standard, and have strong randomness.

Table 1. NIST-800-22 test results.

Statistical Tests	p-Values										Result
	Seq1	Seq2	Seq3	Seq4	Seq5	Seq6	Seq7	Seq8	Seq9	Seq10	
ApproximateEntropy Text	0.8094	0.1941	0.0781	0.3518	0.4390	0.3812	0.4203	0.1690	0.1884	0.0589	Successful
BlockFrequency Text	0.9347	0.2822	0.9547	0.0925	0.6961	0.4518	0.1352	0.4160	0.3816	0.1934	Successful
CumulativeSums Text-1	0.7034	0.9290	0.7701	0.4770	0.0354	0.6270	0.4488	0.2083	0.4378	0.5493	Successful
CumulativeSums Text-2	0.8561	0.9968	0.8754	0.7377	0.0426	0.2912	0.2621	0.1019	0.3783	0.1853	Successful
FFT Text	0.9732	0.9066	0.4508	0.2911	0.4921	0.1912	0.8145	0.4508	0.0226	0.1359	Successful
Frequency Text	0.8666	0.8408	0.9040	0.4541	0.0235	0.6507	0.7674	0.1743	0.9330	0.5541	Successful
LinearComplexity Text	0.2833	0.8136	0.5262	0.2415	0.6749	0.4776	0.9849	0.2676	0.8014	0.3305	Successful
LongestRun Text	0.3615	0.2823	0.5065	0.4150	0.7894	0.7386	0.0683	0.1561	0.5800	0.2138	Successful
OverlappingTemplate Text	0.2713	0.8537	0.8457	0.6464	0.2555	0.1803	0.4144	0.9091	0.7819	0.7349	Successful
Rank Text	0.6985	0.1675	0.6198	0.2927	0.5757	0.3860	0.3147	0.8761	0.3737	0.2093	Successful
Runs Text	0.6066	0.6691	0.6771	0.2721	0.3432	0.1041	0.5789	0.7783	0.6718	0.6011	Successful
Serial Text-1	0.0096	0.8837	0.0110	0.5441	0.1669	0.0331	0.8454	0.1955	0.7045	0.6886	Successful
Serial Text-2	0.1784	0.6697	0.2170	0.5832	0.0293	0.3877	0.9621	0.4920	0.7287	0.5582	Successful

4.3. Histogram Analysis

There are three channels—R, G and B—in color images; the abscissa of the histogram containing these three channels reflects the statistical characteristics of the distribution of every pixel [45,46]. Different plain images and cipher images, as well as their relevant histograms, are shown in Figure 5. The experimental results show that the pixel values of the R, G and B channels of color cipher image are almost uniformly distributed, so the influence of statistical analysis is greatly eliminated [47,48].

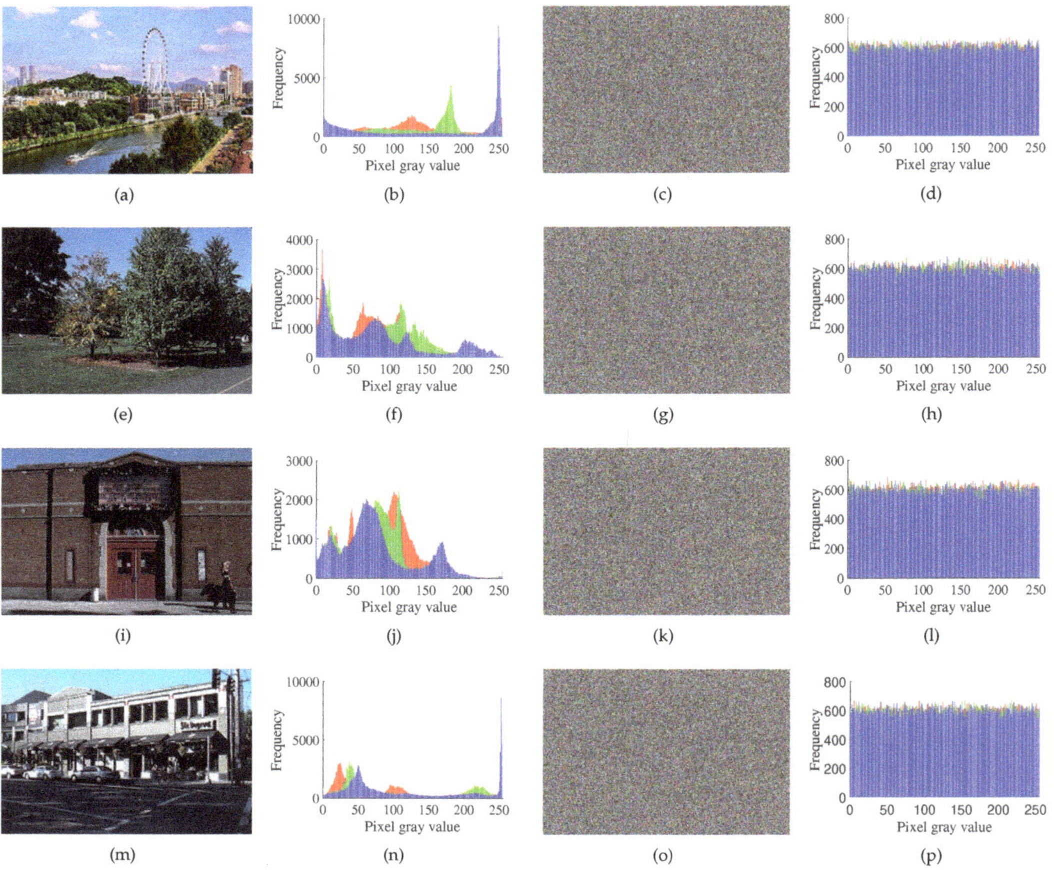

Figure 5. The histograms of images before and after encryption: (**a**) plain image of "Zhong shan"; (**b**) histogram of the plain image of "Zhong shan"; (**c**) cipher image of "Zhong shan"; (**d**) histogram of the cipher image of "Zhong shan"; (**e**) plain image of "Greenlake10"; (**f**) histogram of the plain image of "Greenlake10"; (**g**) cipher image of "Greenlake10"; (**h**) histogram of the cipher image of "Greenlake10"; (**i**) plain image of "Greenlake13"; (**j**) histogram of the plain image of "Greenlake13"; (**k**) cipher image of "Greenlake13"; (**l**) histogram of the cipher image of "Greenlake13"; (**m**) plain image of "Greenlake47"; (**n**) histogram of the plain image of "Greenlake47"; (**o**) cipher image of "Greenlake47"; (**p**) histogram of cipher image of "Greenlake47".

4.4. Correlation Analysis

For the plain image, the correlation between adjacent pixels is strong [49,50]. Gray value of a pixel tends to be close to the gray values of its adjacent pixels. Therefore, the attacker can speculate about the gray value of a pixel from the gray value of its adjacent pixels [51,52]. An encryption system with good performance should satisfy the requirement that adjacent pixels of cipher image have low correlation coefficients to each other in order to resist the statistical attack. Correlation coefficients are commonly used to measure the correlation of two pixels and the calculations of it are defined as [53,54]:

$$
\begin{cases}
E(x) = \frac{1}{N} \sum_{i=1}^{N} x_i \\
D(x) = \frac{1}{N} \sum_{i=1}^{N} (x_i - E(x))^2 \\
\mathrm{cov}(x,y) = \frac{1}{N} \sum_{i=1}^{N} (x_i - E(x))(y_i - E(y)) \\
\gamma_{xy} = \frac{\mathrm{cov}(x,y)}{\sqrt{D(x)} \times \sqrt{D(y)}}
\end{cases}
\tag{12}
$$

where the gray value of every pixel is represented by x and y, while $E(x)$ represents the mean value, $D(x)$ represents the variance, $\mathrm{cov}(x,y)$ represents the covariance and γ_{xy} represents the correlation coefficients.

The correlation coefficients before and after encryption of the selected image are shown in Table 2 where "Anti-Diag". represents the correlation coefficient in the anti-diagonal direction. Figure 6 shows the correlation of plain image and cipher image in horizontal, vertical, diagonal and anti-diagonal directions. It can be seen that there is no obvious correlation between adjacent pixels of a cipher image. Therefore, the cipher images encrypted by the algorithm designed in this paper have high security and can resist the statistical analysis [55].

Table 2. Correlation coefficients of two adjacent pixels.

Pictures	Plain Image				Cipher Image			
	Vert.	Horiz.	Diag.	Anti-Diag.	Vert.	Horiz.	Diag.	Anti-Diag.
7.1.02.tiff	0.9480	0.9429	0.9113	0.9456	−0.0021	0.0303	0.0087	−0.0002
7.1.09.tiff	0.9309	0.9654	0.9208	0.9207	−0.0083	−0.0257	−0.0354	−0.0225
5.1.12.tiff	0.9709	0.9608	0.9429	0.9403	−0.0256	−0.0035	0.0040	−0.0157
5.2.10.tiff	0.9415	0.9364	0.9032	0.9015	0.0032	0.0163	−0.0069	−0.0107

4.5. Sensitivity Analysis

Key sensitivity is an essential indicator of the security of the encryption system. It represents the difference in the decryption results when the same cipher image is decrypted with slightly different keys. For the sake of detecting the susceptibility of the scheme to the key, the first three sequences generated by the initial key are superimposed and combined into a color map, and the minimum precision of $x_1(0)$ is 10^{-15}. The initial key $x_1(0)$ is perturbed with the minimum precision to generate four new sequences, and the first three new sequences are superimposed and combined into a new color map. The two color images are differentiated to get the difference image and the histogram corresponding to the difference image. The initial key $x_2(0)$ is processed in the same way, as shown in Figure 7. By adding 10^{-3} to the initial key $x_1(0)$, four sequences are obtained through cellular neural chaos, and these four sequences are compared with the four sequences generated by no change of $x_1(0)$, as shown in Figure 8. It can be seen from Figures 7 and 8 that the encryption system designed in this paper has high security and strong sensitivity to keys, which increases the difficulty for attackers to decipher the cipher image.

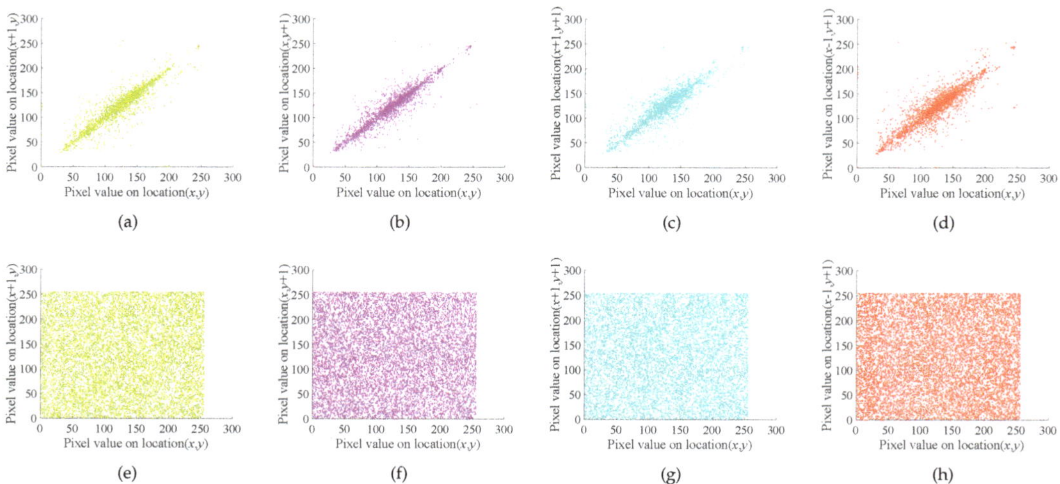

Figure 6. Correlation coefficients distribution map of plain image and cipher image of "7.1.02.tiff": (**a**) "7.1.02.tiff" plain image horizontal correlation; (**b**) "7.1.02.tiff" plain image is vertical correlation; (**c**) "7.1.02.tiff" plain image diagonal correlation; (**d**) "7.1.02.tiff" plain image against angular direction correlation; (**e**) "7.1.02.tiff" cipher image horizontal correlation; (**f**) "7.1.02.tiff" cipher image vertical correlation; (**g**) "7.1.02.tiff" cipher image diagonal correlation; (**h**) "7.1.02.tiff" cipher image inverse diagonal correlation.

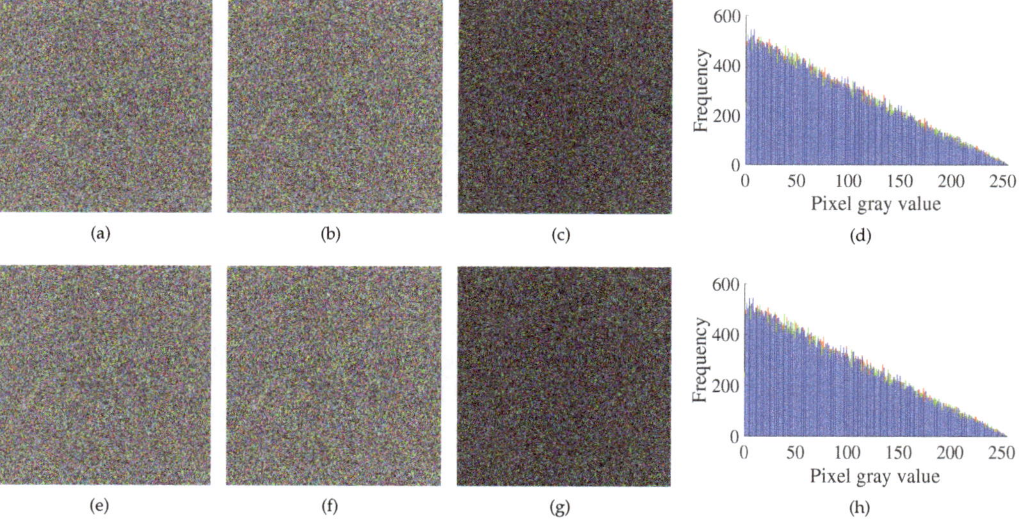

Figure 7. The key sensitivity test: (**a**) $x_1(0), x_2(0), x_3(0), x_4(0)$; (**b**) $x_1(0) + 10^{-15}, x_2(0), x_3(0), x_4(0)$; (**c**) Difference image after key perturbation; (**d**) Difference histogram after key perturbation; (**e**) $x_1(0), x_2(0), x_3(0), x_4(0)$; (**f**) $x_1(0), x_2(0) + 10^{-15}, x_3(0), x_4(0)$; (**g**) Difference image after key perturbation; (**h**) Difference histogram after key perturbation.

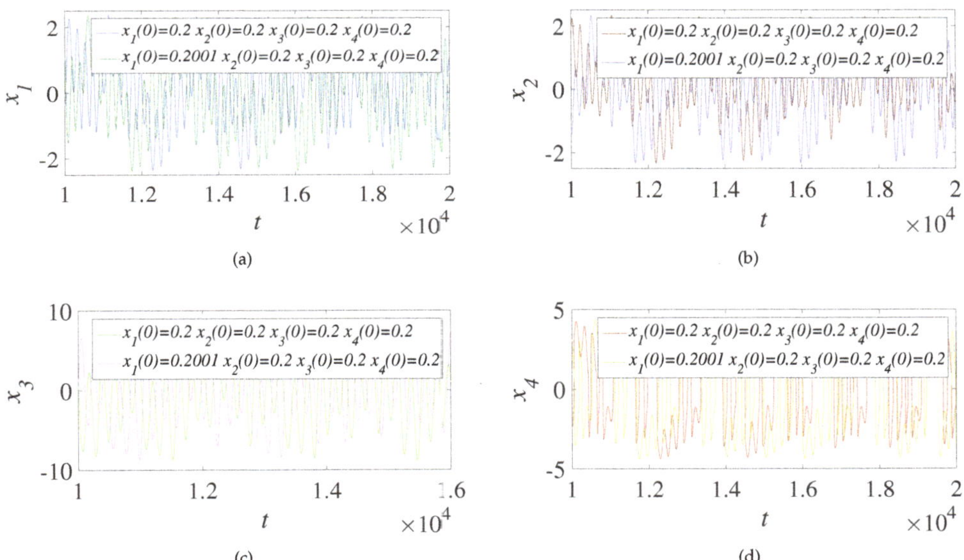

Figure 8. Comparison of four sequences (**a–d**) before and after key $x_1(0)$ perturbation.

Plaintext sensitivity is also one of the important indexes of encryption system security, which indicates the ability of encryption system to resist the differential attack. A secure encryption system should be highly sensitive to plain image. The Number of Pixels Change Rate (NPCR) and Unified Average Changing Intensity (UACI) can be used to represent the difference between two plain images with one pixel difference. The calculation formula is [56]:

$$\begin{cases} NPCR = \frac{1}{H \times W} \times \sum\limits_{i=1}^{H} \sum\limits_{j=1}^{W} D(i,j) \times 100\% \\ UACI = \frac{1}{H \times W} \times \sum\limits_{i=1}^{H} \sum\limits_{j=1}^{W} \frac{|v_1(i,j) - v_2(i,j)|}{255} \times 100\% \end{cases} \tag{13}$$

where $D(i,j) = \begin{cases} 0, v_1(i,j) = v_2(i,j) \\ 1, v_1(i,j) \neq v_2(i,j) \end{cases}$. $v_1(i,j)$ and $v_2(i,j)$ denote the pixel values at positions v_1 and v_2. For a digital image with a gray level of 256, 99.6094% and 33.4635% are ideal values of the NPCR and UACI, respectively.

Firstly, select a pixel from the "Lena" gray image randomly so that we can obtain a new image by changing its pixel value. Then, the two gray images which differ by only one pixel are each encrypted to obtain two ciphertext images. Finally, the NPCR and UACI values of the two encrypted images are obtained and the above operations will be repeated 50 times to obtain 50 groups of NPCR and UACI values. The NPCR and UACI average values of the gray images are shown in Table 3.

Table 3. NPCR and UACI.

Pictures	NPCR (99.6094%)	UACI (33.4635%)
1.2.04.tiff	99.6093%	33.5974%
1.2.07.tiff	99.6078%	33.5580%
1.2.08.tiff	99.6154%	33.5209%
5.1.11.tiff	99.5544%	33.4018%

The NPCR and UACI values obtained each time are shown in Figure 9. The NPCR and UACI average values are very close to the theoretical value. Therefore, the encryption system designed in this paper is extremely sensitive to both plain images and keys. The encryption algorithm designed in this study is safer and can resist the differential attack.

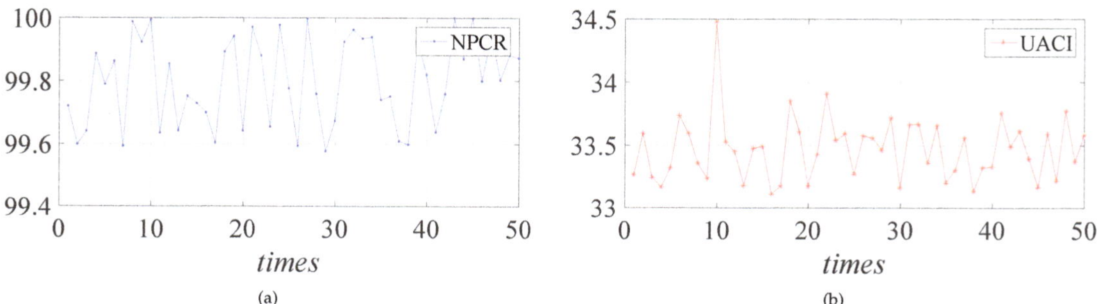

Figure 9. NPCR (**a**) and UACI (**b**).

4.6. Information Entropy Analysis

The degree of the randomness of the system can be expressed by information entropy. The information entropy of the image is positively correlated with the encryption effect. The larger the information entropy is, the better effect the encryption will have. The formula of information entropy is defined as [57]:

$$H(n) = -\sum_{i=0}^{G-1} -1 P(n_i) \log_2 P(n_i) \tag{14}$$

where G represents the number of gray level values of the image and $P(n_i)$ the frequency of pixels with gray value i. The range of gray value of an image with a gray level of 256 is $[0, 255]$, and 8 is its ideal information entropy. When the value of information entropy is closer to 8, the image encryption has better effect [58].

Table 4 shows the information entropy before and after image encryption. The information entropy of the cipher image is very close to the theoretical value of information entropy. It is proven that the pixel value distribution of the cipher image is highly random and the encryption effect is better. Therefore, the algorithm can effectively resist the information entropy attack [33].

Table 4. Information entropy of the plain image and cipher image.

Pictures	Plain Image	Cipher Image
7.1.02.tiff	4.0045	7.9993
5.1.11.tiff	6.4523	7.9970
5.1.12.tiff	6.7057	7.9972
5.2.10.tiff	5.7056	7.9992

4.7. Psnr and Ssim

Peak Signal-to-Noise Ratio (PSNR) and Structural SIMilarity (SSIM) are often used to reflect the encryption quality. PSNR is essentially the same as the Mean Square Error (MSE) and can be obtained by MSE. The calculation formula is [59]:

$$\begin{cases} MSE = \frac{1}{H \times W} \sum\limits_{i=1}^{H} \sum\limits_{j=1}^{W} \left(P(i,j) - C(i,j)\right)^2 \\ PSNR = 10 \times \log_{10}\left(\frac{Q^2}{MSE}\right) \end{cases} \tag{15}$$

where the height and width of the image are represented by H and W, respectively, the pixel level of the image is represented by Q, the plain image pixels are represented by $P(i,j)$, and the cipher image pixels are represented by $C(i,j)$. SSIM is defined as [59]:

$$SSIM(p,c) = \frac{\left(2\mu_p\mu_c + (0.01L)^2\right)\left(2\sigma_{pc} + (0.03L)^2\right)}{\left(u_p^2 + u_c^2 + (0.01L)^2\right)\left(\sigma_p^2 + \sigma_c^2 + (0.03L)^2\right)} \tag{16}$$

where the average values of the plain image P and the cipher image C are denoted by u_p and u_c, respectively. The variance of the plain image and the cipher image denoted by σ_p^2 and σ_c^2 indicates that the covariance of the plain image and the cipher image represented by σ_{pc}. $(0.01L)^2$ and $(0.03L)^2$ are used as constant numbers to maintain stability. L represents the dynamic range of pixel values.

The range of SSIM is from -1 to 1. When the two images are the same, SSIM is 1. The smaller the PSNR and SSIM are, the better the encryption quality is. Tables 5 and 6 show the encryption quality of the proposed scheme and the classic encryption schemes in recent years.

Table 5. PSNR of cipher image with different algorithms.

Pictures	This Paper	Ref. [1]	Ref. [60]	Ref. [28]
7.1.02.tiff	8.9518	9.1033	8.9731	8.9801
5.2.10.tiff	8.7620	8.7684	8.7660	8.7621
5.1.13.tiff	4.9032	4.9585	4.9168	4.9141
5.2.08.tiff	9.6225	9.6389	9.6378	9.6198

Table 6. SSIM of cipher image based on different algorithms.

Pictures	This Paper	Ref. [1]	Ref. [60]	Ref. [28]
7.1.02.tiff	0.0102	0.0108	0.0103	0.0109
5.1.11.tiff	0.0101	0.0099	0.0101	0.0109
5.2.10.tiff	0.0087	0.0098	0.0100	0.0091
5.1.13.tiff	0.0037	0.0057	0.0085	0.0067

The experimental results show that the PSNR and SSIM values obtained by the proposed algorithm are lower than those of other proposed approaches. Therefore, this encryption scheme has certain advantages, and the image encryption quality is high.

4.8. Robust Noise Analysis

Robustness means that the system still has certain performance under interference or at random. Image robustness refers to the fact that the image still has a certain degree of fidelity after undergoing various signal processing or attacks. The image can still be recognized, with low distortion. Add 20% salt-and-pepper noise and 80 × 80 occlusion noise to the cipher image "Figure 5a". The experimental results are shown in the figure below [34,60,61].

It can be seen from Figure 10 that the decrypted images can still be easily identified with high fidelity after noise is added to the cipher image, which indicates the robustness of the image encryption system that can resist noise attacks.

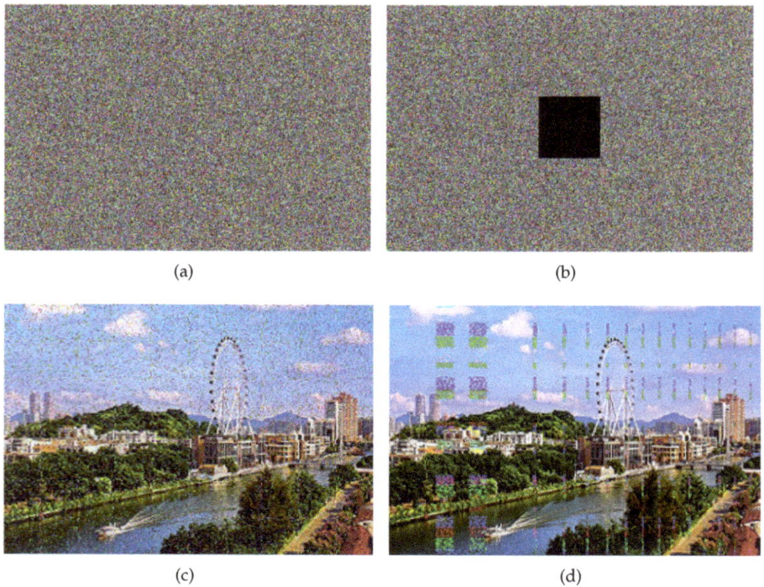

Figure 10. (a) Salt-and-pepper noise cipher image; (b) Occlusion noise cipher image; (c) Decryption of cipher image with salt-and-pepper noise; (d) Decryption of cipher image with occlusion noise.

5. Conclusions

This paper proposes a security-enhanced image communication scheme based on CNN under the cryptanalysis. First, the complex characteristics of CNN are used to generate some sequences. Then, a plain image and these CNN-based sequences are confused, permuted and diffused to get the cipher image. Utilizing the complex dynamics of CNN can effectively enhance the confusion, diffusion and avalanche of encryption. Theoretical analysis and experimental results both demonstrate its safety performance. From the perspective of cryptanalysis, the structure of an image cipher can effectively resist various common attacks. Therefore, the image communication scheme based on CNN proposed in this paper is a competitive security technology method.

Author Contributions: Methodology, H.W.; Project administration, H.W. and C.Z.; Software, J.X., R.C. and D.S.; Supervision, C.Z.; Validation, J.X., Y.L. (Yunlong Liao), R.C., L.W., Y.S., Q.L., Z.L., S.Z., Y.L. (Yuxuan Liu), A.H., T.L., C.C. and J.W.; Writing—original draft, J.X.; Writing—review & editing, H.W. All authors have read and agreed to the published version of the manuscript.

Funding: This work was supported in part by the National Science Foundation of China under Grant 62071088 and Grant 61571092; in part by the Project for Innovation Team of Guangdong University under Grant 2018KCXTD033; in part by the Project for Zhongshan Social Public Welfare Science and Technology under Grant 2019B2007; in part by the Science and Technology Projects of Guangdong Province under Grant 2021A0101180005; in part by the Research Project for Talent of UESTC Zhongshan Institute under Grant 418YKQN07 and Grant 419YKQN23; and in part by the Opening Project Guangdong Province Key Laboratory of Information Security Technology under Grant 2020B1212060078.

Data Availability Statement: Not applicable.

Conflicts of Interest: The authors declare no conflict of interest.

References

1.	Chunyan, S.; Yulong, Q. A Novel Image Encryption Algorithm Based on DNA Encoding and Spatiotemporal Chaos. *Entropy* **2015**, *17*, 6954–6968.
2.	Gopalakrishnan, T.; Ramakrishnan, S. Chaotic Image Encryption with Hash Keying as Key Generator. *IETE J. Res.* **2017**, *63*, 172–187. [CrossRef]
3.	Li, A.; Belazi, A.; Kharbech, S.; Talha, M.; Xiang, W. Fourth Order MCA and Chaos-Based Image Encryption Scheme. *IEEE Access* **2019**, *7*, 66395–66409. [CrossRef]
4.	Chai, X.; Chen, Y.; Broyde, L. A novel chaos-based image encryption algorithm using DNA sequence operations. *Opt. Lasers Eng.* **2017**, *88*, 197–213. [CrossRef]
5.	Kalpana, M.; Ratnavelu, K.; Balasubramaniam, P.; Kamali, M. Synchronization of chaotic-type delayed neural networks and its application. *Nonlinear Dyn.* **2018**, *93*, 543–555. [CrossRef]
6.	Li, M.; Guo, Y.; Huang, J.; Li, Y. Cryptanalysis of a chaotic image encryption scheme based on permutation-diffusion structure. *Signal Process. Image Commun.* **2018**, *62*, 164–172. [CrossRef]
7.	Zhang, X.; Wang, L.; Zhou, Z.; Niu, Y. A chaos-based image encryption technique utilizing hilbert curves and h-fractals. *IEEE Access* **2019**, *7*, 74734–74746. [CrossRef]
8.	Xie, E.Y.; Li, C.; Yu, S.; Lu, J. On the cryptanalysis of Fridrich's chaotic image encryption scheme. *Signal Process.* **2017**, *132*, 150–154. [CrossRef]
9.	Panna, B.; Kumar, S.; Jha, R.K. Image Encryption Based on Block-wise Fractional Fourier Transform with Wavelet Transform. *IETE Tech. Rev.* **2019**, *36*, 600–613. [CrossRef]
10.	Noshadian, S.; Ebrahimzade, A.; Kazemitabar, S. Optimizing chaos based image encryption. *Multimed. Tools Appl.* **2018**, *77*, 25569–25590. [CrossRef]
11.	Musanna, F.; Dangwal, D.; Kumar, S.; Malik, V. A chaos-based image encryption algorithm based on multiresolution singular value decomposition and a symmetric attractor. *Imaging Sci. J.* **2020**, *68*, 24–40. [CrossRef]
12.	Li, Y.; Wang, C.; Chen, H. A hyper-chaos-based image encryption algorithm using pixel-level permutation and bit-level permutation. *Opt. Lasers Eng.* **2017**, *90*, 238–246. [CrossRef]
13.	El-Khamy, S.; Korany, N.; ElSherif, M. A security enhanced robust audio steganography algorithm for image hiding using sample comparison in discrete wavelet transform domain and RSA encryption. *Multimed. Tools Appl.* **2017**, *76*, 24091–24106. [CrossRef]
14.	Feng, W.; Zhang, J. Cryptanalzing a Novel Hyper-Chaotic Image Encryption Scheme Based on Pixel-Level Filtering and DNA-Level Diffusion. *IEEE Access* **2020**, *8*, 209471–209482. [CrossRef]
15.	Tsafack, N.; Sankar, S.; Abd-El-Atty, B.; Kengne, J.; Jithin, K.C.; Belazi, A.; Mehmood, I.; Bashir, A.; Song, O.Y.; Abd El-Latif, A. A New Chaotic Map With Dynamic Analysis and Encryption Application in Internet of Health Things. *IEEE Access* **2020**, *8*, 137731–137744. [CrossRef]
16.	Wang, N.; Li, Q.; Abd El-Latif, A.; Peng, J.; Yan, X.; Niu, X. A novel template protection scheme for multibiometrics based on fuzzy commitment and chaotic system. *Signal Image Video Process.* **2015**, *9*, 99–109. [CrossRef]

17. Abd EL-Latif, A.A.; Abd-El-Atty, B.; Abou-Nassar, E.M.; Venegas-Andraca, S.E. Controlled alternate quantum walks based privacy preserving healthcare images in Internet of Things. *Opt. Laser Technol.* **2020**, *124*, 105942. [CrossRef]
18. Wu, T.; Zhang, C.; Chen, Y.; Cui, M.; Huang, H.; Zhang, Z.; Wen, H.; Zhao, X.; Qiu, K. Compressive sensing chaotic encryption algorithms for OFDM-PON data transmission. *Opt. Express* **2021**, *29*, 3669–3684. [CrossRef]
19. Hua, Z.; Zhou, Y.; Huang, H. Cosine-transform-based chaotic system for image encryption. *Inf. Sci.* **2019**, *480*, 403–419. [CrossRef]
20. Liu, Y.; Zhang, J.; Han, D.; Wu, P.; Moon, Y.S. A multidimensional chaotic image encryption algorithm based on the region of interest. *Multimed. Tools Appl.* **2020**, *79*, 17669–17705. [CrossRef]
21. Li, C.; Luo, G.; Qin, K.; Li, C. An image encryption scheme based on chaotic tent map. *Nonlinear Dyn.* **2017**, *87*, 127–133. [CrossRef]
22. Ozkaynak, F. Brief review on application of nonlinear dynamics in image encryption. *Nonlinear Dyn.* **2018**, *92*, 305–313. [CrossRef]
23. Ouannas, A.; Karouma, A.; Grassi, G.; Pham, V.; Luong, V.S. A novel secure communications scheme based on chaotic modulation, recursive encryption and chaotic masking. *Alex. Eng. J.* **2021**, *60*, 1873–1884. [CrossRef]
24. Ratnavelu, K.; Kalpana, M.; Balasubramaniam, P.; Wong, K.; Raveendran, P. Image encryption method based on chaotic fuzzy cellular neural networks. *Signal Process.* **2017**, *140*, 87–96. [CrossRef]
25. Cheng, G.; Wang, C.; Xu, C. A novel hyper-chaotic image encryption scheme based on quantum genetic algorithm and compressive sensing. *Multimed. Tools Appl.* **2020**, *79*, 29243–29263. [CrossRef]
26. Roy, A.; Misra, A.; Banerjee, S. Chaos-based image encryption using vertical-cavity surface-emitting lasers. *Optik* **2019**, *176*, 119–131. [CrossRef]
27. Li, C.; Zhang, Y.; Xie, E.Y. When an attacker meets a cipher-image in 2018: A year in review. *J. Inf. Secur. Appl.* **2019**, *48*, 102361. [CrossRef]
28. Xu, L.; Li, Z.; Li, J.; Hua, W. A novel bit-level image encryption algorithm based on chaotic maps. *Opt. Lasers Eng.* **2016**, *78*, 17–25. [CrossRef]
29. He, C.; Ming, K.; Wang, Y.; Wang, Z. A Deep Learning Based Attack for The Chaos-based Image Encryption. *arXiv* **2019**, arXiv:1907.12245.
30. Li, G.; Yang, B.; Pu, Y.; Xu, W. Synchronization of generalized using to image encryption. *Int. J. Pattern Recognit. Artif. Intell.* **2017**, *31*, 1754009. [CrossRef]
31. Norouzi, B.; Mirzakuchaki, S. An image encryption algorithm based on DNA sequence operations and cellular neural network. *Multimed. Tools Appl.* **2017**, *76*, 13681–13701. [CrossRef]
32. Zhang, L.; Zhang, X. Multiple-image encryption algorithm based on bit planes and chaos. *Multimed. Tools Appl.* **2020**, *79*, 20753–20771. [CrossRef]
33. Li, M.; Fan, H.; Xiang, Y.; Li, Y.; Zhang, Y. Cryptanalysis and Improvement of a Chaotic Image Encryption by First-Order Time-Delay System. *IEEE Multimed.* **2018**, *25*, 92–101. [CrossRef]
34. Zhang, X.; Liu, W.; Dundar, M.; Badve, S.; Zhang, S. Towards large-scale histopathological image analysis: Hashing-based image retrieval. *IEEE Trans. Med. Imaging* **2015**, *34*, 496–506. [CrossRef]
35. Zhang, X.; Wang, C.; Zheng, Z. An efficient chaotic image encryption algorithm based on self-adaptive model and feedback mechanism. *KSII Trans. Internet Inf. Syst.* **2017**, *11*, 1785–1801.
36. Musanna, F.; Kumar, S. A novel fractional order chaos-based image encryption using Fisher Yates algorithm and 3-D cat map. *Multimed. Tools Appl.* **2019**, *78*, 14867–14895. [CrossRef]
37. Wang, J.; Zhi, X.; Chai, X.; Lu, Y. Chaos-based image encryption strategy based on random number embedding and DNA-level self-adaptive permutation and diffusion. *Multimed. Tools Appl.* **2021**, *80*, 16087–16122. [CrossRef]
38. Lin, M.; Long, F.; Guo, L. Grayscale image encryption based on Latin square and cellular neural network. In Proceedings of the 2016 Chinese Control and Decision Conference (CCDC), Yinchuan, China, 28–30 May 2016; pp. 2787–2793.
39. Alawida, M.; Samsudin, A.; Sen Teh, J.; Alkhawaldeh, R.S. A new hybrid digital chaotic system with applications in image encryption. *Signal Process.* **2019**, *160*, 45–58. [CrossRef]
40. Preishuber, M.; Huetter, T.; Katzenbeisser, S.; Uhl, A. Depreciating Motivation and Empirical Security Analysis of Chaos-Based Image and Video Encryption. *IEEE Trans. Inf. Forensics Secur.* **2018**, *13*, 2137–2150. [CrossRef]
41. The USC-SIPI Image Database. Available online: http://sipi.usc.edu/database (accessed on 23 June 2021).
42. The Ground Truth Database. Available online: http://www.cs.washington.edu/research/imagedatabase (accessed on 23 June 2021).
43. Chen, G.; Mao, Y.; Chui, C. A symmetric image encryption scheme based on 3D chaotic cat maps. *Chaos Solitons Fractals* **2004**, *21*, 749–761. [CrossRef]
44. Wen, H.; Yu, S.; Luuml, J. Breaking an Image Encryption Algorithm Based on DNA Encoding and Spatiotemporal Chaos. *Entropy* **2019**, *21*, 246. [CrossRef] [PubMed]
45. Sasikaladevi, N.; Geetha, K.; Sriharshini, K.; Durga Aruna, M. RADIANT - hybrid multilayered chaotic image encryption system for color images. *Multimed. Tools Appl.* **2019**, *78*, 11675–11700. [CrossRef]
46. Wen, H.; Zhang, C.; Huang, L.; Ke, J.; Xiong, D. Security Analysis of a Color Image Encryption Algorithm Using a Fractional-Order Chaos. *Entropy* **2021**, *23*, 258. [CrossRef] [PubMed]
47. Khan, M.; Ahmad, J.; Javaid, Q.; Saqib, N. An efficient and secure partial image encryption for wireless multimedia sensor networks using discrete wavelet transform, chaotic maps and substitution box. *J. Mod. Opt.* **2017**, *64*, 531–540. [CrossRef]

48. Weng, H.; Zhang, C.; Chen, P.; Chen, R.; Xu, J.; Liao, Y.; Liang, Z.; Shen, D.; Zhou, L.; Ke, J. A Quantum Chaotic Image Cryptosystem and Its Application in IoT Secure Communication. *IEEE Access* **2021**, *9*, 20481–20492.
49. Faragallah, O.S.; Afifi, A.; ElShafai, W.; ElSayed, H.S.; Naeem, E.A.; Alzain, M.A.; AlAmri, J.F.; Soh, B.; ElSamie, F.E.A. Investigation of Chaotic Image Encryption in Spatial and FrFT Domains for Cybersecurity Applications. *IEEE Access* **2020**, *8*, 42491–42503. [CrossRef]
50. Wu, T.; Zhang, C.; Huang, H.; Zhang, Z.; Wei, H.; Wen, H.; Qiu, K. Security Improvement for OFDM-PON via DNA Extension Code and Chaotic Systems. *IEEE Access* **2020**, *8*, 75119–75126. [CrossRef]
51. Mani, P.; Rajan, R.; Shanmugam, L.; Hoon Joo, Y. Adaptive control for fractional order induced chaotic fuzzy cellular neural networks and its application to image encryption. *Inf. Sci.* **2019**, *491*, 74–89. [CrossRef]
52. Meng, L.; Yin, S.; Zhao, C.; Li, H.; Sun, Y. An improved image encryption algorithm based on chaotic mapping and discrete wavelet transform domain. *Int. J. Netw. Secur.* **2020**, *22*, 155–160.
53. Luo, Y.; Yu, J.; Lai, W.; Liu, L. A novel chaotic image encryption algorithm based on improved baker map and logistic map. *Multimed. Tools Appl.* **2019**, *78*, 22023–22043. [CrossRef]
54. Wen, H.; Yu, S. Cryptanalysis of an image encryption cryptosystem based on binary bit planes extraction and multiple chaotic maps. *Eur. Phys. J. Plus* **2019**, *134*, 337. [CrossRef]
55. Pan, X.; Wu, J.; Li, Z.; Zhang, C.; Deng, C.; Zhang, Z.; Wen, H.; Gao, Q.; Yang, J.; Yi, Z.; et al. Laguerre-Gaussian mode purity of Gaussian vortex beams. *Optik* **2021**, *230*, 166320. [CrossRef]
56. Yan, X.; Wang, X.; Xian, Y. Chaotic Image Encryption Algorithm Based on Fractional Order Scrambling Wavelet Transform and 3D Cyclic Displacement Operation. *IEEE Access* **2020**, *8*, 208718–208736. [CrossRef]
57. Li, C.; Lin, D.; Feng, B.; Lu, J.; Hao, F. Cryptanalysis of a Chaotic Image Encryption Algorithm Based on Information Entropy. *IEEE J. Transl. Eng. Health Med.* **2018**, *6*, 75834–75842. [CrossRef]
58. Joshi, A.B.; Kumar, D.; Mishra, D.; Guleria, V. Colour-image encryption based on 2D discrete wavelet transform and 3D logistic chaotic map. *J. Mod. Opt.* **2020**, *67*, 933–949. [CrossRef]
59. Li, G.; Wang, L. Double chaotic image encryption algorithm based on optimal sequence solution and fractional transform. *Vis. Comput.* **2019**, *35*, 1267–1277. [CrossRef]
60. Yin, Q.; Wang, C. A New Chaotic Image Encryption Scheme Using Breadth-First Search and Dynamic Diffusion. *Int. J. Bifurc. Chaos* **2018**, *28*, 1850047. [CrossRef]
61. Lai, H.; Yan, P.; Shu, X.; Wei, Y.; Yan, S. Instance-aware hashing for multi-label image retrieval. *IEEE Trans. Image Process.* **2016**, *25*, 2469–2479. [CrossRef]

Article

Modified Hilbert Curve for Rectangles and Cuboids and Its Application in Entropy Coding for Image and Video Compression

Yibiao Rong [1,2,3], Xia Zhang [1,2,3] and Jianyu Lin [1,2,3,*]

[1] Department of Electronic Engineering, Shantou University, Shantou 515063, China; ybrong@stu.edu.cn (Y.R.); 18xzhang8@stu.edu.cn (X Z.)

[2] Guangdong Provincial Key Laboratory of Digital Signal and Image Processing, Shantou University, Shantou 515063, China

[3] Key Laboratory of Intelligent Manufacturing Technology, Shantou University, Ministry of Education, Shantou 515063, China

* Correspondence: jianyulin@stu.edu.cn

Abstract: In our previous work, by combining the Hilbert scan with the symbol grouping method, efficient run-length-based entropy coding was developed, and high-efficiency image compression algorithms based on the entropy coding were obtained. However, the 2-D Hilbert curves, which are a critical part of the above-mentioned entropy coding, are defined on squares with the side length being the powers of 2, i.e., 2^n, while a subband is normally a rectangle of arbitrary sizes. It is not straightforward to modify the Hilbert curve from squares of side lengths of 2^n to an arbitrary rectangle. In this short article, we provide the details of constructing the modified 2-D Hilbert curve of arbitrary rectangle sizes. Furthermore, we extend the method from a 2-D rectangle to a 3-D cuboid. The 3-D modified Hilbert curves are used in a novel 3-D transform video compression algorithm that employs the run-length-based entropy coding. Additionally, the modified 2-D and 3-D Hilbert curves introduced in this short article could be useful for some unknown applications in the future.

Keywords: scan route; Hilbert curve; run-length-based entropy coding; image and video compression

Citation: Rong, Y.; Zhang, X.; Lin, J. Modified Hilbert Curve for Rectangles and Cuboids and Its Application in Entropy Coding for Image and Video Compression. *Entropy* **2021**, *23*, 836. https://doi.org/10.3390/e23070836

Academic Editor: Amelia Carolina Sparavigna

Received: 28 May 2021
Accepted: 23 June 2021
Published: 29 June 2021

Publisher's Note: MDPI stays neutral with regard to jurisdictional claims in published maps and institutional affiliations.

1. Introduction

Entropy coding plays a critical role in data compression, such as image and video compression, etc. Two commonly used algorithms for entropy coding are Huffman coding [1] and arithmetic coding [2]. In terms of compression efficiency, arithmetic coding is preferred. However, arithmetic coding has a higher computational complexity because it requires multiplication and division during the coding process. To resolve the complexity issue, approximations are used in binary arithmetic coding algorithms, such as [3–5], etc. These binary arithmetic coding algorithms are practical algorithms because the coding of a multiple symbol source can always be converted to coding of a sequence of binary symbol sources. For example, in image compression, the bit-plane coding method [6] and the symbol grouping coding method [7,8] eventually convert the quantized coefficients to binaries to code.

Although the existing binary arithmetic coding algorithms solved the computational complexity issue, it is still not computationally efficient in extremely low entropy conditions because arithmetic coding algorithms encode symbols one by one. For example, to code a binary source with the probabilities $p = 0.999$ for the symbol "0" and $1 - p = 0.001$ for the symbol "1", arithmetic coding needs to code 999 "0"s on average before it codes a "1". On the other hand, run-length-based entropy coding [7,8] is much more computationally efficient for low-entropy coding situations, as it does not need to code the "0"s one by one. Note, low-entropy sources are very common in compression. For example, in subband image compression, most of the quantized coefficients in a subband are zeros. The positions of the non-zero coefficients in a subband normally form an extremely low-entropy source.

For non-stationary binary sources, the binary arithmetic coding algorithms use probability estimators to adapt to the probability variations; whereas the run-length-based entropy coding uses the symbol grouping method to handle non-stationary binary sources [7]. For coding 2-dimensional (and higher-dimensional) subband coefficient arrays, binary arithmetic coding can estimate the probabilities from the coded adjacent coefficients in different spatial directions (context modeling). However, for the run-length-based binary entropy coding, the 2-D coefficient array needs to be scanned into a 1-D array before the run-length coding can be performed. As a result, for run-length-based binary entropy coding, exploiting probability estimations in different spatial directions on the 2-D array before scanning is very difficult. Thus, to achieving coding efficiency, variations in the original 2-D signal need to be maximumly kept into the scanned 1-D array, which requires that nearby elements in the 2-D array are still nearby in the 1-D scanned array. Ideally, adjacent elements in the 2-D array are required to be adjacent elements in the 1-D scanned array, which is impossible, as can be easily shown. However, different scan routes lead to different scatterings of the 2-D nearby elements. Thus, scan routes with small scattering are desired. The Hilbert curve [9,10] is such a route.

Figure 1 shows a 2-D Hilbert curve. As can be seen, the Hilbert curve tries to keep the $2^k \times 2^k$ ($k = 1, 2, \ldots$) elements in the 2-D array together in the scanned 1-D array. In fact, the locality-preserving feature of the Hilbert curve has been extensively studied [10–13]. Thus, using the Hilbert curve scan route, the variations within a 2-D subband coefficient array are greatly kept into the 1-D scanned coefficient array. Indeed, combining the Hilbert scan with the symbol grouping method, an efficient entropy coding was achieved, and high-efficiency image compression algorithms were obtained [7,8]. However, 2-D Hilbert curves are defined on a square of sizes $2^i \times 2^i$ ($i = 1, 2, \ldots$) [9,14]. In other words, not only the array shape is a square, but also the side length of the square can only be the powers of 2, i.e., 2^i. Yet, a subband is normally a rectangle of arbitrary sizes. It is not straightforward to modify the Hilbert curve from the $2^i \times 2^i$ squares to an arbitrary rectangle. In [7,8], details of this modification are not provided.

In this short article, we provide the details of constructing the modified 2-D Hilbert curve of arbitrary rectangle sizes. Furthermore, we extend the method from a 2-D rectangle to a 3-D cuboid. The entropy coding in the 3-D transform video compression algorithm introduced in [15] uses the 3-D Hilbert curve. Test results show that the algorithm is promising. However, the original 3-D Hilbert curve is defined on a cube of side length of power of 2, i.e., size $2^i \times 2^i \times 2^i$ ($i = 1, 2, \ldots$) [16]. Because the 3-D modified Hilbert curves for cuboids were not available at the time, videos were cropped to the size of 1024×1024 for testing the algorithm prototype in [15]. The extension from a 2-D rectangle to a 3-D cuboid makes the prototype proposed in [15] a practical video compression algorithm that accommodates arbitrary rectangle video sizes. Further, Hilbert curves have been widely used in many applications, such as image data encryption, query, and retrieval [17,18], etc. Extending the original Hilbert curves to arbitrary size rectangles and cuboids could be useful for some unknown applications in the future.

2. Two- and Three-Dimensional Modified Hilbert Curves

2.1. The 2-D Modified Hilbert Curve

The original 2-D Hilbert curve connects a square array of the size of $2^i \times 2^i$. We denote it as the ith order Hilbert curve. Hilbert curves of orders $i = 1$, $i = 2$, and $i = 3$ are respectively shown in Figure 1a–c. There is an important property of the Hilbert curve. As can be seen from Figure 1a–c, the starting and the ending points (green and red points in the graphs) are always on one side of the $2^i \times 2^i$ square. Since the starting and the ending points are at the ends of one side of the Hilbert curve square, a Hilbert curve of any order i can easily be represented by a square with labeled starting and ending points like Figure 1d, when the internal structure does not need to be shown.

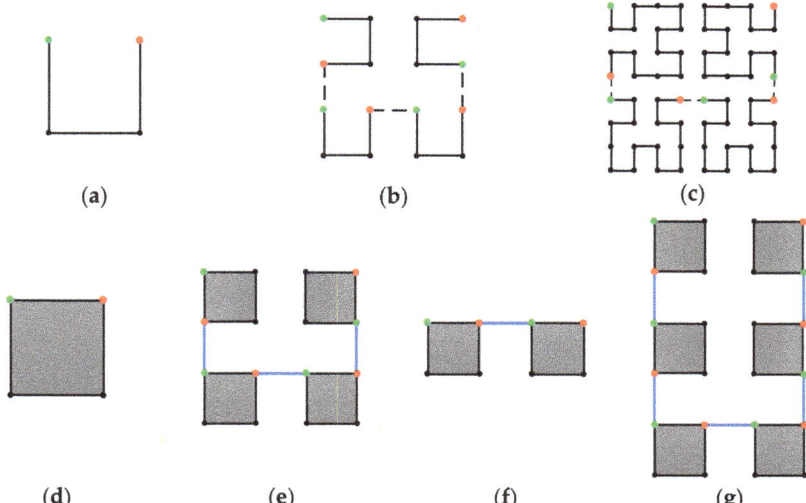

Figure 1. 2-D Hilbert curve properties. (**a–c**) are, respectively, the 1st order, the 2nd order, and the 3rd order Hilbert curves. (**d**) A simple notation to represent the ith order 2-D Hilbert curve. (**e**) Construction of the $(i-1)^{\text{th}}$ order Hilbert curve from the ith order Hilbert curve. (**f**) Reduction of 2^i points on the height H for the $(i+1)^{\text{th}}$ order Hilbert curve. (**g**) Increasing of 2^i points on the height H for the $(i+1)^{\text{th}}$ order Hilbert curve.

Now, we can show that an $(i+1)^{\text{th}}$ order Hilbert curve can be easily constructed from four i^{th} order Hilbert curves. First, replace the 4 points in the 1st order Hilbert curve in Figure 1a with four i^{th} order Hilbert curves whose starting and ending points are arranged as indicated in Figure 1e; then, connect the ending points and starting points of the ith Hilbert curves orderly as indicated in Figure 1e, and the $(i+1)^{\text{th}}$ order Hilbert curve is constructed. With the 1st order Hilbert curve provided by Figure 1a and the method of constructing the $(i+1)^{\text{th}}$ order Hilbert curve from the ith Hilbert curve, we can construct the Hilbert curve of any order by mathematical induction.

Now, our task is to construct a scanning route that is close to the Hilbert curve for a 2-D array of size $W \times H$, where W and H are the element numbers along the vertical and horizontal direction, respectively.

The basic idea is to divide the $W \times H$ rectangle array into N small square arrays of the size $2^{i_n} \times 2^{i_n}$, $n = 1, 2, \ldots, N$, and $i_1 \geq i_2 \geq \ldots \geq i_N$. For example, a 12×8 rectangle array can be divided into three $2^{i_n} \times 2^{i_n}$ square arrays with $i_1 = 3$ and $i_2 = i_3 = 2$. Apparently, within each square, an $(i_n)^{\text{th}}$ order Hilbert curve can be easily constructed as shown in Figure 2a. By appropriately arranging the directions of each Hilbert curve, the 3 Hilbert curves are connected to form the desired route, as shown in Figure 2b.

To form a route close to the Hilbert curve, there are two requirements. First, the largest i_n needs to be selected in order, i.e., select the largest i_1 first, then the largest $i_2, \ldots,$ etc. Without this restriction, the constructed curves may deviate from the Hilbert curve significantly. As an extreme example, for the $12 \times 8 = 96$ points in Figure 2, one can simply choose $i_1 = i_2 = \ldots = i_{96} = 0$, i.e., use the 96 points as 96 small square arrays. In this case, there are too many possible routes. Most of them are not close to the Hilbert curve at all, for example, the raster scan. Secondly, the ending point of the Hilbert curve in the nth square must be adjacent to the starting point of the Hilbert curve of the $(n+1)^{\text{th}}$ square, like the example shown in Figure 2b. For design convenience, the first requirement may not be satisfied strictly sometimes. However, the second requirement must be satisfied.

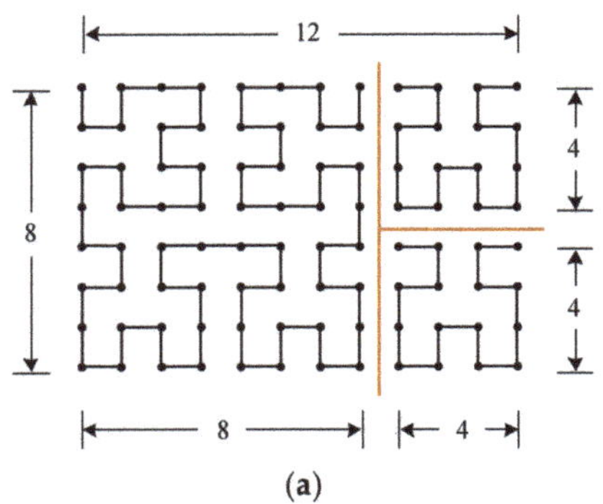

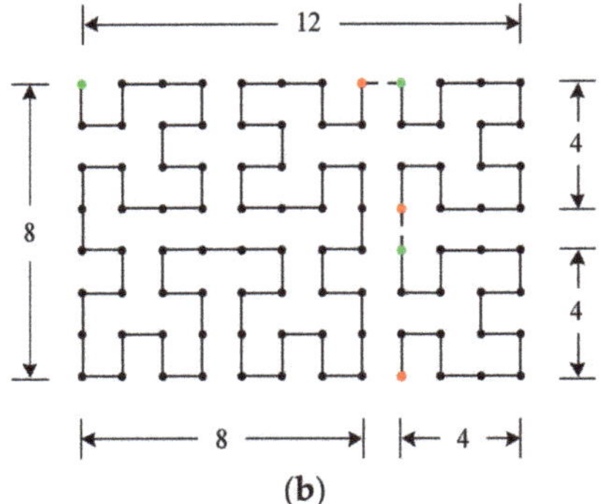

Figure 2. Construction of the 2-D modified Hilbert curve for the array size 12×8. (**a**) Three Hilbert curves are constructed for the 3 divided $2^{i_n} \times 2^{i_n}$ sub-square arrays. (**b**) By selecting appropriate directions for each of the 3 Hilbert curves and connecting the 3 Hilbert curves, the modified 2-D Hilbert curve is constructed.

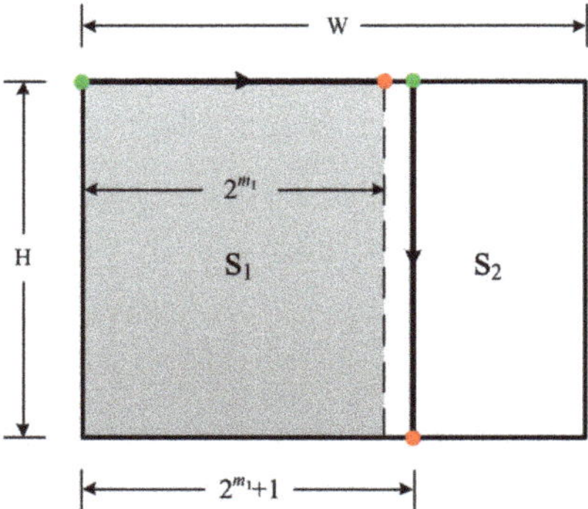

Figure 3. Illustration of the 2-D Modified Hilbert curve construction procedures on a $W \times H$ rectangle array, where $W \geq H$.

Based on the above observations, the procedures of our design method are provided as follows (note, the method is not unique). Without loss of generality, we consider rectangle arrays with $W \geq H$:

1. Find integer m_1 such that $2 \times 2^{m_1} > W$ and $2^{m_1} \leq W$.
2. Construct a modified Hilbert curve within the sub-rectangle S_1 of size $2^{m_1} \times H$. S_1 is a special rectangle with the width being 2^{m_1}. The starting and the ending points of S_1 need to be at the ends of the S_1's top width, as indicated in Figure 3. The construction details for this step is provided shortly. (Note, we use "width" and "height" to represent the number of elements in the two orthogonal directions of a rectangle array throughout the paper. They are not the geometric lengths. Do not get confused with the illustrating diagrams used in the paper.)
3. Once the modified Hilbert curve for the rectangle array S_1 is constructed, the construction of the remaining rectangle array S_2 goes back to step (1) with the starting and the ending points indicated in Figure 3. However, the array size of S_2 is less than half of the original $W \times H$ rectangle.
4. By iterating steps 1 to 3, the subsequent remaining S_2's become smaller and smaller very quickly, with speed faster than 0.5^k, where k is the iteration number. The iteration stops when the remaining S_2 is of size $2^l \times H'$, and the construction is complete. Note, a size of $2^l \times H'$ can always be achieved because the smallest H' can be 1.

Note, when iterating the three steps 1 to 3, if for S_2, the height H is larger than $W - 2^{m_1}$ like the situation shown in Figure 3, then H plays the role of the width W for the new iteration on S_2 because we assumed the initial condition of $W \geq H$ for each iteration. In this case, the design route changes direction because it is always along the width direction, see Figure 3. If for S_2 the height H is still smaller than $W - 2^{m_1}$, the iteration continues along the same design route direction. In the following example, we show a specific design to provide a more intuitive understanding of the procedures.

Suppose we want to design a modified 2-D Hilbert curve for a practical size of $W = 1920/8 = 240$ and $H = 1080/8 = 135$, which is the subband size from the 8×8 subband decomposition on a 1920×1080 image, the standard HDTV size. Following the design procedures:

Iteration 1: $m_1 = 7$, and sub-rectangle arrays S_1 and S_2 for the 1st iteration are: $S_1^{1\text{st iteration}} = 128 \times 135$, $S_2^{1\text{st iteration}} = 112 \times 135$, as shown in Figure 4a.

Iteration 2: Because $H = 135 > 112 = W - 2^{m_1}$, the 135 side of $S_2^{1\text{st iteration}}$ needs to be the width for the 2nd iteration. Then, for the second iteration on the 112×135 array, $m_2 = 7$, the starting (green) point and the ending (red) point are aligned vertically, changing the design route direction, see Figure 4b. The resulting sub-rectangle arrays S_1 and S_2 for the 2nd iteration are $S_1^{2\text{nd}} = 112 \times 128$, $S_2^{2\text{nd}} = 112 \times 7$, as shown in Figure 4b.

Iteration 3: Because $112 > 7 = 135 - 2^{m_2}$, the 112 side of $S_2^{2\text{nd}}$ needs to be the width in the 3rd iteration. Then, for the 3rd iteration on the 112×7 array, $m_3 = 6$, and the starting (green) point and the ending (red) point are aligned horizontally, changing the design route direction again, see Figure 4c. The resulting sub-rectangles S_1 and S_2 for the 3rd iteration are: $S_1^{3\text{rd}} = 64 \times 7$, $S_2^{3\text{rd}} = 48 \times 7$, see Figure 4c. Note, $S_2^{3\text{rd}} = S_1^{4\text{th}} + S_1^{5\text{th}}$ in Figure 4c.

Iteration 4: Because $7 < 48 = 112 - 2^{m_3}$, the 48 side of $S_2^{3\text{rd}}$ is still the width for the 4th iteration. Thus, for the 4th iteration on $S_2^{3\text{rd}} = 48 \times 7$, $m_4 = 5$, the starting (green) point and the ending (red) point are still aligned horizontally, with no design route direction change. The resulting sub-rectangles S_1 and S_2 for the 4th iteration are: $S_1^{4\text{th}} = 32 \times 7$, $S_2^{4\text{th}} = 16 \times 7$, see Figure 4c. Note, $S_2^{4\text{th}} = S_1^{5\text{th}}$ in Figure 4c.

Iteration 5: Similar to Iteration 4, no change of the design route direction is needed. However, the width for the 5th iteration is $16 = 2^4$. By selecting $m_4 = 4$, $S_1^{5\text{th}} = 16 \times 7$, and $S_2^{5\text{th}} = 0 \times 7$, the design completes as indicated in Figure 4c.

Now we go back to provide the details of step 2 of the design procedures. In other words, we need to design a modified Hilbert curve for a rectangle with the width being 2^{m_1} and the height being H. There are three possible situations, (A) $H < 2^{m_1}$, (B) $H > 2^{m_1}$, and (C) $H = 2^{m_1}$. Condition (C) is trivial, where the sub-rectangle is a $2^{m_1} \times 2^{m_1}$ square, and the construction is simply the original Hilbert curve.

Condition (A) $H < 2^{m_1}$:

We start from the $(m_1)^{\text{th}}$ order Hilbert curve, whose height is 2^{m_1}. Thus, we need to reduce the height by $\Delta H = 2^{m_1} - H$. Recall that an integer B can be converted into its binary format $b_s b_{s-1} \ldots b_1 b_0$, i.e., B can be decomposed as

$$B = b_s 2^s + b_{s-1} 2^{s-1} + \ldots + b_1 2^1 + b_0 2^0, \tag{1}$$

where the b_i's are either 0 or 1. Decomposing ΔH using (1), we can perform a reduction of ΔH step by step, with each step achieving a reduction of 2^i points. For example, suppose $\Delta H = 13$. Then, from (1), we have $\Delta H = 8 + 4 + 1$, i.e., $b_3 = 1$, $b_2 = 1$, $b_1 = 0$, and $b_0 = 1$. Thus, we need to reduce 8 points, 4 points, and 1 point on H to achieve the total reduction of $\Delta H = 13$ points.

To perform a reduction of 2^i points, first, a reduction of 2^i points on the $(i+1)^{\text{th}}$ order Hilbert curve is straightforward. By inspection, it can be seen that the top two sub-squares in Figure 1e can be removed, leading to Figure 1f. Then, a reduction of 2^i points on the $(i+1)^{\text{th}}$ order Hilbert curve is achieved.

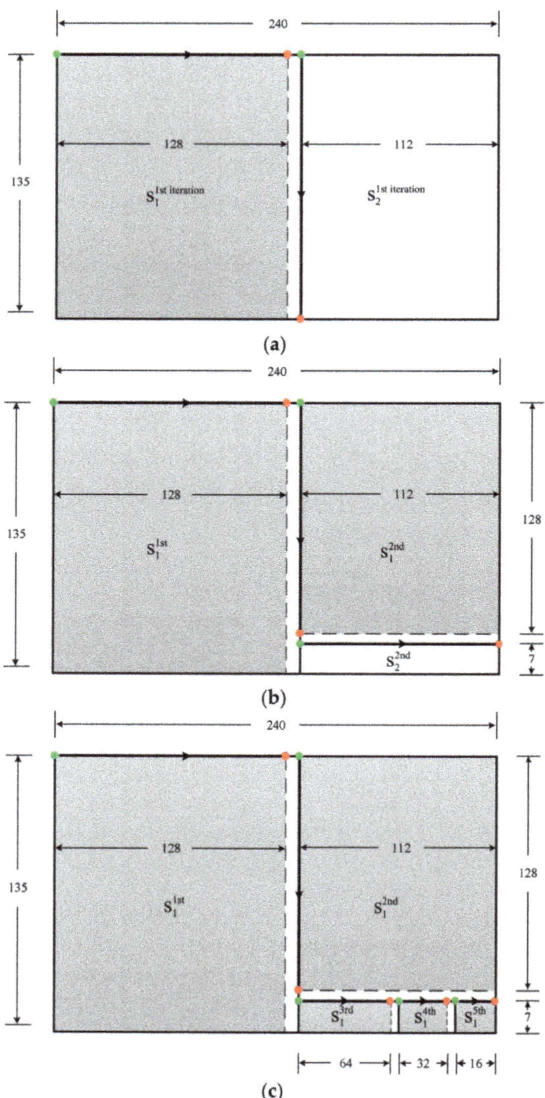

Figure 4. The example of designing the 2-D modified Hilbert curve for the 240×135 array: (**a**) the 1st iteration, (**b**) the 2nd iteration, and (**c**) the 3rd, 4th, and the 5th iteration.

Next, we observe the following property: For an opening-toward-top Hilbert curve of any order, the bottom sub-Hilbert curves always have the same opening-toward-top orientation. As an example, in Figure 5a, a 4th order opening-toward-top Hilbert curve is plotted. The 4th order Hilbert curve can be represented using the structure of Figure 5b, i.e., the main structure is an opening-toward-top 1st order Hilbert curve with 4 sub-curves, which are four 3rd order Hilbert curves denoted by four small shaded squares. As just described above, the removal of the top two sub-squares, or, equivalently, a reduction of 8 points in H in this case, can be easily achieved. Now, it is important to observe from Figure 5b that the bottom two shaded squares, i.e., the bottom two 3rd order sub-Hilbert curves, are also opening-toward-top Hilbert curves. When the original 4th order Hilbert curve is represented using the structure of Figure 5c, for each of the bottom two opening-toward-up 3rd order sub-Hilbert curves, the removal of the top two sub-squares, or equivalently a reduction of 4 points on H in this case, can be achieved. Similarly, it can be seen from Figure 5a,d that reductions of 2 points and 1 point on H can be achieved.

Figure 6 shows a specific example of how the 4th order Hilbert curve is reduced by $\Delta H = 13 = 8 + 4 + 1$. In Figure 5a, the original 4th order $H = 16$ Hilbert curve is shown. Reductions on H by 8, 4, and 1 in each step are respectively shown in Figure 6a–c. After all the sub-reductions are completed, the total reduction of $\Delta H = 13$ is achieved in Figure 6d.

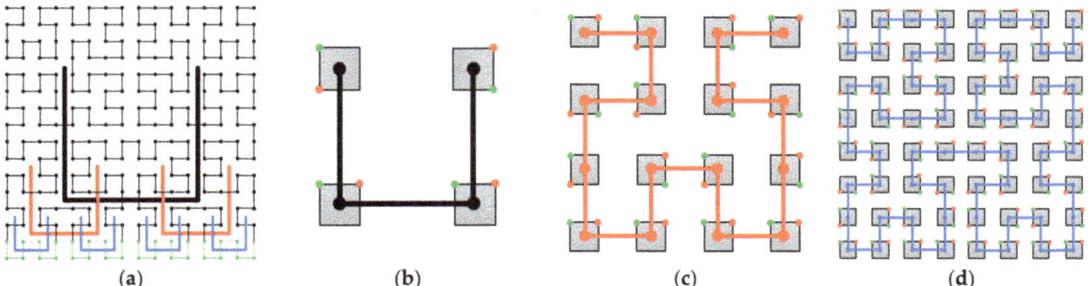

Figure 5. (a) The 4th order 2-D Hilbert curve; The 4th order 2-D Hilbert curve represented using (b) four 3rd order sub-curves; (c) sixteen 2nd order sub-curves; and (d) sixty-four 1st order sub-curves.

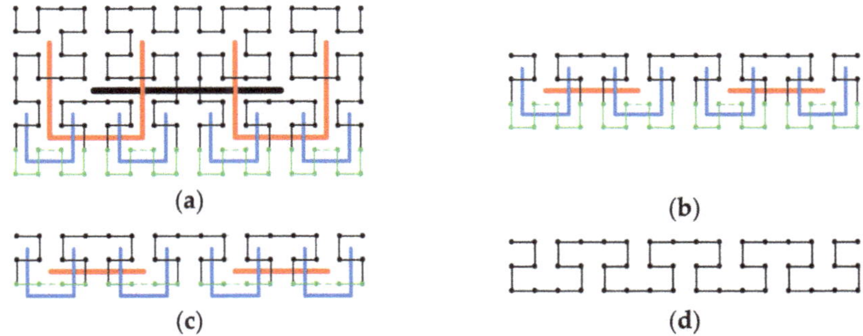

Figure 6. The specific example of reducing H from 16 to 3, i.e., $\Delta H = 13$. (a) The modified curve by an 8-point reduction on H of the 4th order Hilbert curve shown in Figure 5a; (b) a further reduction of 4 points on (a); (c) a further reduction of 1 point on (b); (d) the final result of a total reduction of $\Delta H = 13$ points on the 4th order Hilbert curve is achieved.

<u>Condition (B)</u> $H > 2^{m_1}$:

We need to increase the height by $\Delta H = H - 2^{m_1}$. Note, since $2 \times 2^{m_1} > W \geq H$, $\Delta H = H - 2^{m_1} < 2 \times 2^{m_1} - 2^{m_1} = 2^{m_1} < H$, i.e., $\Delta H < H$. Thus, similar to condition (A), ΔH is decomposed by (1), and we can increase H by 2^i points in each step because the increase in height by 2^i points on the $(i+1)^{\text{th}}$ order Hilbert curve can be achieved using the modification from Figure 1e–g.

With the height of the $(m_1)^{\text{th}}$ order Hilbert curve reduced or increased to H, procedure 2 of the iterative design method described earlier is performed. The modified 2-D Hilbert curve on a $W \times H$ rectangle array using the iterative design procedures is completed.

To provide more intuitions, the 2-D modified Hilbert curves of sizes 27×17, 27×18, and 27×19, constructed using the proposed method, are shown in Figure 7. In addition, the MATLAB codes implementing the proposed method are available in [19]. As can be easily checked, the algorithm runs reasonably fast, and the results can be obtained instantly.

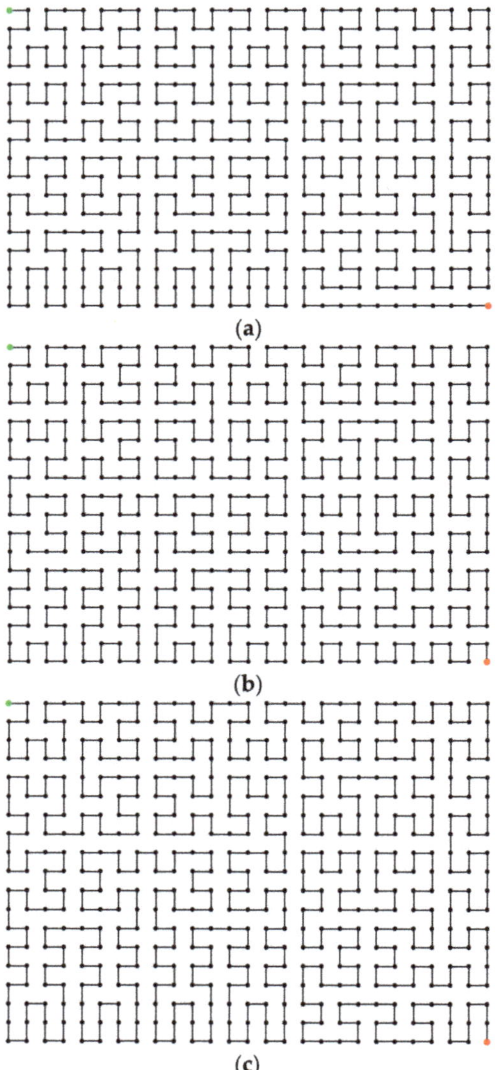

Figure 7. The modified 2-D Hilbert curves of sizes (**a**) 27 × 17, (**b**) 27 × 18, and (**c**) 27 × 19, constructed using the proposed method.

2.2. The 3-D Modified Hilbert Curve

In [15], the binary run-length-based symbol grouping entropy coding method is used in video compression. For the 3-D transform video compression algorithm introduced in [15], conventional motion compensation is not used in order to improve the computational complexity. Instead, a 4-band SCWP transform [20] is performed along the time dimension. In other words, the first step of the video compression algorithm is a 3-D transform. Thus, the transformed coefficients are 3-D subband arrays.

In entropy coding of the quantized 3-D subband coefficient arrays, the 3-D Hilbert curve scan was used to maximally keep the correlations in the 3-D subband into the 1-D scanned array. Because the original 3-D Hilbert curves are for cubes of side length 2^i, in [15], the 1920 × 1080 test videos were cropped to a size of 1024 × 1024 for testing. Apparently, to accommodate an arbitrary rectangle video size, the original 3-D Hilbert curve needs to be modified. Below, we extend the modification method introduced in Section 2.1 for 2-D arrays to 3-D conditions. The 3-D arrays are of size $W \times H \times D$, where W and H are the width and height of the cuboid array, D is the third dimension denoting the depth here, which corresponds to the time dimension of the input video.

The depth D of the 3-D decomposed subband is determined by the parameter "group of pictures" (GOP), which is normally selected to be the powers of 2. As a result, the depths D are also the powers of 2, i.e., $D = 2^d$, $d = 0, 1, 2, \ldots$ For example, the GOP used in [15] was 32, which leads to the depths D of the 3-D subbands being 8, 4, 2, or 1 (for details, please refer to [15]). Furthermore, compared with W and H, D is normally much smaller in our case.

We begin from the original 3-D Hilbert curve. Again, we denote a 3-D Hilbert curve of size $2^i \times 2^i \times 2^i$, the ith order 3-D Hilbert curve. Figure 8a,b, respectively, show the 1st order and the 2nd order 3-D Hilbert curve. Similar to the 2-D situation, the starting and the ending points of any order 3-D Hilbert curves are at the two ends of one side of the cube. Therefore, as shown in Figure 8c, when the internal structure is not needed, a 3-D Hilbert curve of any order i can be represented by a cube with the starting and ending points labeled. With this simple and intuitive representation, the construction of the $(i + 1)^{\text{th}}$ order 3-D Hilbert curve from the ith order 3-D Hilbert curve can be easily demonstrated by Figure 8d. By mathematical induction, given (1) the 1st order 3-D Hilbert curve and (2) the method of constructing the $(i + 1)^{\text{th}}$ order 3-D Hilbert curve from the ith order 3-D Hilbert curve, the 3-D Hilbert curve of any order can be constructed.

Without loss of generality, assume $W \geq H$. As mentioned in our application, D is the powers of 2 and is normally much smaller than W and H. In other words, the modified 3-D Hilbert curve is of size $W \times H \times D = W \times H \times 2^d$, and D is much smaller than W and H. Exploiting these features, the extension to 3-D from 2-D can borrow the 2-D construction procedures introduced in Section 2.1 as follows.

First, consider the situation where W and H are multiples of D, i.e., the cuboid of size $W \times H \times D = (cD) \times (rD) \times D$, where $W = cD$, $H = rD$, c and r are integers. In this case, we can directly extend the 2-D construction method to 3-D construction. To see that, the dth order original 2-D Hilbert curve is compared with the dth order original 3-D Hilbert curve in Figure 9a,b. Because the $D \times D \times D$ cube is the smallest construction block for the 3-D curve and the $D \times D$ square is the smallest construction block for the 2-D curve, the 3-D construction of size $W \times H \times D = (cD) \times (rD) \times D$ can directly borrow the 2-D construction structure of size $W \times H = (cD) \times (rD)$. Figure 9c,d intuitively show the 2-D to 3-D extension by comparing the 2-D Hilbert curve of size $2D \times 2D$ with the modified 3-D Hilbert curve of size $2D \times 2D \times D$.

Next, we consider the situation where W and H are not multiples of D, but $W \geq D$. In this case, we can still borrow the 2-D construction structure, i.e., use the 2-D iterative route similar to Figures 3 and 4c for the height-width (W-H) surface of the 3-D cuboid. This is similar to what we performed in Figure 9c,d, where W and H are multiples of D. The difference is that, in this case, ΔH is not a multiple of D. In this case, we need to consider the non-zero terms in $\Delta H = \sum b_i 2^i$ that are smaller than D, i.e., the $2^i < D$ terms. The $2^i \geq D$ terms are multiples of D, which is the situation previously considered. The $2^i < D$ terms in ΔH, however, need to be handled on the 3-D cubes at the bottom. For example, if in Figure 9d we want to construct a modified 3-D Hilbert curve of size $2D \times 1.5D \times D$ instead of $2D \times 2D \times D$, then the bottom cubes in Figure 9d need to be reduced by $0.5D = 2^{d-1}$ points to achieve $H = 1.5D$.

Therefore, we need to consider adding or reducing 2^i points on the $D \times D \times D = 2^d \times 2^d \times 2^d$ cube, where $0 \leq i \leq d - 1$. This is not difficult. (1) Similar to the 2-D situation, observe in Figure 10a that the bottom 4 sub-cubes (sub-Hilbert curves) are of the same opening-toward-top orientation as its original 3-D Hilbert curve because they all have the starting and ending points at the top surface of the cube. (2) For the 3-D Hilbert curve of size $2^d \times 2^d \times 2^d$, indicated in Figure 10a, reducing and increasing 2^{d-1} points on H can be achieved by Figure 10b,c respectively. Combining (1) and (2) above, reducing or increasing 2^i ($0 \leq i \leq d-1$) points on H can be achieved on a 3-D Hilbert curve of size $2^d \times 2^d \times 2^d$.

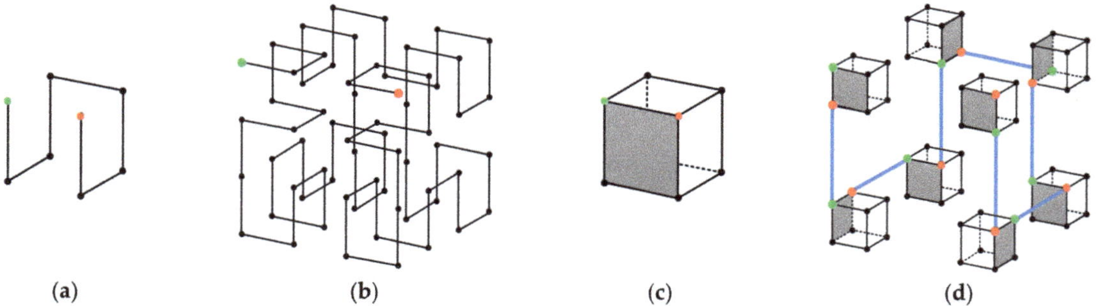

Figure 8. 3-D Hilbert curve properties. (**a**,**b**) are respectively the 1st order and the 2nd order 3-D Hilbert curves. (**c**) A simple notation to represent the ith order 3-D Hilbert curve. (**d**) Construction of the $(i+1)^{\text{th}}$ order Hilbert curve from the ith order Hilbert curve.

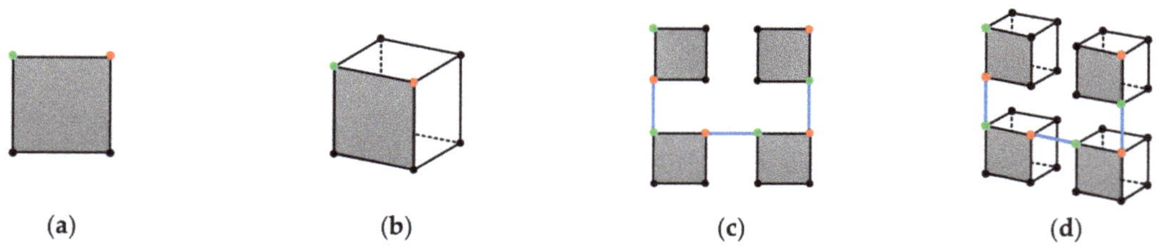

Figure 9. Extension of the 2-D construction to 3-D construction for size $(cD) \times (rD) \times D$, where c and r are integers. (**a**) The smallest construction block for 2-D, the $D \times D$ square. (**b**) The smallest construction block for 3-D, the $D \times D \times D$. cube. (**c**) The $2D \times 2D$ size 2-D Hilbert curve, constructed from the 2-D smallest construction block. (**d**) The $2D \times 2D \times D$ size 3-D Hilbert curve, constructed from the 3-D smallest construction block borrowing the 2-D construction structure of (**c**).

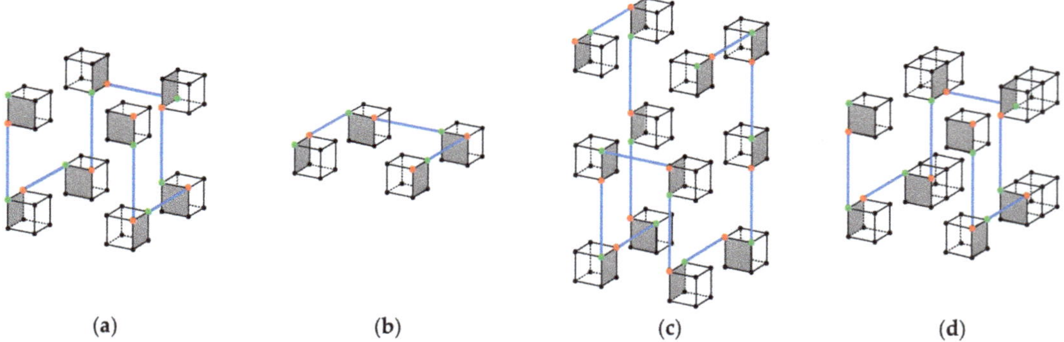

Figure 10. Point increasing and reducing operations on the $(i+1)^{\text{th}}$ order 3-D Hilbert curve. (**a**) The $(i+1)^{\text{th}}$ order 3-D Hilbert curve. (**b**) The i points are reduced on H. (**c**) The i points are increased on H. (**d**) The situation where i points need to be increased in a different direction.

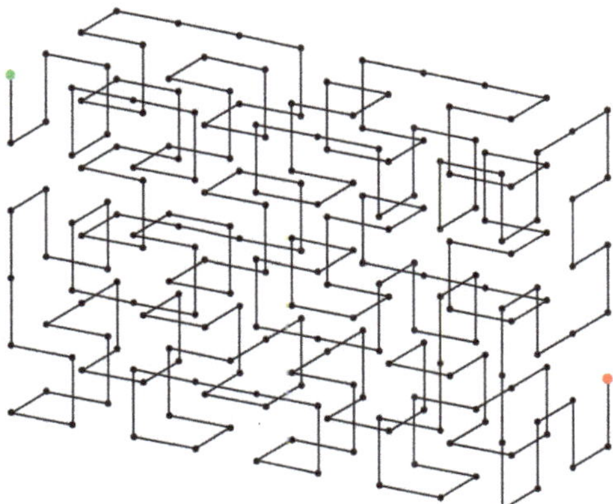

Figure 11. The modified 3-D Hilbert curve of size $9 \times 6 \times 4$, constructed using the proposed method.

Finally, during the 3-D construction described above, for the height-width (W-H) surface of the 3-D cuboid, we can exploit the 2-D iterative route similar to Figures 3 and 4c as long as $W \geq D$ is satisfied for the remaining S_2's in the W-H surface. For example, we can realize the modified 3-D Hilbert curve for the size of $240 \times 135 \times 8$ following exactly the 2-D route shown in Figure 4c, which was used for constructing the 240×135 2-D modified Hilbert curve. Nevertheless, a construction may end up with a residue S_2 on the W-H surface, whose height and width are both smaller than D. For example, using the above method to construct the modified 3-D Hilbert curve of the size $243 \times 135 \times 8$, we end up with a residue cuboid of size $3 \times 7 \times 8$. Then, the 2-D Figure 3 iterative procedure cannot proceed for the W-H surface anymore. We have to consider the situation of constructing a modified 3-D Hilbert curve of the size $W \times H \times D$ with $D = 2^d > W \geq H$.

However, in the 3-D transform video application, D is very small. As mentioned above, in [15], the maximum $D = 8$ even if the GOP used is 32. When $D = 8$, the maximum residue cuboid is only $7 \times 7 \times 8$, which is very small. For such tiny residue cuboids, using some other routes, such as the raster scan, would not lead to any noticeable effect on the final video compression results. On the other hand, the design for the situation of $D = 2^d > W \geq H$ is complex, and thus, for the application of coding the 3-D transformed coefficients in video compression, we can just use a simple scan route for the residue cuboids with $D > W \geq H$. We implemented in MATLAB such 3-D extension with small $D > W \geq H$ residue cuboid connected using the raster scan, which is available at [19]. Figure 11 shows a modified 3-D Hilbert curve of size $9 \times 6 \times 4$ (i.e., $D = 2^2$) produced by MATLAB codes.

We will not lengthily go into the design on the condition $D = 2^d > W \geq H$. For completeness, we only briefly describe that the design is possible using similar ideas we have used up to now. Note, there can be some other methods to handle the $D = 2^d > W \geq H$ situation because the design method is not unique.

For the $D = 2^d > W \geq H$ situation, first, consider the situation where either W or H is a power of 2. Without loss of generality, assume $H = 2^h$. Observe that the sizes $W \times H \times D$, $D \times H \times W, \ldots$, etc., i.e., all the 6 permutations, are the same for our curve construction task. In order to exploit our previously developed construction techniques, we need to change the roles of W, H, and D. Because D is the longest side, we need to use $D = 2^d$ as the width. Since $H = 2^h < D$, use H as the depth. Then, the construction finishes nicely in one step.

For the more difficult situation, where both W and H are not powers of 2, decompose the shortest side H into a sum of 2^i using equation (1): $H = 2^{h_0} + 2^{h_1} + \ldots + 2^{h_n}$, where $h_0 > h_1 \ldots > h_n$ (h_0 corresponds to the most significant bit, i.e., $2^{h_0} > \frac{1}{2}H$). Then, the construction on the $D \times W \times 2^{h_0} = 2^d \times W \times 2^{h_0}$ cuboid is immediately achieved as described above. To increase the thickness from 2^{h_0} to H, the point-increasing operation needs to be along the direction as illustrated in Figure 10d. For the 4 length-increased sub-cubes at the back in Figure 10d, the bottom 2 sub-cubes can use the point-increasing operation we already used, i.e., the one from Figure 10a to Figure 10c, but the top 2 sub-cubes need to use a different point-increasing structure, which is skipped here. We may also need to perform multiple point-increasing operations and then perform a point-decreasing operation to achieve the desired value H, and the operations need to be performed individually for the sub-cubes at the back of the $D \times W \times 2^{h_0}$ cuboid. The sizes of the sub-cubes can be different depending upon the W value. As a result, the implementation is complex. We will not go into the details further since currently, there is no immediate application.

3. Conclusions and the Near Future Work

We have shown the method of modifying the 2-D Hilbert curve to fit an arbitrary $W \times H$ rectangle array and the method of modifying the 3-D Hilbert curve to fit a cuboid array of size $W \times H \times 2^d$. These modified Hilbert curves can be used in entropy coding for image and video compression. Furthermore, since the construction of the modified 2-D and 3-D Hilbert curves is not straightforward, the methods presented in this short article could be useful for some unknown applications in the future.

The 2-D modified Hilbert curve has already been used in the run-length-based symbol grouping entropy coding method for lossy and lossless image compression. High compression efficiency is achieved, as shown in [7,8].

Because of using the 3-D Hilbert curve, the video compression algorithm prototype introduced in [15] only tested videos with the cropped size of 1024×1024, although some promising results were shown. On applying the 3-D modified Hilbert curve for coding to the 3-D subband coefficients so that the algorithm can handle arbitrary video sizes, together with some other fine tunings, we are completing the video compression algorithm very soon. We will systematically compare the performances of the new video compression algorithm with state-of-the-art video compression algorithms, such as HEVC, etc., in terms of compromise between complexity and compression efficiency. From the preliminary test results shown in [15], we expect that the final completed video compression algorithm using the 3-D modified Hilbert curve developed in this paper will be competitive to state-of-the-art video compression algorithms in certain important situations, such as the compression at the high video quality.

Author Contributions: Conceptualization, J.L.; Investigation, J.L., Y.R. and X.Z.; Supervision, J.L.; Writing—original draft, J.L. and X.Z.; Writing—review & editing, J.L. and Y.R. All authors have read and agreed to the published version of the manuscript.

Funding: This research was funded by Shantou University.

Conflicts of Interest: The authors declare no conflict of interest.

References

1. Huffman, D. A method for the construction of minimum redundancy codes. *Proc. Inst. Radio Eng.* **1952**, *40*, 1098–1101. [CrossRef]
2. Witten, I.H.; Neal, R.M.; Cleary, J.G. Arithmetic coding for data compression. *Commun. ACM* **1987**, *30*, 520–540. [CrossRef]
3. Pennebaker, W.B.; Mitchell, J.L.; Langdon, G.G.; Arps, R.B. An overview of the basic principles of the Q-coder adaptive binary arithmetic coder. *IBM J. Res. Dev.* **1988**, *32*, 717–726. [CrossRef]
4. Douglas Withers, W. A Rapid Probability Estimator and Binary Arithmetic Coder. *IEEE Trans. Inform.* **2001**, *47*, 1533–1537. [CrossRef]
5. Belyaev, E.; Forchhammer, S.; Liu, K. An Adaptive Multialphabet Arithmetic Coding Based on Generalized Virtual Sliding Window. *IEEE Signal Process. Lett.* **2017**, *24*, 1034–1038. [CrossRef]
6. Taubman, D.S.; Marcellin, M.W. *JPEG2000: Image Compression Fundamentals, Standards, and Practice*; Kluwer Academic Publishers: Switzerland, Cham, 2002.
7. Lin, J. A New Perspective on Improving the Lossless Compression Efficiency for Initially Acquired Images. *IEEE Access* **2019**, *7*, 144895–144906. [CrossRef]
8. Lin, J. Reversible Integer-to-Integer Wavelet Filter Design for Lossless Image Compression. *IEEE Access* **2020**, *8*, 89117–89129. [CrossRef]

9. Wikipedia. Available online: https://en.wikipedia.org/wiki/Hilbert_curve (accessed on 27 June 2021).
10. Hilbert, D. Über die stetige Abbildung einer Linie auf ein Flächenstück. *Math. Ann.* **1891**, *38*, 459–460. [CrossRef]
11. Moon, B.; Jagadish, H.; Faloutsos, C. Analysis of the Clustering Properties of Hilbert Space-filling Curve. *IEEE Trans. Knowl. Data Eng.* **2001**, *13*, 124–141. [CrossRef]
12. Jafadish, H.V. Analysis of the Hilbert curve for representing two-dimensional space. *Inf. Process. Lett.* **1997**, *62*, 17–22. [CrossRef]
13. Abel, D.J.; Mark, D.M. A Comparative Analysis of Some Two-Dimensional Orderings. *Int. J. Geogr. Inf. Syst.* **1990**, *4*, 21–31. [CrossRef]
14. Liu, X.; Schrack, G.F. Encoding and decoding the Hilbert order. *Softw. Pract. Exper.* **1996**, *26*, 1335–1346. [CrossRef]
15. Lin, J. Improving the Compression Efficiency for Transform Video Coding. In Proceedings of the 2017 4th International Conference on Systems and Informatics (ICSAI), Hangzhou, China, 11–13 November 2017.
16. Liu, X.; Schrack, G.F. An algorithm for encoding and decoding the 3-D Hilbert order. *IEEE Trans. Image Process.* **1997**, *6*, 1333–1337. [PubMed]
17. Bourbakis, N.; Alexopoulos, C. Picture data encryption using scan patterns. *Pattern Recognit.* **1992**, *25*, 567–581. [CrossRef]
18. Hu, F.C.; Tsai, Y.H.; Chung, K.L. Space-filling approach for fast window query on compressed images. *IEEE Trans. Image Process.* **2000**, *9*, 2109–2116.
19. Available online: https://stumail-my.sharepoint.cn/:u:/g/personal/jianyulin_stu_edu_cn/EZxWBMLZ9ndCiKCI-QwtzM4BxL0IS6MtaEyZ2A-LpLl3dA (accessed on 27 June 2021).
20. Lin, J.; Smith, M.J.T. Spectrum Decomposition for Image/Signal Coding. *IEEE Trans. Signal Process.* **2013**, *61*, 1065–1071. [CrossRef]

Article

Retinex-Based Fast Algorithm for Low-Light Image Enhancement

Shouxin Liu, Wei Long, Lei He, Yanyan Li * and Wei Ding

School of Mechanical Engineering, Sichuan University, Chengdu 610065, China; liushouxin@stu.scu.edu.cn (S.L.); long_wei@scu.edu.cn (W.L.); markushe_scu@163.com (L.H.); dingwei1995@stu.scu.edu.cn (W.D.)
* Correspondence: yyl_scu@163.com; Tel.: +86-15002820593

Abstract: We proposed the Retinex-based fast algorithm (RBFA) to achieve low-light image enhancement in this paper, which can restore information that is covered by low illuminance. The proposed algorithm consists of the following parts. Firstly, we convert the low-light image from the RGB (red, green, blue) color space to the HSV (hue, saturation, value) color space and use the linear function to stretch the original gray level dynamic range of the V component. Then, we estimate the illumination image via adaptive gamma correction and use the Retinex model to achieve the brightness enhancement. After that, we further stretch the gray level dynamic range to avoid low image contrast. Finally, we design another mapping function to achieve color saturation correction and convert the enhanced image from the HSV color space to the RGB color space after which we can obtain the clear image. The experimental results show that the enhanced images with the proposed method have better qualitative and quantitative evaluations and lower computational complexity than other state-of-the-art methods.

Keywords: Retinex; image enhancement; gamma correction; low-light image; HSV color space

Citation: Liu, S.; Long, W.; He, L.; Li, Y.; Ding, W. Retinex-Based Fast Algorithm for Low-Light Image Enhancement. *Entropy* **2021**, *23*, 746. https://doi.org/10.3390/e23060746

Academic Editor: Amelia Carolina Sparavigna

Received: 15 May 2021
Accepted: 8 June 2021
Published: 13 June 2021

Publisher's Note: MDPI stays neutral with regard to jurisdictional claims in published maps and institutional affiliations.

1. Introduction

Images captured with a camera in weakly illuminated environments are often degraded. For example, these types of images with low contrast and low light, reduce visibility. The object and detail information cannot be captured, which can reduce the performance of image-based analysis systems, such as computer vision systems, image processing systems and intelligent traffic analysis systems [1–3].

In order to address the above problems, a great number of low-light image enhancement methods have been proposed. Generally, the existing methods can be divided into three categories, namely the HE-based (histogram equalization) algorithm, Retinex-based algorithm and non-linear transformation [4–6]. The HE-based algorithm is the simplest method; the main idea of this method is to adjust illuminance by equalizing the histogram of the input low-light image. To address the shortage of conventional HE algorithms, over enhancement and loss of detail information, a great number of improved and HE-based methods have been proposed, such as contrast-limited equalization (CLAHE), bi-histogram equalization with a plateau limit (BHE), exposure-based sub-image histogram equalization (ESIHE) and exposure-based multi-histogram equalization contrast enhancement for non-uniform illumination images (EMHE) [7–12]. However, HE-based methods neglect the noise hidden in the dark region of low-light images. The Retinex model is a color perception model of human vision, which consists of illumination and reflectance [13,14]. The aim of Retinex-based algorithms is to estimate the right illumination image or reflectance image from its degraded image by different filters to achieve low brightness enhancement [15,16]. Some classic algorithms are single-scale Retinex (SSR) and multi-scale Retinex (MSR). In order to solve color distortion, multi-scale Retinex with color restoration (MSRCR) was proposed, which introduced color restoration in multi-scale Retinex. After that, some improved algorithms introduced different types of filters to replace the traditional Gaussian filter, such as the improved Gaussian filter, improved guided filter, bright-pass filter and so on [17–19]. Even though image texture details can be restored well via the Retinex-based method, the halo effect is introduced into enhanced images. Common non-linear functions are gamma correction, sigmoid transfer function and logarithmic transfer function [20–22]; these types of methods are pixel-wise operations for natural low-light images. Compared with other non-linear functions, the gamma transfer function is wildly used in the field of image processing, but the limitation of gamma correction is that if the parameter γ is too small, it will amplify the noise of the target image; by contrast, if the parameter γ is close to 1, satisfactory enhanced results will not be obtained. Therefore, estimating a suitable γ value is the key to obtaining satisfactory enhanced results.

In this paper, we utilize the gamma transfer function to estimate the illumination and achieve brightness enhancement via the Retinex model. The enhanced image achieves satisfactory light enhancement and global brightness equalization; thus, our method can restore more information than other methods. The final experimental results show that compared with other state-of-the-art methods, the enhanced images through our algorithm have better qualitative and quantitative evaluations. Some examples of natural low-light images and enhanced images with the proposed RBFA method are shown in Figure 1. All low-light images in Figure 1 were captured by the authors of this paper.

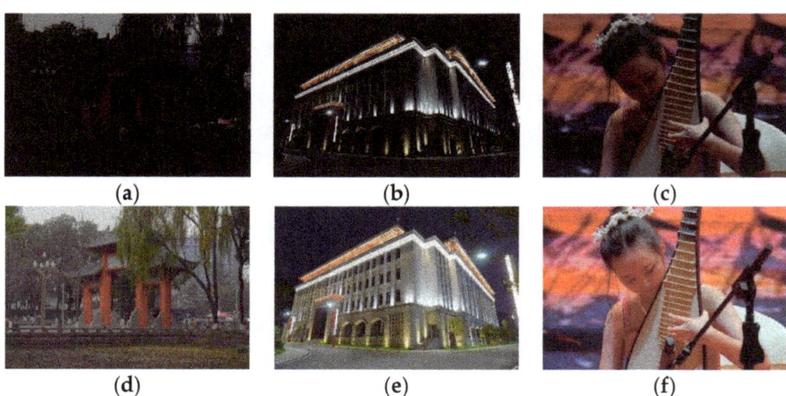

Figure 1. Top row (**a–c**): natural low-light images, bottom row (**d–f**): enhanced images with our proposed RBFA method.

The rest of this paper is organized as follows: Section 2 describes the corresponding works of the proposed algorithm in this paper. In Section 3, the details of the proposed method are introduced. Section 4 presents the comparative experiment results with other state-of-the-art methods and describes the computational complexity comparison. The work is concluded in Section 5.

2. Related Work

We introduce the Retinex model, gamma correction and HSV color space in this section, which construct the basis of our method.

2.1. Retinex Model

The classical Retinex model assumes that the observed image consists of reflectance and illumination. The Retinex model can be expressed as follows [23].

$$\mathbf{H} = \mathbf{R} \bullet \mathbf{L} \tag{1}$$

where H is the observed image, R and L represent the reflectance and the illumination of the image, respectively. The operator '•' denotes the multiplication. In this paper, we utilize the logarithmic transformation to reduce computational complexity. We can obtain the following expression.

$$log(\mathbf{H}) = log(\mathbf{R}\bullet\mathbf{L}) \tag{2}$$

Finally, we can obtain Equation (3) to estimate the reflectance in the HSV color space.

$$log(\mathbf{R}) = log(\mathbf{V}) - log(\mathbf{L}) \tag{3}$$

2.2. Gamma Correction

The gamma transfer function is wildly used in the field of image processing, and the corresponding gamma transfer function can be expressed as follows [24,25].

$$g(x,y) = u(x,y)^{\gamma} \tag{4}$$

where $g(x,y)$ denotes the gray level of the enhanced image at pixel location (x,y), $u(x,y)$ is the gray level of the input low-light image at pixel location (x,y), and γ represents the parameter of the gamma transfer function. The shape of the gamma transfer function can be affected by parameter γ; the influence of different values of γ is shown in Figure 2.

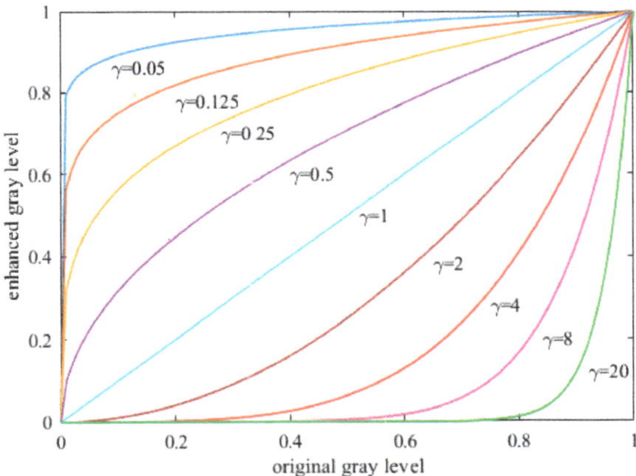

Figure 2. The shapes of gamma functions with different γ values.

According to the Figure 2, we can see that the enhanced gray level increases monotonically with decreased parameter γ; if we want to achieve a higher value of the gray level, we have to let the size of parameter γ fall within the range from 0 to 1. Contrastingly, the enhanced gray level decreases monotonically with increased parameter γ.

2.3. HSV Color Space

The HSV color space consists of a hue component (H), saturation component (S) and value component (V) [26,27]. The value component represents the brightness intensity of the image. The advantage of the HSV color space is that any component can be adjusted without affecting each other [28]; more specifically, the input image is transferred from the RGB (red, green, blue) color space to the HSV color space, which can eliminate the strong color correlation of the image in the RGB color space. Therefore, this work is based on the HSV color space [29]. Commonly, image enhancement in RGB color space need to process R, G and B, three components, but we only need to process the V component in this work. Therefore, this will greatly reduce the image processing time.

3. Our Approach

The details of proposed algorithm are described in this section. Based on the descriptions in Section 2.2, in this work we only focus on the V component to adjust the brightness of the low-light image; the flowchart of the proposed method is shown in Figure 3. We choose an image named "Arno" to illustrate the enhancement process of the proposed method, the processing of image enhancement and corresponding histograms are shown in Figure 4.

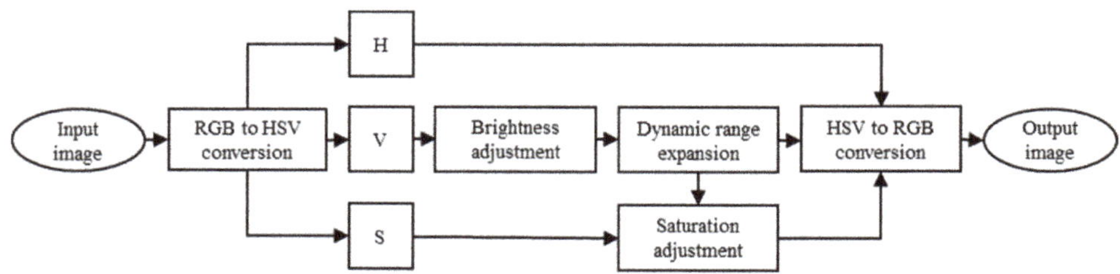

Figure 3. The flowchart of the proposed method.

(**a**) input image

(**b**) original V component

(**c**) brightness adjustment

(**d**) dynamic range expansion

(**e**) original S component

(**f**) adjusted S component

(**g**) enhanced image

(**h**) reference image

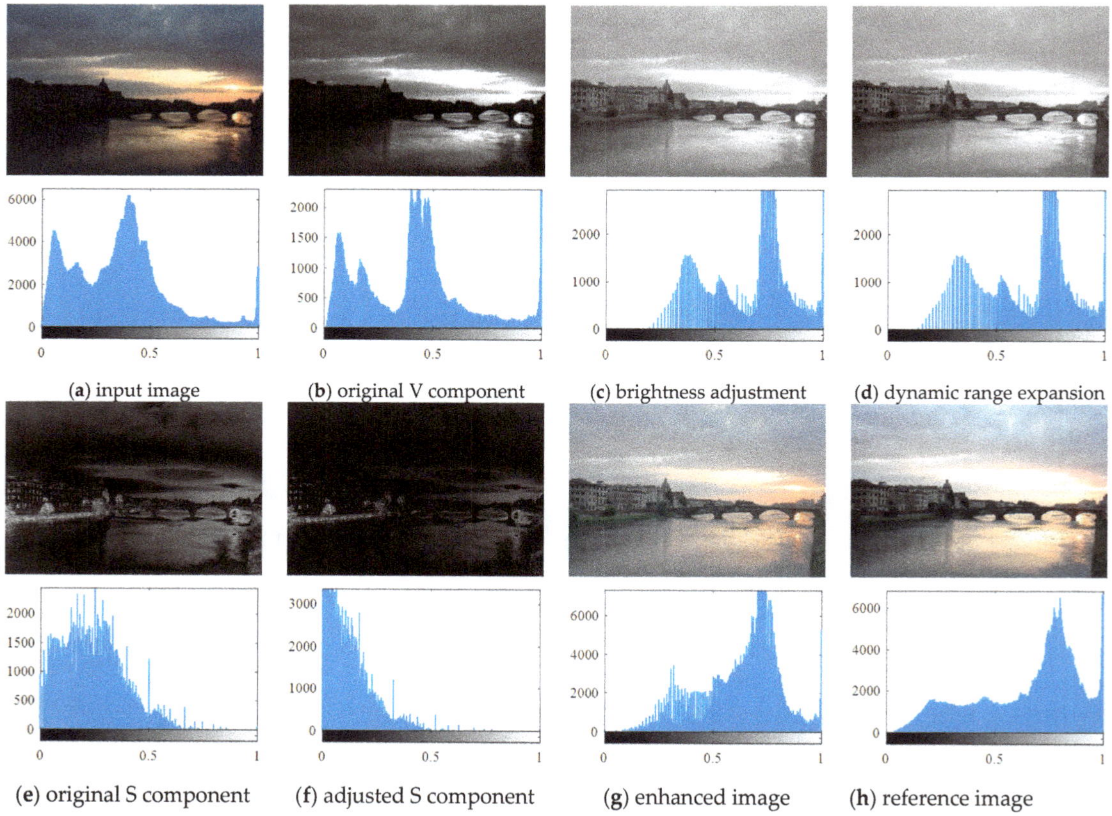

Figure 4. Low-light image enhancement process and corresponding grayscale histograms.

In our method, we use gamma correction to estimate the illumination and the Retinex model to achieve brightness enhancement. Compared with using filters to estimate the illumination, using gamma correction to estimate the illumination can effectively reduce the computational time. The key to gamma correction is to compute the value of the gamma parameter; the details of the gamma parameter determined are described as follows.

3.1. Brightness Enhancement

The gray levels of a low-light image are mainly concentrated in the low gray level area, and the dynamic range of low gray levels is very narrow. Combing Figure 2, we can see that the higher the gray level dynamic range of the input image, the higher the gray level dynamic range of the output image. Therefore, we use linear enhancement to stretch the gray level dynamic range before gamma correction, and we make the value of the stretched gray level fall within the range of (0, 1) to prevent over-enhancement. The used linear function in this paper can be expressed as follows.

$$Vmax = \max(V(x,y)) \tag{5}$$

$$V1(x,y) = \frac{1}{Vmax} * V(x,y) \tag{6}$$

where $Vmax$ denotes the maximum pixel value of V component, max(.) denotes take the maximum value of $V(x,y)$, $V(x,y)$ is the pixel value of the original V component at location (x,y), $V1(x,y)$ is the enhanced pixel value at location (x,y) and '$*$' represents the multiplication.

The maximum value of the low-light image is usually lower than 1; we can infer that $\frac{1}{Vmax} > 1$, so this linear function can stretch the dynamic range of the low-light image, and we also can obtain that $V1(x,y) \leq 1$.

After the gray level dynamic range is stretched, we adopt gamma correction to estimate illumination. For a low-light image, the lower the brightness intensity, the lower the gray level. Therefore, we take this feature into consideration. First, based on the global histogram, we compute the mean gray level value, which can reflect the overall brightness level to a certain extent. The corresponding computational formula is expressed as Equation (7), and we can obtain the mean gray level value via this equation.

$$m = \frac{\sum_{i=0}^{L} P(i) * i}{\sum_{i=0}^{L} P(i)} \tag{7}$$

where m is the mean value of gray levels, L denotes the maximum value of gray levels of an image and $P(i)$ is the histogram of gray level i.

In this paper, we assume that the gray levels more than zero and less than $m + 1$ are the extreme low gray levels. In fact, this part of the gray level is the key to determine the mean gray level of the low-light image. Based on the above descriptions, we design a formula to convert the gray level of this part into a constant, and use this constant to compute the gamma value. The corresponding transfer formula is expressed as Equation (8).

$$c = \frac{\sum_{i=1}^{m} P(i) * i}{128 * \sum_{i=1}^{m} P(i)} \tag{8}$$

where c is the value of conversion result and c is a positive number. Low-light images may have similar mean values, which will lead to similar c values. In order to enlarge the difference of c values among different images, we use the following expression to enlarge c values.

$$c1 = \frac{1}{1 + e^{-c}} \tag{9}$$

where $c1$ represents the enlarged c value. In addition, we also think that the focus of brightness enhancement lies in the low gray level area rather than the high gray level area. Therefore, we take the distribution of the low gray level as one of the important bases for estimating the gamma value. In order to calculate the distribution of the low gray level, the cumulative distribution function (CDF) is used to calculate the distribution of the gray level in this part. In this paper, we consider the gray level less than 128 to be the low gray level area.

$$cdf(j) = \sum_{0}^{j} pdf(i) \tag{10}$$

$$pdf(i) = \frac{p(i)}{M * N} \tag{11}$$

where $p(i)$ is the number of pixels that have gray level i, M and N are the length and width of the image, j is the threshold point of CDF and we set j equals to 128. Then we weigh the CDF value with the $c1$ value to obtain the gamma parameter value.

$$\gamma = w * c1 + (1 - w) * cdf \tag{12}$$

where γ represents the gamma parameter, w is the weighted value and equals to 0.48. Combining Equations (4), (6) and (12), we can get the final expression as follows.

$$VL(x,y) = V1(x,y)^{w*c1+(1-w)*cdf} \tag{13}$$

where $VL(x,y)$ denotes the pixel location (x,y) of illumination image. Combing Equations (3) and (13), we can get the reflectance, and it is shown as follows.

$$log(R) = log(V) - log(VL) \tag{14}$$

We get the enhanced V component as follows:

$$VE = \exp(log(V) - log(VL)) \tag{15}$$

The enhanced V component and corresponding histogram are shown in Figure 4c.

3.2. Dynamic Range Expansion

After brightness enhancement, the pixel values are easily concentrated in the higher gray level range, which leads to the grayscale dynamic range becoming narrow with low contrast in the enhanced image. We can adjust the contrast of the image by enlarging the V component gray level [30,31]. In order to avoid pixels values concentrated in the higher gray level range, we use a piecewise function to further stretch the gray level dynamic range to achieve dynamic range expansion. The corresponding expression can be expressed as follows.

$$VE'(x,y) = \begin{cases} VE(x,y), & VE(x,y) \geq 0.5 \\ 2 * (VE(x,y))^2, & VE(x,y) < 0.5 \end{cases} \tag{16}$$

The dynamic range enlarged V component and corresponding histogram are shown in Figure 4d.

3.3. Saturation Adjustment

In addition to brightness, the color saturation also directly affects the visual experience. In the HSV color space, the mean value of the S component and V component of a clear image should be approximately equal [32,33]. However, with the adjustment of brightness, the mean value of the V component changes greatly, which affects the image color. Based on the mean difference between the V component and the S component, Formula (20) is designed to adjust the S component. The details of our method are described as follows. Firstly, we use Equation (17) to compute the mean difference between the V component and S component.

$$VES = VE'mean - Smean \tag{17}$$

where VES is the mean difference, $VE'mean$ is the mean value of enhanced V component and $Smean$ is the mean value of S component. The expression used to compute $VE'mean$ is shown below.

$$VE'mean = \frac{\sum_0^i VE'(i) * i}{M * N} \tag{18}$$

where i denotes the gray level, and $VE'(i)$ is the number of pixels that have gray level i. M and N are the length and width of the image. Similiarly, we can get Equation (19) to compute the $Smean$.

$$Smean = \frac{\sum_0^i S(i) * i}{M * N} \tag{19}$$

where i denotes the gray level, $S(i)$ is the number of pixels that have gray level i. From the above description, we adjust the S component value to reduce the mean difference value between the VE' component and S component to achieve the purpose of color saturation adjustment. After VES is obtained, we use it to adjust the S component. According to Section 2.2, if we want to enlarge the value of the S component, we have to ensure that the gamma parameter lies in the range (0,1). On the contrary, we need to ensure that the parameter value is greater than 1 to reduce the value of the S component. Therefore, we use Equation (20) to achieve this step.

$$S1(x,y) = S(x,y)^{1+(-1)^{2-n}*(|VES|^2+|VES|)}, \; n = \left\{ \begin{array}{l} 0 \; VES < 0 \\ 1 \; VES \geq 0 \end{array} \right. \tag{20}$$

where $S1(x,y)$ denotes the pixel location (x,y) of the adjusted S component, and $S(x,y)$ is the pixel location (x,y) of the original S component. According to Equation (17), we can see that if $VES < 0$, we know that $VEmean < Smean$, so we need to reduce the value of the S component. Meanwhile, from Equation (20) we know that $n = 0$ and $1 + (-1)^{2-n} * \left(|VES|^2 + |VES| \right) > 1$, then we get $S1(x,y) < S(x,y)$. Similarly, we can see that when $VES > 0$, we also can get $S1(x,y) > S(x,y)$. The original S component and corresponding histogram are shown in Figure 4e and the adjusted S component and corresponding histogram are shown in Figure 4f.

4. Comparative Experiment and Discussion

This section describes the comparative experiment with the existing methods and experimental results. The comparative methods used include the LECARM algorithm [34], FFM algorithm [7], LIME algorithm [17], AFEM algorithm [1], JIEP algorithm [15] and SDD algorithm [35]. All comparative experiments are performed in MATLAB R2020b on a PC running Windows 10 with an Intel (R) Core (TM) i7-10875H CPU @ 2.30 GHz and 16 GB of RAM. Due to the length limitation of this paper, we use 10 images to illustrate the comparative results; the reference images are shown in Figure 5. All test images and reference images come from the public MEF dataset [36], which, in total, include 24 low-light images.

| (a) | (b) | (c) | (d) | (e) |

Figure 5. *Cont.*

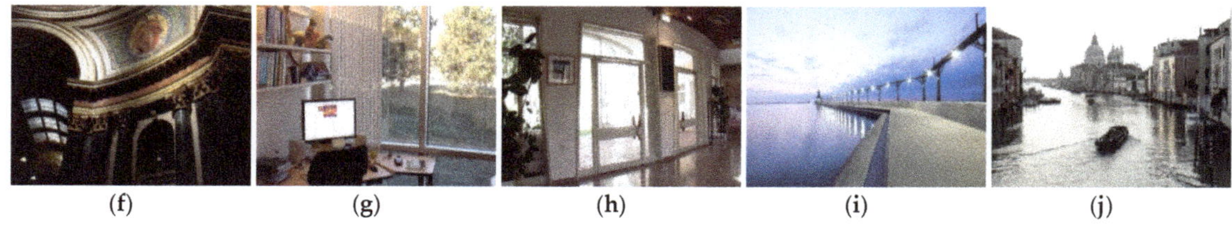

(f)   (g)   (h)   (i)   (j)

Figure 5. Reference images (**a–j**).

4.1. Computational Time Comparison

We test the time consumed for different algorithms to process different size images, and the test results are shown in Table 1.

Table 1. Time cost of different methods.

Image Size	100 × 100	700 × 700	1300 × 1300	1900 × 1900	2500 × 2500	3100 × 3100	3700 × 3700	4300 × 4300
LECARM	0.151	0.396	0.707	1.234	1.934	2.823	3.951	5.398
AFEM	0.048	0.204	0.566	1.136	2.014	3.075	4.674	5.959
LIME	0.030	0.124	0.394	0.825	1.437	2.203	3.209	4.363
FFM	0.182	5.043	17.071	36.819	65.744	95.577	142.183	197.190
SDD	0.222	8.882	34.930	79.808	139.754	209.301	345.587	526.162
JIEP	0.079	3.565	13.332	29.159	45.297	55.327	82.677	120.519
Proposed	0.013	0.071	0.249	0.519	0.909	1.419	2.076	2.804

In Table 1, the shortest times are highlighted in bold case values, and the second-shortest times are highlighted with underlined values. Table 1 shows that the proposed method takes the shortest time for processing each image due to the lowest computational complexity, in comparison to both the FFM method and SDD method, which consume the longer time. We also can learn that JIEP's time consumption is higher than the AFEM method and less than the FFM method. The time consumptions of AFEM, LECARM and LIME are similar because of the same computational complexity. Generally, the proposed RBFA algorithm consumes the least time on average, and the processing speed of the image is the fastest.

We made the data in Table 1 into a line chart to analyze the computational comparison of different methods as shown in Figure 6. Figure 6 shows that the computational complexity of the proposed method RBFA is $O(N)$, and it is the lowest among all the methods, in comparison to SDD's computational complexity, which is the highest. The computational complexity of the SDD method is $O(N^2)$, which results in the SDD method costing more time on image processing. The computational complexity of both the FFM method and JIEP method are $O(N\log N)$, but the time increment of the FFM method is higher than the JIEP method. The computation complexity of AFEM, LECARM, LIME and the proposed method MFGC are $O(N)$; although the computational complexity is the same, the time increment of the proposed algorithm is the smallest with the same amount of data, proving that the proposed method has the lowest computational complexity.

4.2. Visual Comparison

Although the results of the LECARM method preserve the original hue and saturation and have a higher brightness intensity of each image, this algorithm easily results in non-uniform global light, further decreasing the visual experience, such as in the mid area of Figures 7–10, which have higher illumination than other areas. The SDD method results show that there are some regions that become blurred, such as in the middle area of Figures 10 and 11. From the enhanced images, we can see that the performance of FFM is unstable, because some results of the FFM method are inadequate brightness enhancement, for instance, the whole area of Figures 12 and 13. From the results of the JIEP method, we can see that this algorithm is focused on normal under exposure and does not perform well in extreme low illuminance regions, such as in the middle region of Figure 8 and the bottom region of Figure 14. It is clear that the results of the LIME method show uneven brightness and over-enhancement in some areas, such as the lower middle area of Figure 10, middle area of Figures 14 and 15 and the wall in the Figure 16. The results of the AFEM algorithm are not satisfactory because the brightness increment is too small to restore the details covered with dark regions, such as the bottom area of Figures 14 and 16. As we can see, the results of the proposed RBFA method achieved the global brightness balance after enhancement via the proposed method; the color retained is more natural than the other methods.

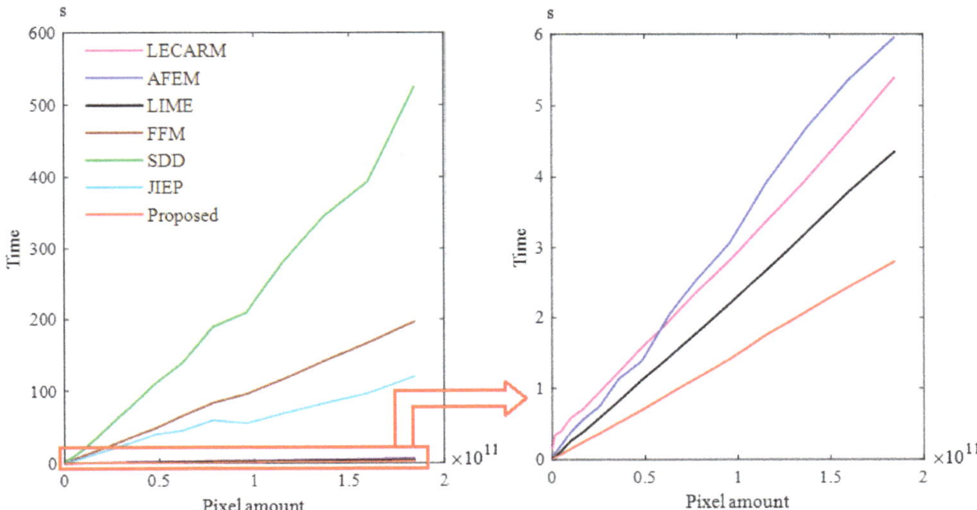

Figure 6. Result of computational complexity comparison.

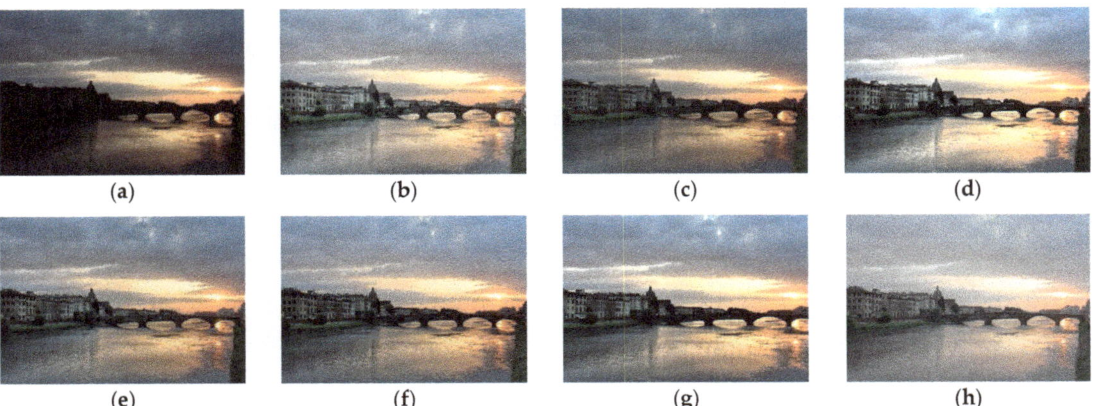

Figure 7. Comparing enhanced results of Arno with different methods. (**a**) Input image, (**b**) enhanced with LECARM, (**c**) enhanced with FFM, (**d**) enhanced with LIME, (**e**) enhanced with AFEM, (**f**) enhanced with JIEP, (**g**) enhanced with SDD, (**h**) enhanced with proposed RBFA method.

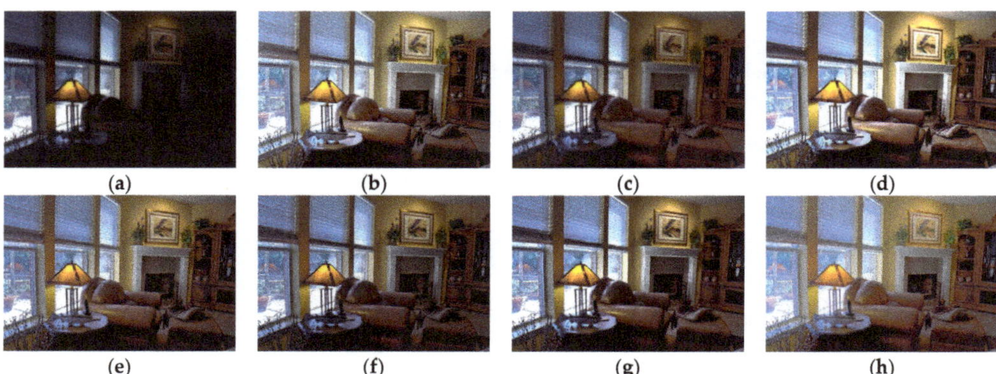

Figure 8. Comparing the enhanced results of Room with different methods. (**a**) Input image, (**b**) enhanced with LECARM, (**c**) enhanced with FFM, (**d**) enhanced with LIME, (**e**) enhanced with AFEM, (**f**) enhanced with JIEP, (**g**) enhanced with SDD, (**h**) enhanced with proposed RBFA method.

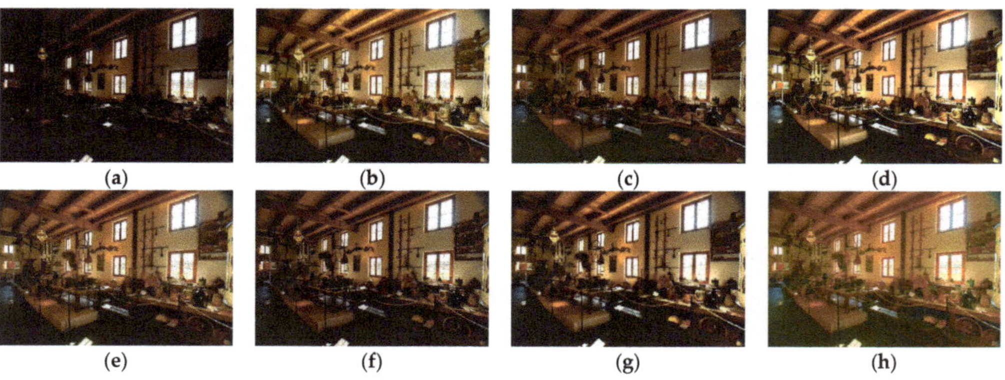

Figure 9. Comparing enhanced results of Farmhouse with different methods. (**a**) Input image, (**b**) enhanced with LECARM, (**c**) enhanced with FFM, (**d**) enhanced with LIME, (**e**) enhanced with AFEM, (**f**) enhanced with JIEP, (**g**) enhanced with SDD, (**h**) enhanced with proposed RBFA method.

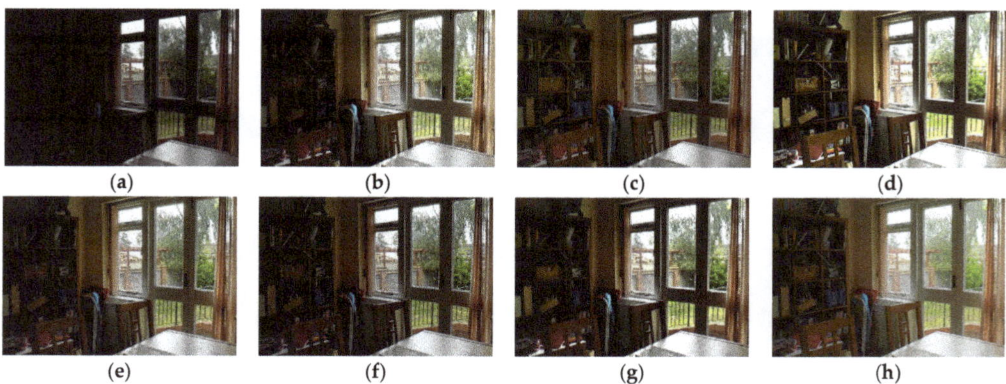

Figure 10. Comparing enhanced results of House with different methods. (**a**) Input image, (**b**) enhanced with LECARM, (**c**) enhanced with FFM, (**d**) enhanced with LIME, (**e**) enhanced with AFEM, (**f**) enhanced with JIEP, (**g**) enhanced with SDD, (**h**) enhanced with the proposed RBFA method.

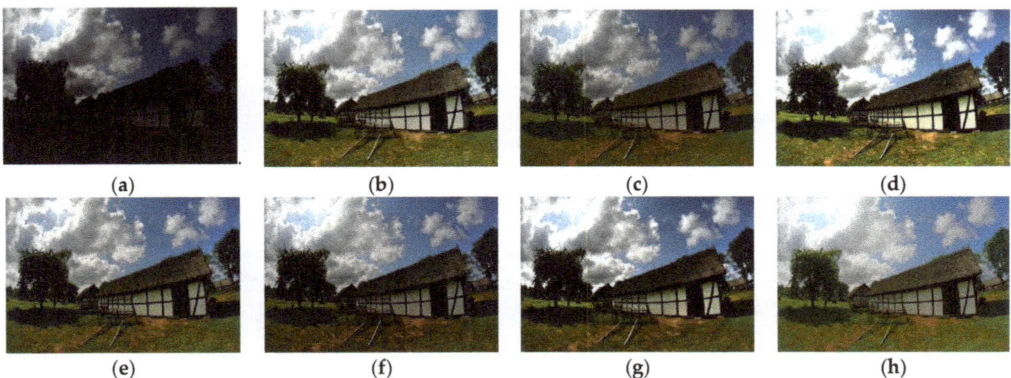

Figure 11. Comparing enhanced results of Cru with different methods. (**a**) Input image, (**b**) enhanced with LECARM, (**c**) enhanced with FFM, (**d**) enhanced with LIME, (**e**) enhanced with AFEM, (**f**) enhanced with JIEP, (**g**) enhanced with SDD, (**h**) enhanced with proposed RBFA method.

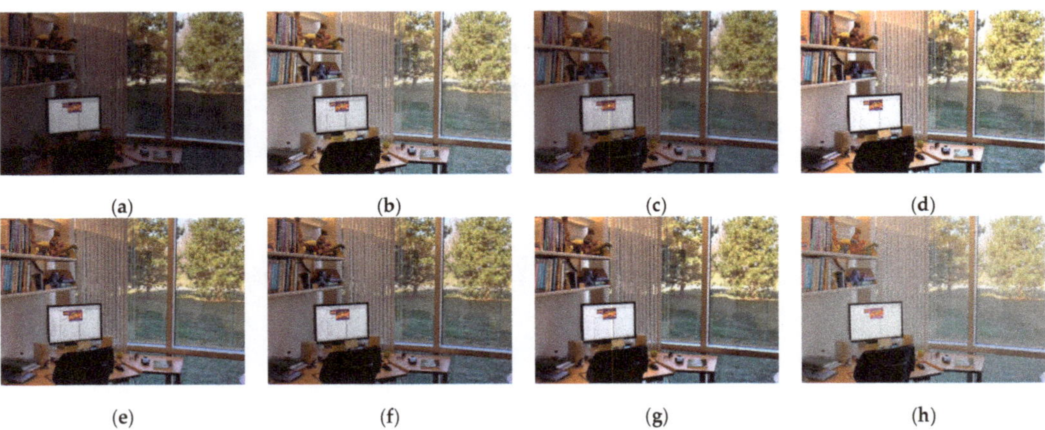

Figure 12. Comparing enhanced results of Office with different methods. (**a**) Input image, (**b**) enhanced with LECARM, (**c**) enhanced with FFM, (**d**) enhanced with LIME, (**e**) enhanced with AFEM, (**f**) enhanced with JIEP, (**g**) enhanced with SDD, (**h**) enhanced with proposed RBFA method.

4.3. Objective Assessment

Because human eyes often lose some details when we observe a picture, we choose one no-reference image quality assessment metric (perception-based image quality evaluator (PIQE)), three full reference image quality measure metrics (mean-squared error (MSE), structural similarity (SSIM), and peak signal-to-noise ratio (PSNR)) and lightness order error (LOE) to measure the quality of the enhanced images. The results of the different quality measure methods are shown in Table 2; these values represent the average value. The best scores are highlighted in bold case values, and the second-best scores are highlighted with underline values.

We can see from Table 2 that the proposed method obtained the best score four times and second-best score once. As shown in Table 2, the PIQE values of different methods fall within the range from 38.601 to 51.457, which means that the quality of all enhanced images is very similar and close, and the enhanced images via the proposed method obtained the best score. The smaller the LOE value, the more natural the enhancement effect. We can see that the LOE value of the proposed method is the best. This also means that the naturalness of the preservation of the proposed method is efficient. MSE is calculated by taking the average of the square of the difference between the reference image and enhanced image; the smaller the value is, the higher the similarity between the reference image and the enhanced image. The result of the proposed method is only 4.59 lower than the best LIME result. SSIM assesses the visual impact of three characteristics of an image: luminance, contrast and structure. The bigger the SSIM value, the higher the image quality; we see that the enhanced image via the proposed method preserved the highest similarity to the reference image. We know that the PSNR value of the proposed method is also the highest, which means that our method is useful for low-light image enhancement. Generally, the image quality enhanced by the proposed method is better than other comparative methods.

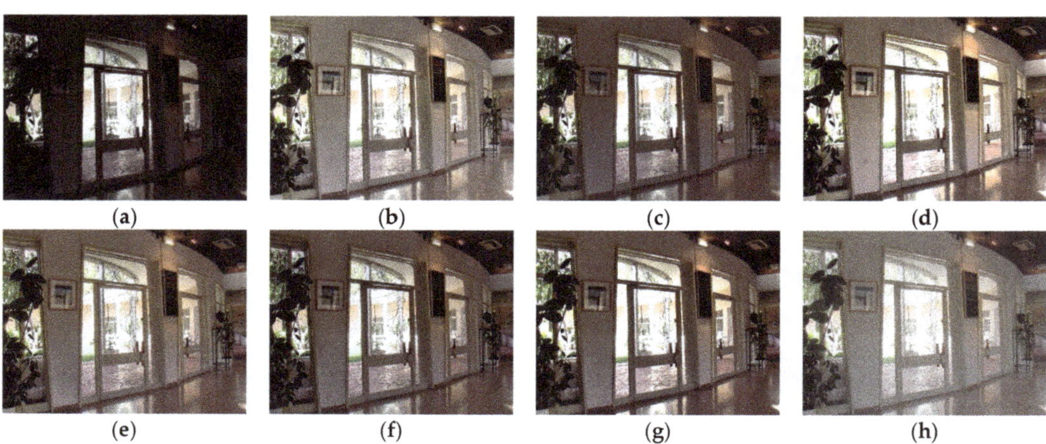

Figure 13. Comparing enhanced results of Door with different methods. (**a**) Input image, (**b**) enhanced with LECARM, (**c**) enhanced with FFM, (**d**) enhanced with LIME, (**e**) enhanced with AFEM, (**f**) enhanced with JIEP, (**g**) enhanced with SDD, (**h**) enhanced with proposed RBFA method.

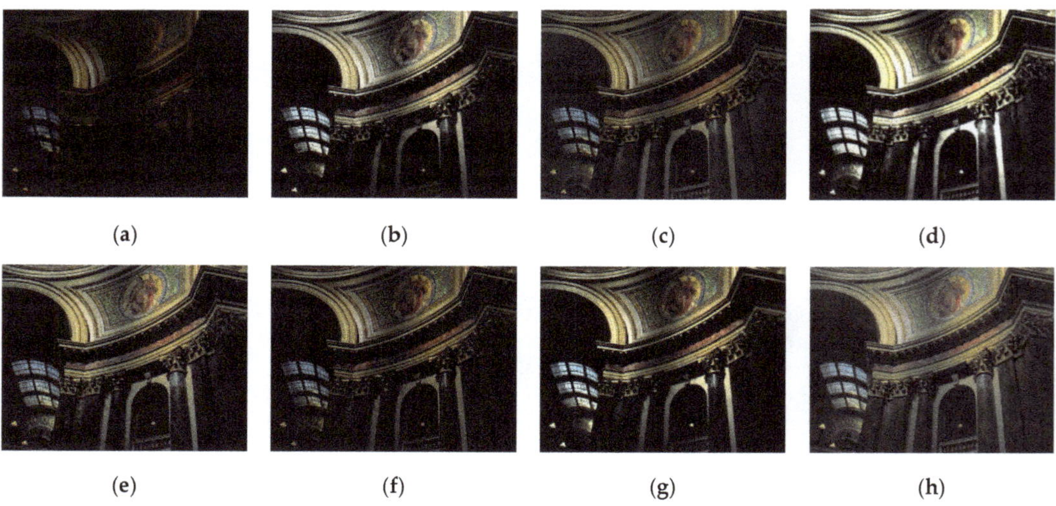

Figure 14. Comparing enhanced results of Capitol with different methods. (**a**) Input image, (**b**) enhanced with LECARM, (**c**) enhanced with FFM, (**d**) enhanced with LIME, (**e**) enhanced with AFEM, (**f**) enhanced with JIEP, (**g**) enhanced with SDD, (**h**) enhanced with proposed RBFA method.

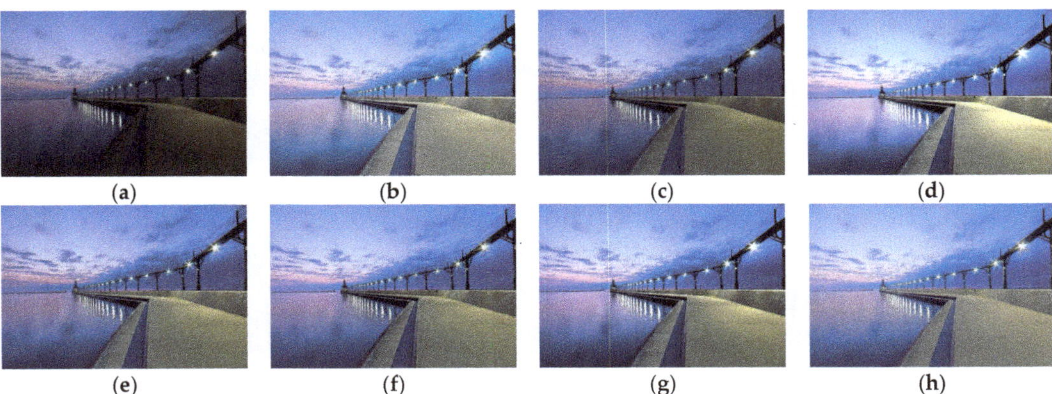

Figure 15. Comparing enhanced results of Venice with different methods. (**a**) Input image, (**b**) enhanced with LECARM, (**c**) enhanced with FFM, (**d**) enhanced with LIME, (**e**) enhanced with AFEM, (**f**) enhanced with JIEP, (**g**) enhanced with SDD, (**h**) enhanced with proposed RBFA method.

Figure 16. Comparing enhanced results of Venice with different methods. (**a**) Input image, (**b**) enhanced with LECARM, (**c**) Enhanced with FFM, (**d**) enhanced with LIME, (**e**) enhanced with AFEM, (**f**) enhanced with JIEP, (**g**) enhanced with SDD, (**h**) enhanced with proposed RBFA method.

Table 2. Results of image quality measure metrics with different methods.

Metrics	LECARM	AFEM	FFM	JIEP	LIME	SDD	Proposed
PIQE	39.818	39.809	42.884	40.072	42.705	51.457	**38.601**
LOE	415.594	253.646	291.906	296.568	749.862	493.806	**7.660**
MSE	3777.2175	2021.305	2823.849	2241.768	**1153.584**	1617.479	1158.174
SSIM	0.531	0.747	0.709	0.732	0.739	0.751	**0.753**
PSNR	12.504	16.350	14.464	15.847	18.136	17.511	**18.258**

5. Conclusions

We proposed the Retinex-based fast enhancement method in this paper. This method can address uneven brightness and greatly improve the brightness of low-light areas. The proposed method is more efficient. In general, the proposed RBFA algorithm performance is better than other state-of-the-art methods, combining the results of the comparative experiment, computational complexity comparison and quality assessment. In other words, the proposed RBFA method is a simple and efficient low-light image-enhancement algorithm.

Author Contributions: Conceptualization, S.L.; methodology, S.L.; software, W.L.; validation, L.H.; formal analysis, W.D.; investigation, L.H.; resources, Y.L.; data curation, L.H.; writing—original draft preparation, S.L.; writing—review and editing, S.L.; visualization, W.L.; supervision, W.L.; project administration, Y.L.; funding acquisition, Y.L. All authors have read and agreed to the published version of the manuscript.

Funding: This research was funded by Science and Technology Department of Sichuan Province, grant number 2020JDRC0026.

Acknowledgments: This work was supported by Science and Technology Department of Sichuan Province, People's Republic of China.

Conflicts of Interest: We declare that we do not have any commercial or associative interest that represents a conflict of interest in connection with the manuscript entitled 'Retinex-Based Fast Algorithm for Low-Light Image Enhancement'.

References

1. Fu, X.; Zeng, D.; Huang, Y.; Liao, Y.; Ding, X.; Paisley, J. A fusion-based enhancing method for weakly illuminated images. *Signal Process.* **2016**, *129*, 82–96. [CrossRef]
2. Wang, Y.F.; Liu, H.M.; Fu, Z.W. Low-Light Image Enhancement via the Absorption Light Scattering Model. (in English). *IEEE Trans. Image Process.* **2019**, *28*, 5679–5690. [CrossRef] [PubMed]
3. Wang, W.; Wu, X.; Yuan, X.; Gao, Z. An Experiment-Based Review of Low-Light Image Enhancement Methods. *IEEE Access.* **2020**, *8*, 87884–87917. [CrossRef]
4. Bora, D.J.; Bania, R.K.; Che, N. A Local Type-2 Fuzzy Set Based Technique for the Stain Image Enhancement. *Ing. Solidar.* **2019**, *15*, 1–22. [CrossRef]
5. Yun, H.J.; Wu, Z.Y.; Wang, G.J.; Tong, G.; Yang, H. A Novel Enhancement Algorithm Combined with Improved Fuzzy Set Theory for Low Illumination Images. *Math. Probl. Eng.* **2016**, *2016*, 1–9. [CrossRef]
6. Rahman, S.; Rahman, M.M.; Abdullah-Al-Wadud, M.; Al-Quaderi, G.D.; Shoyaib, M. An adaptive gamma correction for image enhancement. *Eurasip. J. Image Vide* **2016**, *2016*, 35–48. [CrossRef]
7. Dai, Q.; Pu, Y.F.; Rahman, Z.; Aamir, M. Fractional-Order Fusion Model for Low-Light Image Enhancement. *Symmetry* **2019**, *11*, 574. [CrossRef]
8. Reddy, E.; Reddy, R. Dynamic Clipped Histogram Equalization Technique for Enhancing Low Contrast Images. *Proc. Natl. Acad. Sci. India Sect. A Phys. Sci.* **2019**, *89*, 673–698. [CrossRef]
9. Ooi, C.H.; Kong, N.S.P.; Ibrahim, H. Bi-Histogram Equalization with a Plateau Limit for Digital Image Enhancement. *IEEE T Consum. Electr.* **2009**, *55*, 2072–2080. [CrossRef]
10. Singh, K.; Kapoor, R. Image enhancement using Exposure based Sub Image Histogram Equalization. *Pattern Recogn Lett.* **2014**, *36*, 10–14. [CrossRef]
11. Tan, S.F.; Isa, N.A.M. Exposure Based Multi-Histogram Equalization Contrast Enhancement for Non-Uniform Illumination Images. *IEEE Access* **2019**, *7*, 70842–70861. [CrossRef]
12. Zuiderveld, K. Contrast limited adaptive histogram equalization. *Graph. Gems Iv* **1994**, 474–485. [CrossRef]
13. Li, M.D.; Liu, J.Y.; Yang, W.H.; Sun, X.Y.; Guo, Z.M. Structure-Revealing Low-Light Image Enhancement Via Robust Retinex Model. *IEEE T Image Process* **2018**, *27*, 2828–2841. [CrossRef]
14. Zhou, Z.Y.; Feng, Z.; Liu, J.L.; Hao, S.J. Single-image low-light enhancement via generating and fusing multiple sources. *Neural Comput. Appl.* **2020**, *32*, 6455–6465. [CrossRef]
15. Cai, B.; Xu, X.; Guo, K.; Jia, K.; Hu, B.; Tao, D. A Joint Intrinsic-Extrinsic Prior Model for Retinex. In Proceedings of the 2017 IEEE International Conference on Computer Vision (ICCV), Venice, Italy, 22–29 October 2017; pp. 4020–4029.
16. Li, Z.; Xiaochen, H.; Jiafeng, L.; Jing, Z.; Xiaoguang, L. A Naturalness-Preserved Low-Light Enhancement Algorithm for Intelligent Analysis. *Chin. J. Electron.* **2019**, *28*, 316–324.
17. Guo, X.J.; Li, Y.; Ling, H.B. LIME: Low-Light Image Enhancement via Illumination Map Estimation. *IEEE T Image Process* **2017**, *26*, 982–993. [CrossRef] [PubMed]
18. Kim, W.; Lee, R.; Park, M.; Lee, S.H. Low-Light Image Enhancement Based on Maximal Diffusion Values. *IEEE Access* **2019**, *7*, 129150–129163. [CrossRef]
19. Wang, W.C.; Chen, Z.X.; Yuan, X.H.; Wu, X.J. Adaptive image enhancement method for correcting low-illumination images. *Inf. Sci.* **2019**, *496*, 25–41. [CrossRef]
20. Chang, Y.; Jung, C.; Ke, P.; Song, H.; Hwang, J. Automatic Contrast-Limited Adaptive Histogram Equalization With Dual Gamma Correction. *IEEE Access* **2018**, *6*, 11782–11792. [CrossRef]
21. Srinivas, K.; Bhandari, A.K. Low light image enhancement with adaptive sigmoid transfer function. *IET Image Process* **2020**, *14*, 668–678. [CrossRef]
22. Kansal, S.; Tripathi, R.K. Adaptive gamma correction for contrast enhancement of remote sensing images. *Multimed Tools Appl* **2019**, *78*, 25241–25258. [CrossRef]

23. Al-Hashim, M.A.; Al-Ameen, Z. Retinex-based multiphase algorithm for low-light image enhancement. *Traitement Du Signal* **2020**, *37*, 733–743. [CrossRef]

24. Ashiba, M.I.; Tolba, M.S.; El-Fishawy, A.S.; El-Samie, F.E.A. Gamma correction enhancement of infrared night vision images using histogram processing. *Multimed Tools Appl* **2019**, *78*, 27771–27783. [CrossRef]

25. Kallel, F.; Ben Hamida, A. A New Adaptive Gamma Correction Based Algorithm Using DWT-SVD for Non-Contrast CT Image Enhancement. *IEEE Trans Nanobioscience* **2017**, *16*, 666–675. [CrossRef]

26. Chandrasekharan, R.; Sasikumar, M. Fuzzy Transform for Contrast Enhancement of Nonuniform Illumination Images. *IEEE Signal Proc Let* **2018**, *25*, 813–817. [CrossRef]

27. Li, Z.; Jia, Z.; Yang, J.; Kasabov, N. Low Illumination Video Image Enhancement. *IEEE Photonics J.* **2020**, *12*, 1–13. [CrossRef]

28. Dhal, K.G.; Ray, S.; Das, S.; Biswas, A.; Ghosh, S. Hue-Preserving and Gamut Problem-Free Histopathology Image Enhancement. *Iran. J. Sci. Technol. Trans. Electr. Eng.* **2019**, *43*, 645–672. [CrossRef]

29. Lyu, W.J.; Lu, W.; Ma, M. No-reference quality metric for contrast-distorted image based on gradient domain and HSV space. *J. Vis. Commun. Image Represent.* **2020**, *69*, 102797–102806. [CrossRef]

30. Deng, H.; Sun, X.; Liu, M.; Ye, C.; Zhou, X. Image enhancement based on intuitionistic fuzzy sets theory. *Iet Image Process* **2016**, *10*, 701–709. [CrossRef]

31. Wang, Z.; Wang, K.; Liu, Z.; Zeng, Z. Study on Denoising and Enhancement Method in SAR Image based on Wavelet Packet and Fuzzy Set. In Proceedings of the 2019 IEEE 4th Advanced Information Technology, Electronic and Automation Control Conference (IAEAC), Chengdu, China, 20–22 December 2019; pp. 1541–1544.

32. Zhu, Q.; Mai, J.; Shao, L. A Fast Single Image Haze Removal Algorithm Using Color Attenuation Prior. *IEEE T Image Process* **2015**, *24*, 3522–3533. [CrossRef]

33. Gupta, R.; Khari, M.; Gupta, V.; Verdu, E.; Wu, X.; Herrera-Viedma, E.; Crespo, R.G. Fast Single Image Haze Removal Method for Inhomogeneous Environment Using Variable Scattering Coefficient. *Cmes-Comput. Modeling Eng. Sci.* **2020**, *123*, 1175–1192. [CrossRef]

34. Ren, Y.; Ying, Z.; Li, T.H.; Li, G. LECARM: Low-Light Image Enhancement Using the Camera Response Model. *IEEE T Circ Syst Vid* **2019**, *29*, 968–981. [CrossRef]

35. Hao, S.; Han, X.; Guo, Y.; Xu, X.; Wang, M. Low-Light Image Enhancement with Semi-Decoupled Decomposition. *IEEE T Multimed.* **2020**. [CrossRef]

36. Ma, K.; Duanmu, Z.; Yeganeh, H.; Wang, Z. Multi-Exposure Image Fusion by Optimizing A Structural Similarity Index. *IEEE Trans. Comput. Imaging* **2018**, *4*, 60–72. [CrossRef]

Article

Multi-Person Tracking and Crowd Behavior Detection via Particles Gradient Motion Descriptor and Improved Entropy Classifier

Faisal Abdullah [1], Yazeed Yasin Ghadi [2], Munkhjargal Gochoo [3], Ahmad Jalal [1] and Kibum Kim [4],*

[1] Department of Computer Science, Air University, Islamabad 44000, Pakistan; 191633@students.au.edu.pk (F.A.); ahmadjalal@mail.au.edu.pk (A.J.)
[2] Department of Computer Science and Software Engineering, Al Ain University, Abu Dhabi 122612, United Arab Emirates; Yazeed.ghadi@aau.ac.ae
[3] Department of Computer Science and Software Engineering, United Arab Emirates University, Al Ain 15551, United Arab Emirates; mgochoo@uaeu.ac.ae
[4] Department of Human-Computer Interaction, Hanyang University, Ansan 15588, Korea
* Correspondence: kibum@hanyang.ac.kr

Abstract: To prevent disasters and to control and supervise crowds, automated video surveillance has become indispensable. In today's complex and crowded environments, manual surveillance and monitoring systems are inefficient, labor intensive, and unwieldy. Automated video surveillance systems offer promising solutions, but challenges remain. One of the major challenges is the extraction of true foregrounds of pixels representing humans only. Furthermore, to accurately understand and interpret crowd behavior, human crowd behavior (HCB) systems require robust feature extraction methods, along with powerful and reliable decision-making classifiers. In this paper, we describe our approach to these issues by presenting a novel Particles Force Model for multi-person tracking, a vigorous fusion of global and local descriptors, along with a robust improved entropy classifier for detecting and interpreting crowd behavior. In the proposed model, necessary preprocessing steps are followed by the application of a first distance algorithm for the removal of background clutter; true-foreground elements are then extracted via a Particles Force Model. The detected human forms are then counted by labeling and performing cluster estimation, using a K-nearest neighbors search algorithm. After that, the location of all the human silhouettes is fixed and, using the Jaccard similarity index and normalized cross-correlation as a cost function, multi-person tracking is performed. For HCB detection, we introduced human crowd contour extraction as a global feature and a particles gradient motion (PGD) descriptor, along with geometrical and speeded up robust features (SURF) for local features. After features were extracted, we applied bat optimization for optimal features, which also works as a pre-classifier. Finally, we introduced a robust improved entropy classifier for decision making and automated crowd behavior detection in smart surveillance systems. We evaluated the performance of our proposed system on a publicly available benchmark PETS2009 and UMN dataset. Experimental results show that our system performed better compared to existing well-known state-of-the-art methods by achieving higher accuracy rates. The proposed system can be deployed to great benefit in numerous public places, such as airports, shopping malls, city centers, and train stations to control, supervise, and protect crowds.

Keywords: bat optimization; human crowd behavior (HCB); improved entropy (IE); Jaccard similarity; multi-person counting; particles gradient motion (PGM); speeded up robust features (SURF)

Citation: Abdullah, F.; Ghadi, Y.Y.; Gochoo, M.; Jalal, A.; Kim, K. Multi-Person Tracking and Crowd Behavior Detection via Particles Gradient Motion Descriptor and Improved Entropy Classifier. *Entropy* **2021**, *23*, 628. https://doi.org/10.3390/e23050628

Academic Editor: Amelia Carolina Sparavigna

Received: 1 April 2021
Accepted: 12 May 2021
Published: 18 May 2021

Publisher's Note: MDPI stays neutral with regard to jurisdictional claims in published maps and institutional affiliations.

1. Introduction

Multi-person tracking is currently one of the most essential and challenging research topics in the computer vision community [1–9]. Because of the common availability of high-quality low-cost video cameras and considering the inefficiency of manual surveillance and monitoring systems, automated video surveillance is now essential for today's crowded and complex environments. To monitor, control, and protect crowds, accurate information about numbers plays a vital role in operational and security efficiencies [10–16]. The counting and tracking of many persons is a challenging problem [17–25] due to occlusions, the constant displacement of people, different perspectives and behaviors, varying illumination levels, and because, as the crowd gets bigger, the allocation of pixels per person decreases.

A primary concern in surveillance and monitoring systems is to identify human crowd behaviors and supervise the crowd to prevent disasters and unforeseen events [26–34]. The analysis of human behavior in crowded scenes is one of the most important and challenging areas in current research [35–43]. Traditional visual surveillance systems that depend purely on manpower to analyze videos is inefficient because of the enormous number of cameras and screens that require monitoring, human fatigue due to time spent on lengthy monitoring periods, paucity of essential fore-knowledge and training in what to look for, and also because of the colossal amount of video data that is generated per day. Such issues necessitate an automated visual surveillance system that can reliably detect, isolate, analyze, identify, and alert responders to unusual events in real time. Automated surveillance systems seek to detect human behaviors automatically in crowded scenes, and it has many potential applications, such as security, care of the elderly and infirm, traffic monitoring, inspection tasks, military applications, robotic vision, sports analysis, video surveillance, and pedestrian traffic monitoring [44–52].

In this research article, we propose a robust new particles-based approach for multi-person counting and tracking, which addresses the problematic fact that, as the density of a crowd increases, the number of pixels allocated per human decreases. By using our particles-based approach, we were able to count and track multiple persons in crowded scenes and efficiently deal with occlusions, arbitrary movements, and overlaps. We also propose a new approach for crowd behavior detection using an improved entropy classifier based on the fusion of global and local descriptors extraction. First of all, we applied pre-processing steps on extracted video frames for noise removal, edge detection, and contrast adjustment, then human/non-human detection was performed using multi-level thresholding and morphological operations. We applied a distance algorithm for human silhouette extraction. After that, our work involved two facets: (i) multi-people tracking and (ii) crowd behavior detection. In the multi-person tracking phase, we first verified the extracted silhouettes by a particles force model, then we converted extracted foreground objects into particles, and, using physics phenomena of the mutually interacting particles force model, non-human objects were discarded. As every extracted human silhouette is a collection of particles, by treating groups of particles that make one silhouette as a cluster, we performed labeling and cluster estimation using a K-nearest neighbors search algorithm to count the persons. We then fixed the human silhouettes with a unique integer ID, and, using normalized cross correlation as a cost function and the Jaccard similarity index, multi-person tracking was performed. However, for crowd behavior detection, we used a fusion of global and local descriptors, that is, after foreground extraction, we extracted a human crowd contour as a global descriptor and a particles gradient motion (PGM) descriptor, along with geometric and speeded up robust features (SURF) as local descriptors. Using this fusion of global and local descriptors, bat optimization was then applied for optimal descriptors. Finally, by using Shannon's information entropy theory [53], we introduced an improved entropy classifier to detect crowd behavior.

Experimental results show that our proposed system performed better compared to existing well-known state-of-the-art methods. The proposed system has huge potential applications, such as crowd density estimation, security, care of the elderly and vulnerable, sports analysis, inspection tasks, military applications, robotic vision, video surveillance, and pedestrian traffic monitoring. The major contributions of this paper can be highlighted as follows:

1. We propose a new particles force model for human silhouettes verification, which is a necessary step for accurate counting and tracking of multiple persons in crowded scenes.
2. We developed a novel particles gradient motion local descriptor and human crowd contour as a global descriptor, while the fusion of global and local features was used for crowd behavior detection.
3. We designed an improved entropy classifier to analyze contextual information and classify crowd behavior in a more efficient manner.
4. We evaluated the performance of our proposed multi-person tracking approach on a publicly available benchmark PETS2009 dataset while crowd behavior detection performance was evaluated on the publicly available benchmark UMN dataset and the proposed method was fully validated for efficacy, surpassing other state-of-the-art methods, including deep learning.

The remaining structure of this paper was arranged as follows: Section 2 describes related work. A detailed overview of the proposed model for multi-person tracking and crowd behavior detection is mentioned in Section 3, which includes preprocessing, human silhouettes extraction, the particles force model, multi-person counting, multi-person tracking, global and local features extraction, bat optimization, and an improved entropy classifier. In Section 4, we evaluate the performance of our proposed approach on a publicly available benchmark dataset and give a detailed comparison of our proposed approach with other state-of-the-art methods. Lastly, in Section 5, we sum up the paper and outline future directions.

2. Related Work

During the last few years, several algorithms and systems have been developed by different researchers for crowd counting, tracking, and human behavior detection [54–62]. Here, we divide the related work into two parts, namely, human crowd behavior detection systems and multi-person counting and tracking systems.

2.1. Crowd Behavior Detection Systems

Many contributions have been proposed to describe crowd behavior using various models [63–69]. Crowd behavior detection is a challenging problem due to the arbitrary movements of individuals and groups, partial or full occlusions, different outlooks and behaviors, posture changes, and composite backgrounds [70–76]. To detect human behaviors automatically in crowded areas, S. Wu et al. in [77] constructed a density function of optical flow based on class-conditional probability and described the motion of crowds using divergent centers and potential destinations so that anomalies can be detected on the basis of a Bayesian framework. However, the system is not effective for arbitrary movements or overlaps. S. Choudhary et al. in [78] proposed a SIFT feature extraction technique, along with a Genetic Algorithm for optimal feature extraction; anomalies were detected by checking feature set movement behaviors. Their proposed system has a very high computational processing demand. Direkoglu et al. in [79] used a one-class SVM, along with features based on optical flow to detect crowd behavior; their system is limited by the accuracy limitations of optical flow estimation. W. G. Aguilar et al. in [80] introduced a moved-pixels density-based statistical modeling approach for detecting abnormal crowd behavior. This system has low computational cost, but the efficiency decreases with increasing complexity of the situation being monitored, e.g., serious occlusions. A. Shehzed et al. in [81] first detected humans and then the gaussian smoothing technique was used to detect anomalous behavior; however, the accuracy of the system decreases with illumination changes and occlusions because thresholding is used for detection. W. Ren et al. in [82] introduced a behavior entropy model for detecting abnormal crowd behavior using spatio-temporal information, along with behavior certainty of pixels, but the system is vulnerable to certain misclassifications due to interclass similarities. G. Wang et al. in [83] addressed the crowd behavior detection problem by using the pyramid Lucas-

Kanade optical flow [84] method based on location estimation of adjacent flow; however, the proposed method is not effective for an unstructured crowd. R Mehran et al. in [85] placed a grid of particles on the image and introduced a social force model for detecting crowd behavior. Bellomo, N. et al. in [86] pursued two specific objectives: the derivation of a general mathematical structure based on appropriate developments of the kinetic theory suitable for capturing the main features of crowd dynamics and the derivation of macroscopic equations from the underlying mesoscopic description. Colombo, R.M. et al. in [87] dealt with macroscopic modelling of crowd movements, particularly how non-local interactions are influenced by walls, obstacles, and exits. An ad hoc numerical algorithm, along with heuristic evaluation of its convergence, was also provided. Khan, S.D. et al. in [88] proposed scale estimation network SENet and head detection network. The SENet takes the input image and predicts the distribution of scales (in terms of histogram) of all heads in the input image, which are later on classified by a detection network.

2.2. Multi-Person Counting and Tracking Systems

True foreground extraction, i.e., human pixels, is only one of the primary steps for accurate counting and tracking of humans in crowded scenes [89–93]. Several approaches and systems have been introduced by many researchers for multi-person counting and tracking. In [94], S. Choudri et al. proposed a pixels-based people counting model using the fusion of a pixel map-based algorithm along with human detection to count only human classified pixels. They applied a depth map, image segmentation, and a human presence map that was updated with a human mask for the purpose of counting people; however, the system has misclassification problems due to interclass similarities. H. Chen et al. in [95] proposed a new color and intensity patch segmentation approach for tracking and detection of human body parts and for the full body. They applied fusion of color space segmentations for the detection of body parts and for the full body. For tracking, based on the velocity of a target, they adaptively selected the track gate size. A target's likely forward position was predicted based on the target's previous velocity and direction. The proposed algorithm achieved satisfactory results only when the count of peoples was limited in the view, i.e., efficiency decreases as the crowd increases. In [96], J. Garcia et al. introduced a head tracking-based directional people counter. Using several circular patterns and preprocessing steps, people's heads were detected. For the tracking application, a Kalman filter was used, and counting was achieved on the bases of head detection and tracking. The effectiveness of the proposed algorithm decreases during serious occlusions, arbitrary movements, and overlaps. M. Vinod et al. in [97] introduced object tracking and counting using new morphological techniques. The frame-difference technique, followed by mor-phological processing and region growing, was used for counting people. Moving objects were extracted by determining their movements, and then tracking was performed using color features. As the illumination of the scene changed, the efficiency of the proposed algorithm decreased. G. Liu et al. in [98] proposed a tracker based on a correlation filter. Kalman filter applications were used for tracking. They designed a tracker that detects numerous positions and alternate templates. However, the system was not efficient in dealing with complex situations, such as occlusions and random movements. E. Ristani et al. in [99] used deep learning to track multi-persons. Using CNN, they extracted features and then introduced a weighted triple loss strategy to assign weights during training. Their system was computationally complex, and a huge dataset was essential for training. H. Xu et al. in [100] located humans by their shoulders and heads, and, for tracking, they used trajectory analysis and the Kalman filter, but the system was not effective for arbitrary movements or overlaps.

3. Proposed System Methodology

This section elaborates our proposed methodology for multi-person tracking and crowd behavior detection. We propose a robust multi-person tracking system based on a particles force model and human crowd behavior detection system using an improved entropy classifier with spatio-temporal and particles gradient motion descriptors. In the proposed system, the first step is the preprocessing of extracted video frames from a static camera. Secondly, object detection is transacted using multi-level thresholding, morphological operations, and labeling. Thirdly, for human silhouette extraction, a distance algorithm is applied, and non-human filtering is performed on all extracted labeled objects. At this stage, we administered our work into two streams: the first was for multi-person counting and tracking, where we first performed a human silhouette verification step by converting extracted objects into particles and a robust particles force model was introduced for human silhouette verification. In the next step, after verification of human silhouettes, as all verified human silhouettes are a collection of particles, by treating each group of particles as a cluster we performed labeling and cluster estimation using a K-nearest neighbors searching algorithm for multi-person counting. After that, for multi-person tracking, the position of each detected human silhouette was then located and locked by assigning an integer ID for temporally fixing each human silhouette in the full video, and detected fixed humans were tracked using a Jaccard Similarity Index. However, in the second facet, for crowd behavior detection, the extracted foreground objects were passed through a feature extraction step and multiple distinguishable global and local features were extracted from every frame. After that, all the extracted features were standardized using the bat optimization algorithm. Lastly, in the classification phase, an improved entropy classifier was proposed for detection of crowd behavior. Figure 1 depicts the synoptic schematics of our proposed system.

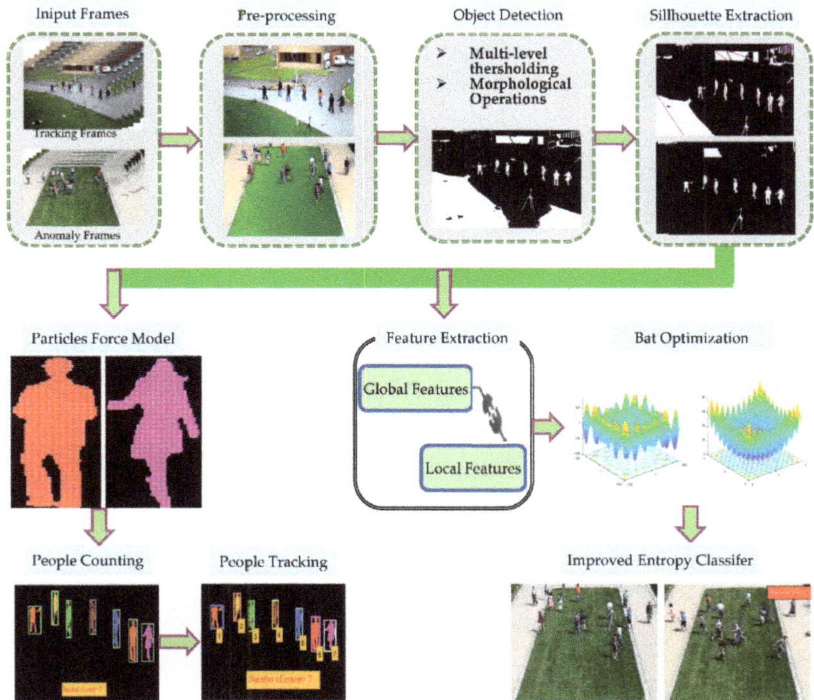

Figure 1. Synoptic schematics of the proposed Multi-Person Tracking and Crowd Behavior Detection system.

3.1. Pre-Processing

During image pre-processing, color frames were extracted from a static video camera $E = [f_1, f_2, f_3, \ldots, f_Z]$, where Z is the total number of frames. These color images were then passed through a Laplacian filter to reduce the noise and sharpen the edges. A Laplacian filter was applied using Equation (1):

$$\nabla^2 f = \frac{\partial^2 f}{\partial^2 x} + \frac{\partial^2 f}{\partial^2 y} \tag{1}$$

where $\nabla^2 f$ is the 2nd order derivative for obtaining the filtered mask. However, a pure Laplacian filter did not produce an enhanced image, thus, to achieve the sharpened enhanced image, we subtracted the Laplacian outcome from the original image using Equation (2):

$$g(x,y) = f(x,y) - \left[\nabla^2 f\right] \tag{2}$$

where the $g(x,y)$ is the sharpened image and $f(x,y)$ is the input image. After obtaining the sharpened image $g(x,y)$, histogram equalization was performed on the sharpened image in order to adjust the contrast of an image using Equation (3):

$$s_k = T(r_k) = (L-1)\sum_{j=0}^{k} p_r(r_j) \quad k = 0,\ 1,\ 2,\ \ldots,\ L-1 \tag{3}$$

where variable r denotes the intensities of an input image to be processed. As usual, we assumed that r is in the range $[0\ L-1]$, with $r = 0$ representing black and $r = L-1$ representing white, while s represents the output intensity level after intensity mapping for every pixel in the input image, having intensity r. However, $p_r(r)$ is the probability density function (PDF) of r, where the subscript on p were used to indicate that it was a PDF of r. Thus, a processed (output) image was achieved using Equation (3) by mapping each pixel in the input image with intensity r_k into a corresponding pixel with level s_k in the output image, as shown in Figure 2.

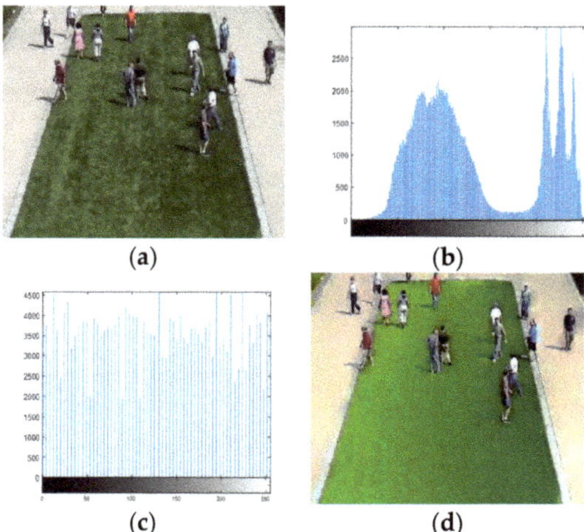

Figure 2. Preprocessing steps. (**a**) Original color frame of a video, (**b**) histogram of original image, (**c**) histogram of enhanced image, and (**d**) enhanced image.

3.2. Human Silhouettes Extraction

After obtaining the preprocessed frames, we performed human/non-human detection by performing multi-level thresholding using Equation (4), as depicted in Figure 3c.

$$th(x,y) = \begin{cases} 1 & \text{if} \quad l(x,y) > t_1, t_2, t_3 \\ 0 & \text{otherwise} \end{cases} \tag{4}$$

where $th(x,y)$ is the threshold image and t_1, t_2, t_3 are the applied thresholds that are defined by Otsu's procedure. In order to extract more useful information, the resultant binary image was inverted using a point processing operation that subtracts every pixel of an image from the maximum level of the image, as shown in Equation (5).

$$C(x,y) = 1 - th(x,y) \tag{5}$$

where $C(x,y)$ is the inverted image, as shown in Figure 3d, and $th(x,y)$ is the binary image with a maximum level of 1. After obtaining the human/non-human binary foreground frames, we performed morphological operations to remove imperfections in the inverted image C. For the removal of small unwanted objects, erosion was performed, and then, to fill small holes while preserving the size and shape of objects, morphological closing was performed. Every object in image C was first eroded using erosion as represented in Equation (6) and then dilated using Equation (7), after which the dilated image was eroded again using the disk-shaped structuring element, as shown in Equation (8).

$$m(x,y) = \begin{cases} 1 & \text{if} \quad S \text{ fits } C \\ 0 & \text{otherwise} \end{cases} \tag{6}$$

$$m(x,y) = \begin{cases} 1 & \text{if} \quad S \text{ hits } C \\ 0 & \text{otherwise} \end{cases} \tag{7}$$

$$Mo = (C \ominus S)((C \oplus S) \ominus S) \tag{8}$$

where C represents the input inverted image and S is the disk-shaped structuring element used for erosion and dilation, while Mo is the resultant image. The erosion of C by S is denoted as $(C \ominus S)$; however, the dilation of C by S is denoted as $(C \oplus S)$. After morphological operations, all the objects in the image were grouped and labeled, which helped in extracting and uniquely analyzing every object that was required for human silhouette extraction.

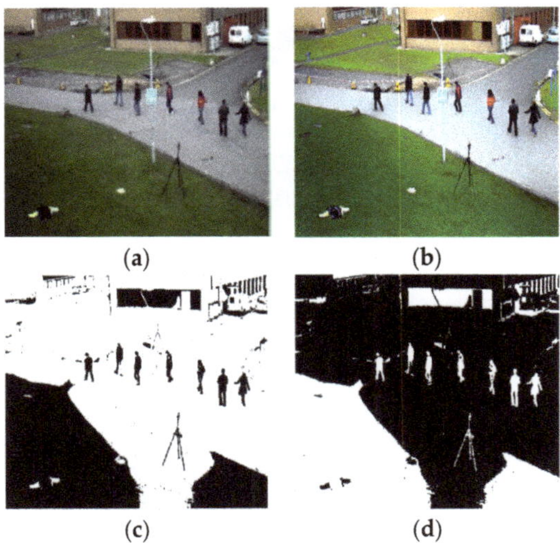

Figure 3. Object detection steps. (**a**) Original color frame of a video, (**b**) enhanced image, (**c**) binary image after multi-level thresholding, and (**d**) inverse of a threshold image.

After human/non-human detection, for human silhouette extraction, we calculated the center and extreme points of each of the labeled objects of M_o, then we extracted each object one by one, and the distance from center to two extreme points was calculated for every object for non-human filtering, as shown in Figure 4. The same procedure was adopted for the frames from frame 1 to frame Z.

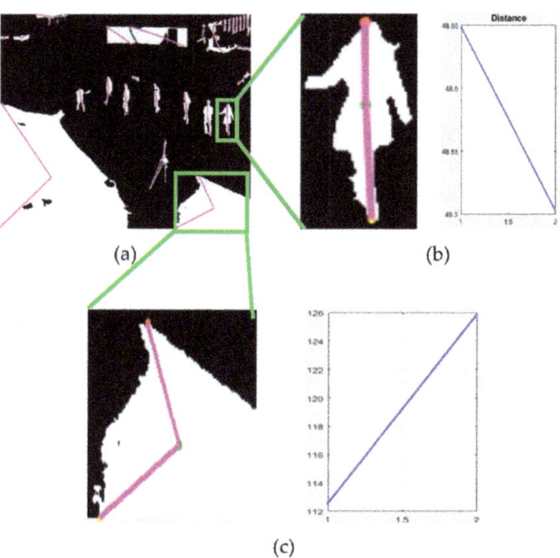

Figure 4. Human silhouette extraction. (**a**) Distance algorithm from the center to two extreme points for every object, (**b**) single silhouette extracted uniquely through labeling, along with its distance graph, and (**c**) a single non-silhouette, along with its distance graph.

After calculating the distances, those objects whose distances were greater than the set threshold were discarded using Equation (9), and only silhouettes resembling humans were retained.

$$E_h = \begin{cases} 0 & \text{if } d_1 > T \cap d_2 > T \\ 1 & \text{otherwise} \end{cases} \tag{9}$$

where the distance from the center to one extreme point is denoted by d_1, the center to the other extreme point distance is represented by d_2, T is the set threshold and E_h is the resultant image. After human silhouette extraction, most of the non-human objects were discarded by the distance algorithm; however, some non-human objects that resembled human objects remained.

3.3. Multi-Person Tracking

For accurate human tracking, the extraction of the true foreground, i.e., human pixels only, is a primary step. Thus, after application of the distance algorithm (mentioned in Section 3.2) for multi-person tracking, we performed the human silhouette verification step using the particles force model, and then the multi-person counting and tracking steps were executed.

3.3.1. Human Silhouettes Verification: Particles Force Model

We present a robust particles force model for human silhouette verification. First of all, every extracted labeled silhouette was converted into particles, as shown in Figure 5a. We treated all pixels as fluid particles, thus, every extracted silhouette was a collection of many particles, as depicted in the magnified view in Figure 5b. Therefore, in our designed method, each silhouette was represented by a set of particles $Q = [p_1, p_2, p_3, \cdots , p_N]$, where N is the total number of particles in one silhouette.

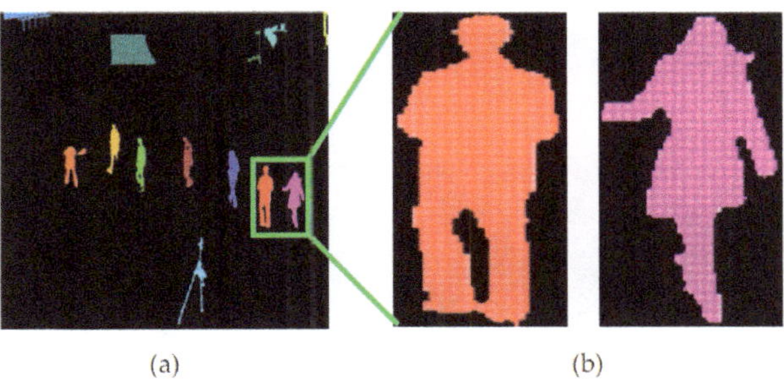

Figure 5. The particles force model. (**a**) Particle conversion of every extracted silhouette and (**b**) magnified view of particle conversion.

We know from physics that, in solids, particles do not have enough kinetic energy to overcome the strong forces of attraction, called bonds, which attract the particles toward each other. Using this physics phenomenon, we found the force of attraction between particles of every extracted silhouette, as shown in Figure 6:

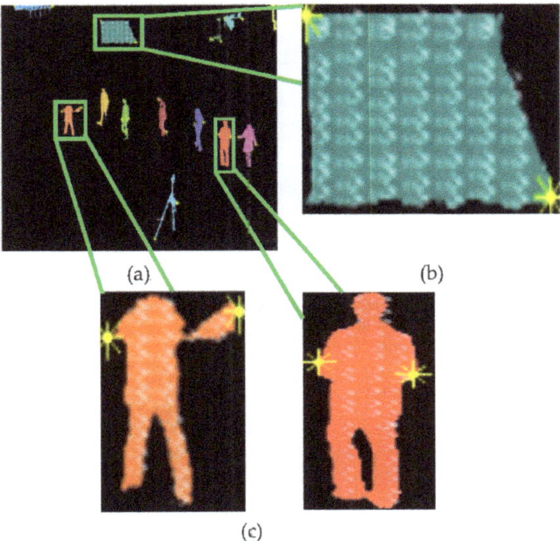

Figure 6. Particles force model. (**a**) Interacting force between two particles (**b**) for non-human silhouettes and (**c**) human silhouettes.

For simplicity, we found the force of attraction between only two mutually interacting particles using Equation (10) in all frames from 1 to Z.

$$F_i = \frac{p_1 p_2}{r^2} \tag{10}$$

where i is in the range $[1\ E]$ with E, representing the maximum number of silhouettes per frame, while F_i is the force of attraction between particle p_1 and p_2 of the ith silhouette and r^2 is the square of Euclidian distance between particles p_1 and p_2. After calculating the force between particles of every silhouette in all video frames, we discarded those silhouettes whose force of attraction was static in frame t and frame $t+1$ using Equation (11):

$$H_s = \begin{cases} 1 & \text{if } \frac{dF_i}{dt} > 0 \\ 0 & \text{otherwise} \end{cases} \tag{11}$$

where $\frac{dF_i}{dt}$ represents the change in attraction force between particles of every ith silhouette, with respect to time between frames t to $t + 1$. After application of the particles force model, we only retained human silhouettes in each frame, as depicted in Figure 7:

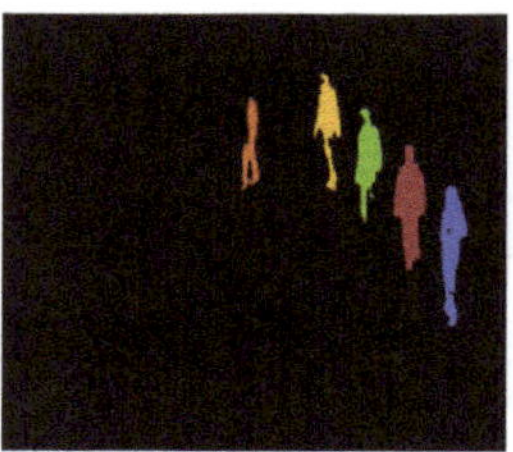

Figure 7. A few examples of verified multi-human silhouettes.

3.3.2. Multi-Person Counting

After extraction of the verified human silhouettes, to count these detected humans silhouettes, which consist of a set of particles, we performed cluster estimation. Since every silhouette is a collection of particles, the group of particles that makes one silhouette was treated as one cluster, and, by using the K-nearest neighbor search algorithm, cluster estimation was performed on every frame, as depicted in Figure 8:

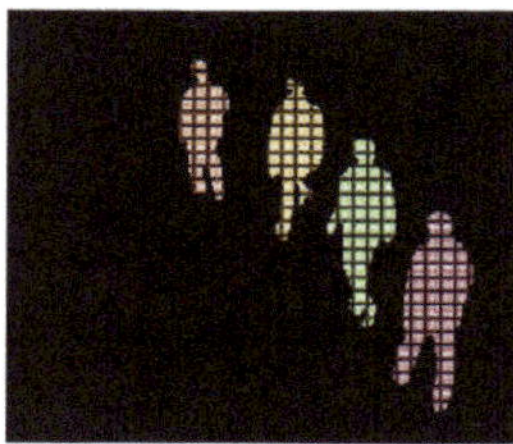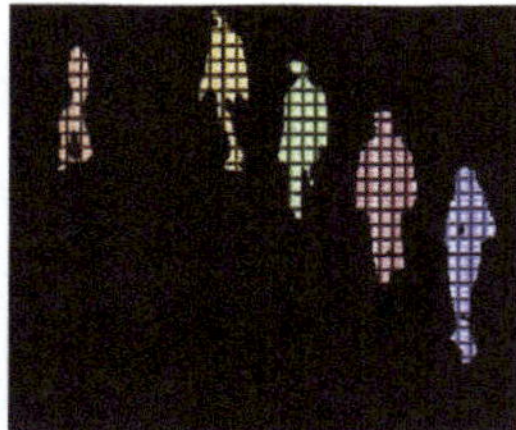

Figure 8. Human contours for cluster estimations.

After that, we labeled clusters in all frames, as shown in Equation (12), and, to make them appear visually, we drew green bounding boxes around each cluster. Thus, by performing cluster estimation and labeling, we counted all the extracted human silhouettes, as shown in Figure 9:

$$I_c = L_m p_N \tag{12}$$

where p_N is the total number of particles in one cluster (the total number of particles in each cluster varies from cluster to cluster and the number of clusters in each frame varies from frame to frame), while L_m represents the label of cluster m and I_c is the resultant extracted labeled cluster that was treated as one silhouette and was considered in counting.

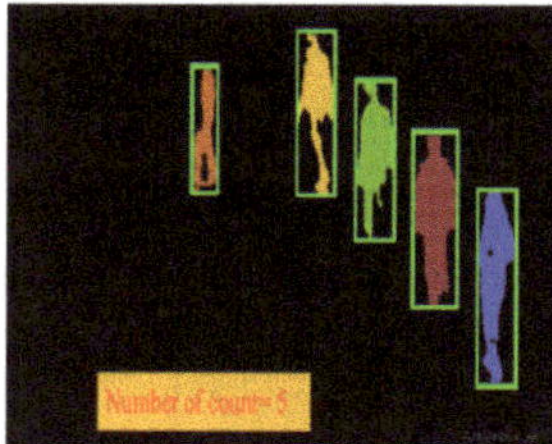

Figure 9. Sample frames of multi-person counts at different time intervals.

3.3.3. Multi-Person Tracking

The goal of person tracking is to establish correspondence between individuals across frames. Thus, to establish correspondence between persons in frame t and frame $t + 1$, we calculated the position and velocity of every detected human silhouette in all frames. In our model, we assumed that people can enter or leave the scene, thus, for temporally fixing of all humans across frames, the position of each human silhouette was located and locked by assigning a unique integer ID, which was fixed to that particular silhouette in all frames. The states of all the predicted persons in frame F_t were stored in a structure and matched with the states of frame F_{t+1}, while the detected fixed human silhouettes were tracked using the Jaccard similarity index.

$$S_t = \sum_{i=1}^{n} I_{Li} \tag{13}$$

While using data association and cross-correlation as a cost function, detected and predicted persons were associated in consecutive frames, as represented in Figure 10. The root steps involved in multi-person tracking are illustrated in Figure 11.

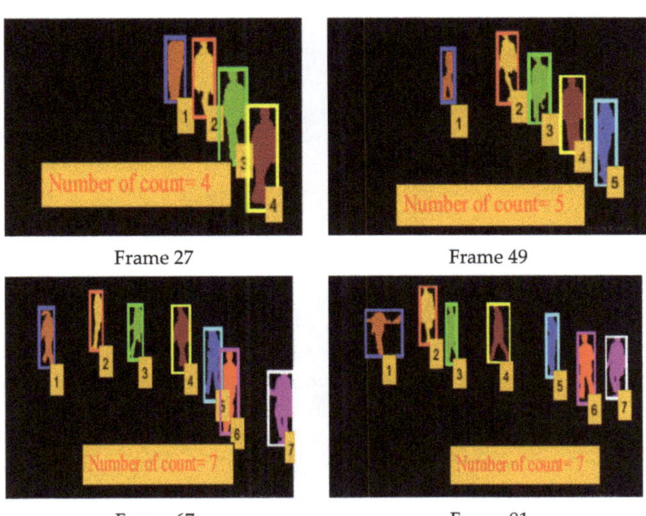

Figure 10. Sample frames of multiple human silhouette-fixing and tracking at different time intervals.

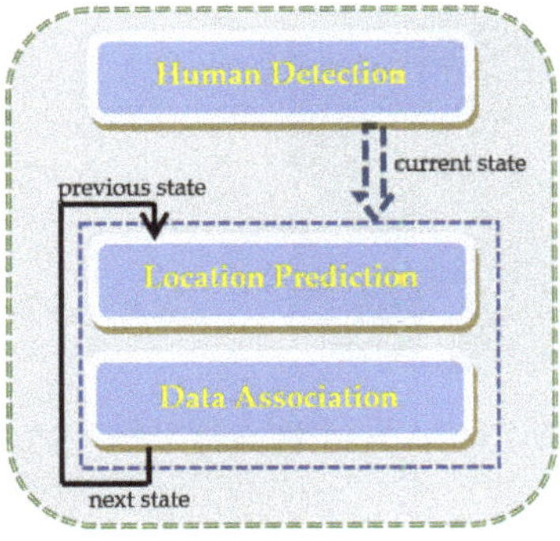

Figure 11. Key steps involved in multi-person tracking.

3.4. Crowd Behavior Detection

Understanding that accurate crowd behavior requires robust global and local feature extraction [101–103], along with a potent decision-making classifier, for crowd behavior detection after applying the distance algorithm (mentioned in Section 3.3), the extracted silhouettes were passed through the feature extraction step and multiple distinguishable global and local features were extracted for every frame. Next, bat optimization was applied for optimal feature extraction and decisions were made by the improved entropy classifier.

3.4.1. Global-Local Descriptors

For the global-local descriptor, we used a fusion of global and local image properties. In global features, we described the visual content of the whole image and we had the ability to represent an image with a single vector. Here, we extracted the crowd contour as a global feature. For local features, we used our newly proposed particles gradient motion features, geometric features, and speeded up robust feature (SURF) [104]. For local features, we extracted interest points and represented them as a set of vectors that respond more vigorously to clutter and occlusions.

Initially, in global features, we found the center of each human and considered all the humans in the scene as a vertex; this can be denoted as $P = \{P1, P2, \ldots, Pn \mid Pi = (Xi, Yi)\}$, where P represents the whole human crowd in the scene, considered as a set of vertices, and (Xi, Yi) are the coordinates of the ith human. We considered only those humans that were at the extreme points and joined them with a line, forming the biggest graph, covering all extreme vertices, as shown in Figure 12. The graph represented the human crowd contour, and thus, the variations in the shape of a graph threw a flash on variations in the outer area of the human crowd, i.e., on global changes. To measure the variations in the crowd contour, we compared the contour temporally by integrating over all of the pixels of the contour. In general, we defined the (p, q) moment of a contour as in Equation (14):

$$m_{p,q} = \sum_x^n \sum_y^n I(x,y) x^p y^q \tag{14}$$

where $I(x,y)$ is the intensity of the pixels in coordinate (x,y). Here, p is the x-order and q is the y-order, whereby, order means the power to which the corresponding component is taken in the sum just displayed. The summation is over all of the pixels of the contour boundary (denoted by n in the equation). It then follows immediately that, if p and q are both equal to 0, then the $m_{0,0}$ moment is actually just the length in pixels of the contour. The moment computation just described gives some rudimentary characteristics of a contour that can be used to compare two contours.

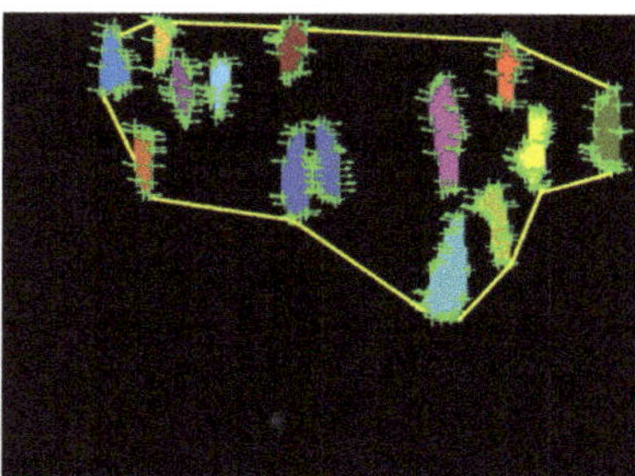

Figure 12. Extraction of human crowd contour as a global feature.

In the SURF descriptor [105], we computed distinctive invariant local features, which detected the interest points and elaborate features that depict some invariance to image noise, rotation, direction, scaling, and changes in illumination. Using SURF, we computed 75 local points for every human silhouette in an image, and thus, for every frame, we had 1050 SURF descriptors in a set of vectors, as shown in Figure 13:

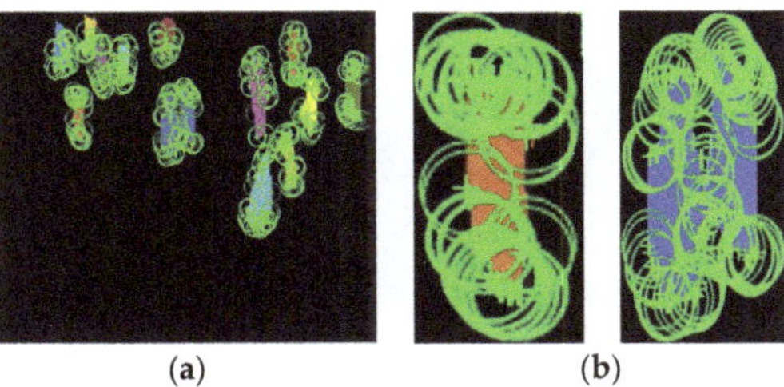

(a)   (b)

Figure 13. (**a**) SURF features for all human silhouettes and (**b**) magnified view of SURF features for two human silhouettes.

In geometric local features, we first identified the skeleton joints of every human silhouette in each frame using a skeleton model, and then, by considering skeleton joints as vertices, we drew poly-shapes and triangles with three or four vertices. By using the left arm, neck, left shoulder, and torso, a left polygon wing was drawn and filled with a color. Similarly, a right polygon wing was drawn and filled with different colors using the right arm, neck, torso, and right shoulder. Additionally, the torso area, lower area, left shoulder triangles, and right shoulder triangles were drawn, as depicted in Figure 14. The areas enclosed under these polygons were analyzed frame by frame, and on the basis of angle differences and area size, normal and abnormal behaviors of human crowds were detected. Algorithm 1 depicts the overall procedure used for the extraction of the strongest body points for human silhouettes.

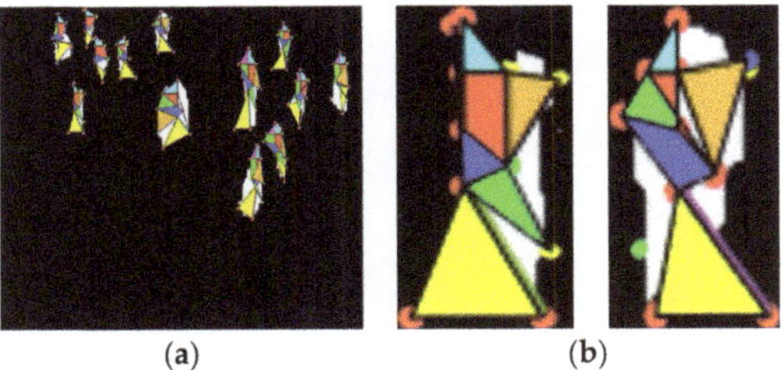

(a)   (b)

Figure 14. (**a**) Geometric features for all human silhouettes. (**b**) Magnified view of geometric features for two human silhouettes.

In particles gradient motion (PGM), we first converted every human silhouette into particles and then only those particles that were on the human contour were considered, and their interaction force was calculated. Generally, every pedestrian in a crowd has a desired direction and velocity v_i^d, calculated using Equation (16). However, in crowded scenes, because of the presence of multiple persons, individual movements are limited, and the actual velocity of each pedestrian v_i is different from their respective expected motion. The actual velocity of particles is calculated using Equation (15).

$$v_i = F_{avg}(x_i, y_i) \tag{15}$$

where $F_{avg}(x_i, y_i)$ is the ith particle average optical flow in the coordinate (x_i, y_i). We calculated the desired velocity v_i^d of particles as:

$$v_i^d = (1 - w_i)\, F(x_i, y_i) + w_i F_{avg}(x_i, y_i) \tag{16}$$

where $F(x_i, y_i)$ represents ith particle optical flow with coordinates (x_i, y_i) and w_i is the panic weight parameter. The pedestrian i displays vanity behaviors as $w_i \to 0$ and collective behaviors as $w_i \to 1$. Linear interpolation was used for the enumeration of efficient optical flow and the adequate average flow field of particles. Thus, on the basis of the actual velocity and the desired velocity, we can calculate the interaction force using Equation (17):

$$F_{\text{int}} = \frac{1}{T}\left(v_i^d - v_i\right) - \frac{dv_i}{dt} \tag{17}$$

where F_{int} is the resultant interaction force, as represented in Figure 15 and T is the relaxation parameter. When the interaction force of particles was greater than the set threshold, it was detected as an abnormal event; otherwise, it was considered to be normal.

Algorithm 1 Extract strongest body points for human silhouettes

---

Input: I: Extracted Human Silhouettes
Output: Strongest body points, i.e., head, shoulders, legs, arms, hips
/* for each connected component, extract body points.
B = bwboundaries(binary_image);
lbl = bwlabel(binary_image);
CC2 = bwconncomp(lbl);
L52 = labelmatrix(CC2);
for objectidx2 = 1:CC2.NumObjects
individualsilheouts2 = bsxfun(@times, closezn, L52 == objectidx2);
[labeledImage2,numberofBlobs2] = bwlabel(individualsilheouts2,4);
end
Aa = individualsilheouts2;
/* Defining a upper, middlle and lower portion for each individual silheouts */
th = thershold;
rps = regionprops(Aa,'Boundingbox', 'Area');
for k = 1 to length(rps) do
w = rps(k). Boundingbox
if height > th and width > th then
upper_region = struct('x',w(1), 'y', w(2), 'width',w(3), 'height', w(4)/5); /* head */
middle_region = struct('x',w(1), 'y', w(2) + w(4)/4, 'width',w(3), 'height', w(4)/4); /* arms */
lower_region = struct('x',w(1), 'y', w(2) + w(4)/2, 'width',w(3), 'height', w(4)/2); /* legs */
j = j+1;
s(j) = w;
end
end
top = [x,max_y]:left = [min_x,y]:bottom = [x,min_y]:right = [max_x,y];
% label the head region%
Head =top pixels of upper region
Right Shoulder = Bottom right pixels of upper region
Left Shoulder = Bottom left pixels of upper region
Right arm = Right Pixels of middle region
Left arm = Left Pixels of middle region
Right foot = Bottom right pixels of lower region
Left foot = Bottom left pixels of lower region
return Head, Shoulders, arms, foots

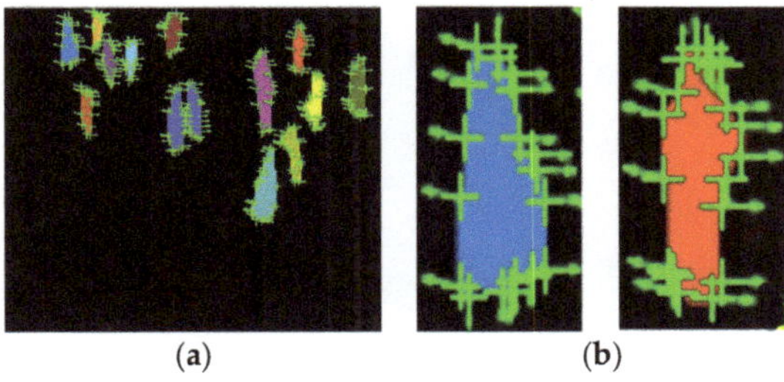

(a)   (b)

Figure 15. (**a**) Particles gradient motion descriptors for all human silhouettes and (**b**) magnified view of PGM for two human silhouettes.

3.4.2. Event Optimization: Bat Optimization

Optimization is a process by which the optimal solutions of a problem that satisfies and objective function are accessed [106–109]. Yang, in [110], introduced an optimization algorithm inspired by a property of bats, known as echolocation. Echolocation is a type of sonar that enables bats to fly and hunt in the dark. The bat optimization (BO) algorithm is composed of multiple variables of a given problem. Using the echolocation capability, bats can detect obstacles in the way and the distance, orientation, type, size, and even the speed of their prey.

BO has multiple agents depicting the parameters of the layout dilemma, as any other metaheuristic mechanism. From real-valued vectors, the initial population is randomly generated with number N and dimension d by considering lower and upper boundaries using Equation (18):

$$X_{ij} = X_{min} + \varphi(X_{max} - X_{min}) \tag{18}$$

where X_{max} and X_{min} are higher and lesser boundaries for dimension j, respectively, $j = 1, 2, \ldots, d$, $i = 1, 2, \ldots$, and N and φ ranged from 0 to 1 is a randomly generated value. After population initialization, we calculated the fitness function and stored the local and the global best. We evaluated the fitness values of all humans, and, on the basis of their movements, new local and global best solutions were obtained; all the humans had velocity Vi^t affected by a predefined frequency f_i, and finally, their new position Xi^t was located temporally, as described in the following Equations:

$$f_i = f_{min} + \beta(f_{max} - f_{min}) \tag{19}$$

$$Vi^t = Vi^{t-1} + (Xi^t - X*)f_i \tag{20}$$

$$Xi^t = Xi^{t-1} + Vi^t \tag{21}$$

where f_i is the frequency of the ith human, f_{min} and f_{max} are lower and higher frequency values, respectively, β represents a randomly generated value, and, after comparison of all solutions, $X*$ illustrates achieved global best location (solution). Figure 16 depicts the flow chart of the algorithm and Figure 17 represents optimization results.

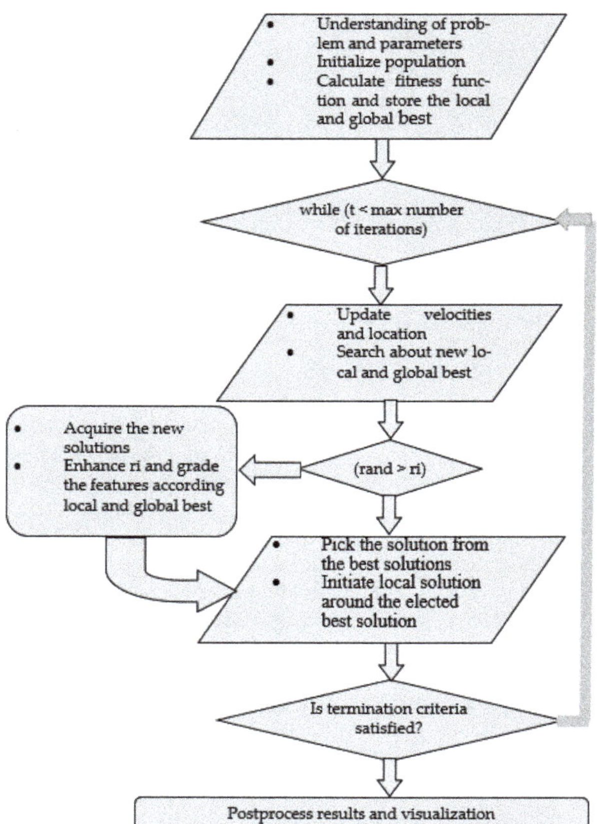

Figure 16. Bat optimization flow chart.

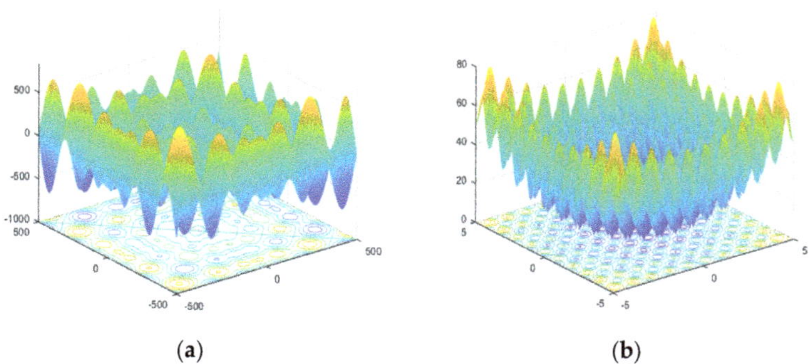

(a)   (b)

Figure 17. Bat optimization results. (a) Normal optimal features; (b) abnormal optimal features.

3.4.3. Improved Entropy Classifier

Using Shannon's information entropy theory [53] to describe the degree of uncertainty, we proposed an improved entropy classifier for the detection of human crowd behavior. First of all, we standardized all the features using Equation (22):

$$X_{ij}{}^* = \frac{X_{ij} - min\{X_j\}}{max\{X_j\} - min\{X_j\}} \tag{22}$$

where X_{ij}^* is the value of the j-th feature for i-th human. $j = 1, 2, \ldots, m$, $i = 1, 2, \ldots, n$, while n is the count of humans and m represents the count of features. After that, the weight of j-th feature for i-th human was calculated using Equation (23):

$$q_{ij} = \frac{X_{ij}^*}{\sum_{i=1}^{n} X_{ij}^*} \tag{23}$$

Thus, the information entropy of each feature was calculated using Equation (24):

$$e_j = -k \sum_{i=1}^{n} \left(q_{ij} \times \ln q_{ij} \right) \tag{24}$$

where $k = \frac{1}{\ln m}$. After calculating the information entropy, we then calculated the difference coefficient and maximum ratio of the difference coefficient using Equations (25) and (26):

$$d_j = 1 - e_j \tag{25}$$

$$D = \frac{\max\left(d_j\right)}{\min\left(d_j\right)}, \qquad (j = 1, 2, \ldots, m) \tag{26}$$

After calculating D, we then built up the scale ratio chart 1–9 using Equation (27):

$$R = \sqrt[a-1]{\frac{D}{a}} \tag{27}$$

where a depicts the highest scale-value worked as an adjustment coefficient by calculating the power $(a - 1)$. The D is allocated to the mapping values from 1 to 9 in the above Equation. After that, from scale 1–9, mapped values were calculated, and judgment matrix R was established with elements r_{ij}, respectively, using Equation (28):

$$r_{ij} = \frac{d_i}{d_j}, \; (d_i > d_j) \tag{28}$$

The obtained judgment matrix satisfied the consistency test because the elements r_{ij} demonstrated the ratio of difference coefficient of two features.

Thus, the consistent weights W_j for each feature were then calculated using an analytical hierarchy process. After that, information entropy was again calculated for each feature, using these weights by utilizing Equation (24). The crowd behavior entropy of the whole system was the summary of all entropies. In this way, for every frame, the entropy value was calculated and utilized as a template. For a small entropy value less than the defined threshold, the behavior was predicted as normal; however, for entropy values higher than the set threshold, the behavior was presumed to be abnormal. A flow chart of the proposed improved entropy classifier is shown in Figure 18. Figure 19 depicts results over event classes.

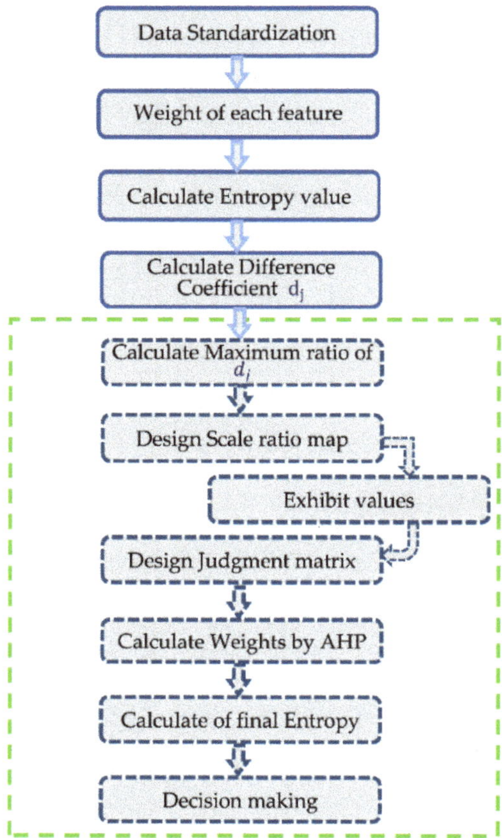

Figure 18. Flow chart of the improved entropy classifier.

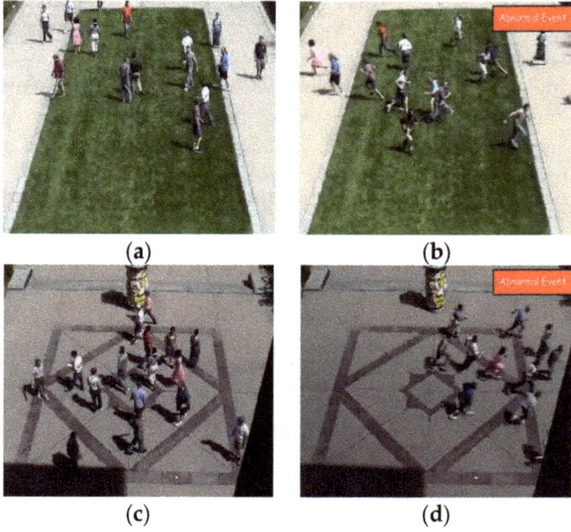

Figure 19. Crowd behavior detection. (**a**,**c**) Normal frames and (**b**,**d**) abnormal frames.

4. Performance Evaluation

In this section, we evaluated the performance of our proposed system. We conducted experiments on two publicly available benchmark datasets to evaluate the accuracy and performance of our proposed model. The PETS2009 dataset was used to evaluate the accuracy of multi-person tracking and the UMN dataset was used to evaluate the accuracy of crowd behavior detection. We started by briefly describing the datasets used, and then the experimental results were discussed. Finally, we showed the mean accuracy of our proposed system. We also compared our proposed model with other state-of-the-art multi-person tracking and crowd behavior detection systems.

4.1. Datasets Description

4.1.1. PETS2009 Dataset

To evaluate different video surveillance challenges, we used PETS2009, one of the publicly available benchmark datasets. The challenges included the S1 dataset for counting persons in a low-density crowd, the S2 dataset for detecting and tracking persons in medium-density crowds, and the S3 dataset for tracking and estimating the number of persons in a high-density crowd. Some sample frames of different synchronized views from PETS2009 dataset are depicted in Figure 20.

Figure 20. Sample frames of different synchronized views from the PETS2009 dataset.

4.1.2. UMN Dataset

To evaluate different video surveillance challenges for crowd behavior detection, UMN is one of the publicly available benchmark datasets. The UMN dataset consists of three different scenes, specifically, two outdoor and one indoor, with videos of 11 various panic scenarios. For the detection of abnormal behavior of a crowd, the UMN dataset is one of the best datasets that is publicly available. There were two outdoor scenes: the lawn scene, consisting of two scenarios with 1453 frames, and the Plaza scene, with three scenarios that had 2142 frames. There were six scenarios in the indoor scene, with 4144 frames. Sample frames of different scenarios of the UMN dataset are shown in Figure 21.

Figure 21. Sample frames of different scenarios of the UMN dataset.

4.2. Experimental Settings and Results

We performed all the experiments on MATLAB, and the hardware system had a 64-bit intel core-i3 2.5 GHz CPU and 6 GB of RAM. Three experimental measures were used to evaluate the performance of the system: (1) mean accuracy of multi-person tracking, (2) the accuracy of human crowd behavior detection, and (3) comparisons between our proposed new system with other current and well-known systems. Experimental results showed that our proposed system produces a higher accuracy rate over existing systems.

4.2.1. Experiment 1: Multi-Person Tracking over the PETS2009 Dataset

Experimental results and mean accuracy of our proposed multi-person counting and tracking model on a publicly available PETS2009 dataset are shown in Tables 1 and 2. The ground truth was obtained by counting the number of persons in every sequence, where one sequence contained 20 frames. Table 1 depicts the mean accuracy of our proposed multi-person counting system on the first 30 sequences. As shown, the mean accuracy of our proposed model was 89.80%.

Table 1. Multi-person counting accuracy over the PETS2009 dataset.

Sequence No (Frame = 20)	Actual Count	Predicted Count	Accuracy
6	3	3	100
12	4	4	100
18	5	4	80
24	6	5	83.33
30	7	6	85.71
	Mean Accuracy = 89.80%		

Table 2. Multi-person tracking accuracy over PETS2009 dataset.

Sequence No (Frame = 20)	Successful	Failure	Accuracy
6	3	0	100
12	4	0	100
18	4	1	80
24	5	1	83.33
30	5	2	71.43
	Mean Accuracy = 86.95%		

Table 2 presents the mean accuracy of our proposed multi-person tracking system. The actual number of humans is the same as for Table 1, while column 2 represents the successful tracking rate of our proposed particles force model and column 3 depicts the failure case. The mean accuracy of our proposed model for multiple person tracking was 86.95%.

4.2.2. Experiment 2: Human Crowd Behavior Detection over the UMN Dataset

Experimental results using the confusion matrix and the mean accuracy of our proposed HCB model on the publicly available UMN dataset are shown in Table 3. The way to evaluate algorithms is to run them throughout a test sequence with initialization from the ground truth position in the first frame.

Table 3. Confusion matrix, showing mean accuracy for human crowd behavior detection on the UMN dataset.

Events	Normal	Abnormal
Normal	88	12
Abnormal	16	84
Mean Accuracy of Event Detection = 86.06%		

4.2.3. Experiment 3: Multi-Person Tracking and HCB Detection Comparisons with State-of-the-Art Methods

We compared our proposed system with other well-known multi-person tracking and human crowd behavior detection methods. As depicted, our system performed better compared to existing well-known state-of-the-art methods. Table 4 shows that, in comparison to other state-of-the-art methods, our proposed system achieved an admirable accuracy rate of 86.06% for crowd behavior detection, which is higher than the accuracy of the force field model (FF) (81.04%) and the social force model (SF) (85.09%). The accuracy of other methods under the same evaluation settings was taken from [77,79].

Table 4. Comparison of the proposed approach with other state-of-the-art methods for human crowd behavior detection on the UMN dataset.

Indoor/Outdoor Scenes	Force Field Model	Social Force Model	Proposed Method
Scene 1	88.69	84.41	87.43
Scene 2	80.00	82.35	83.21
Scene 3	77.92	90.83	90.63
Overall accuracy	81.04%	85.09%	86.06%

Table 5 presents the comparison of our proposed system with other state-of-the-art systems for multi-person counting. Experiment results show that our proposed system achieved a higher accuracy rate of 89.8% over existing methods.

Table 5. Comparison of proposed approach with state-of-the-art multi-person counting methods.

Methods	Counting Accuracy (%)
Pixel-map based algorithm [94]	83.6
Sparsity-driven [111]	86.3
Head Shoulder based detection [100]	86.7
Skin Detection [81]	88.7
Proposed method	89.8

In Table 6, comparisons of multi-person tracking with other state-of-the-art methods show that our proposed system achieved a higher accuracy rate of 86.9% over existing methods.

Table 6. Comparison of the proposed approach with state-of-the-art multi-person tracking methods.

Methods	Tracking Accuracy (%)
Flow Linear Programming [112]	78.8
DDPMO [113]	81.3
Appearance model [114]	83.0
Proposed method	86.9

5. Conclusions

In this paper, we proposed a new robust approach for crowd counting. We introduced and tested tracking and human behavior detection using the idea of a mutually interacting particles force model and an improved entropy classifier with spatio-temporal and particles gradient motion descriptors. Through detailed experiments, we proved the ability of the method to efficiently count, track, and detect the behavior of multiple persons efficiently in crowded scenes. The performance of our new tracking system decreases marginally with increasing numbers of persons in the scene. This is mainly due to full occlusions that occur in the test videos. We achieved promising results on the publicly available benchmark PETS2009 dataset, with an accuracy of 89.80% for multi-person counting and 86.95% for person tracking, as shown in Tables 1 and 2. However, for HCB detection, we achieved promising results on the publicly available benchmark UMN dataset, with an accuracy of 86.06%, as shown in Table 3. Our future work will focus on some occlusion reasoning methods to further tackle the occlusion problems. We will also extend our work to multiple scene detection. We are interested in recognition of different scenes, such as sport scenes, fight scenes, robbery scenes, traffic scenes, and action scenes.

Author Contributions: Conceptualization, F.A., Y.Y.G. and K.K. methodology, F.A. and A.J.; software, F.A.; validation, F.A., Y.Y.G. and K.K.; formal analysis, M.G. and K.K.; resources, Y.Y.G., M.G. and K.K.; writing—review and editing, F.A., A.J. and K.K.; funding acquisition, Y.Y.G. and K.K. All authors have read and agreed to the published version of the manuscript.

Funding: This research was supported by the Basic Science Research Program through the National Research Foundation of Korea (NRF) under Grant 2018R1D1A1A02085645, by the Korea Medical Device Development Fund grant funded by the Korean government (the Ministry of Science and ICT; the Ministry of Trade, Industry and Energy; the Ministry of Health and Welfare; and the Ministry of Food and Drug Safety) (202012D05-02), and by Hanyang University Grant No. 201800000000647.

Data Availability Statement: Data sharing not applicable.

Conflicts of Interest: The authors declare no conflict of interest.

References

1. Fu, Z.; Feng, P.; Angelini, F.; Chambers, J.; Naqvi, S.M. Particle PHD filter based multiple human tracking using online group-structured dictionary learning. *IEEE Access* **2018**, *6*, 14764–14778. [CrossRef]
2. Wen, L.; Lei, Z.; Lyu, S.; Li, S.Z.; Yang, M.H. Exploiting hierarchical dense structures on hypergraphs for multi-object tracking. *IEEE Trans. Pattern Anal. Mach. Intell.* **2015**, *38*, 1983–1996. [CrossRef] [PubMed]
3. Maggio, E.; Taj, M.; Cavallaro, A. Efficient multitarget visual tracking using random finite sets. *IEEE Trans. Circuits Syst. Video Technol.* **2008**, *18*, 1016–1027. [CrossRef]
4. Yilmaz, A.; Javed, O.; Shah, M. Object tracking: A survey. *Acm Comput. Surv. (CSUR)* **2006**, *38*, 13-es. [CrossRef]
5. Marcenaro, L.; Marchesotti, L.; Regazzoni, C.S. Tracking and counting multiple interacting people in indoor scenes. In Proceedings of the Third IEEE International Workshop on Performance Evaluation of Tracking and Surveillance, Copenhagen, Denmark, 1 June 2002; pp. 56–61.
6. Ali, S.; Shah, M. Floor fields for tracking in high density crowd scenes. In Proceedings of the European Conference on Computer Vision, Marseille, France, 12–18 October 2008; Springer: Berlin/Heidelberg, Germany, 2008; pp. 1–14.
7. Jalal, A.; Kim, Y. Dense depth maps-based human pose tracking and recognition in dynamic scenes using ridge data. In Proceedings of the 2014 11th IEEE International Conference on Advanced Video and Signal Based Surveillance (AVSS), Seoul, Korea, 26–29 August 2014; pp. 119–124.
8. Yao, R.; Lin, G.; Xia, S.; Zhao, J.; Zhou, Y. Video object segmentation and tracking: A survey. *ACM Trans. Intell. Syst. Technol. (TIST)* **2020**, *11*, 1–47. [CrossRef]
9. Pervaiz, M.; Jalal, A.; Kim, K. Hybrid Algorithm for Multi People Counting and Tracking for Smart Surveillance. In Proceedings of the 2021 International Bhurban Conference on Applied Sciences and Technologies (IBCAST), Islamabad, Pakistan, 12–16 January 2021; pp. 530–535.
10. Topkaya, I.S.; Erdogan, H.; Porikli, F. Counting people by clustering person detector outputs. In Proceedings of the IEEE International Conference on AVSS, Seoul, Korea, 26–29 August 2014; pp. 313–318.
11. Ren, W.; Wang, X.; Tian, J.; Tang, Y.; Chan, A.B. Tracking-by-Counting: Using Network Flows on Crowd Density Maps for Tracking Multiple Targets. *IEEE Trans. Image Process.* **2020**, *30*, 1439–1452. [CrossRef]
12. Loy, C.C.; Chen, K.; Gong, S.; Xiang, T. Crowd counting and profiling: Methodology and evaluation. In *Modeling, Simulation and Visual Analysis of Crowds*; Springer: New York, NY, USA, 2013; pp. 347–382.
13. Mahmood, M.; Jalal, A.; Kim, K. WHITE STAG model: Wise human interaction tracking and estimation (WHITE) using spatio-temporal and angular-geometric (STAG) descriptors. *Multimed. Tools Appl.* **2010**, *79*, 6919–6950. [CrossRef]
14. Jalal, A.; Kim, K. Wearable inertial sensors for daily activity analysis based on adam optimization and the maximum entropy Markov model. *Entropy* **2020**, *22*, 579.
15. Ryan, D.; Denman, S.; Sridharan, S.; Fookes, C. An evaluation of crowd counting methods, features and regression models. *Comput. Vis. Image Underst.* **2015**, *130*, 1–17. [CrossRef]
16. Idrees, H.; Saleemi, I.; Seibert, C.; Shah, M. Multi-source multi-scale counting in extremely dense crowd images. In Proceedings of the IEEE Conference on Computer Vision and Pattern Recognition, Portland, OR, USA, 23–28 June 2013; pp. 2547–2554.
17. Ekinci, M.; Gedikli, E. Silhouette based human motion detection and analysis for real-time automated video surveillance. *Turk. J. Electr. Eng. Comput. Sci.* **2005**, *13*, 199–229.
18. Younsi, M.; Diaf, M.; Siarry, P. Automatic multiple moving humans detection and tracking in image sequences taken from a stationary thermal infrared camera. *Expert Syst. Appl.* **2020**, *146*, 113171. [CrossRef]

19. Lempitsky, V.; Zisserman, A. Learning to count objects in images. *Adv. Neural Inf. Process. Syst.* **2010**, *23*, 1324–1332.
20. Farooq, M.U.; Saad, M.N.B.; Malik, A.S.; Salih Ali, Y.; Khan, S.D. Motion estimation of high density crowd using fluid dynamics. *Imaging Sci. J.* **2020**, *68*, 141–155. [CrossRef]
21. Sajid, M.; Hassan, A.; Khan, S.A. Crowd counting using adaptive segmentation in a congregation. In Proceedings of the 2016 IEEE International Conference on Signal and Image Processing (ICSIP), Beijing, China, 13–15 August 2016; pp. 745–749.
22. Fehr, D.; Sivalingam, R.; Morellas, V.; Papanikolopoulos, N.; Lotfallah, O.; Park, Y. Counting people in groups. In Proceedings of the 2009 Sixth IEEE International Conference on Advanced Video and Signal Based Surveillance, Genova, Italy, 2–4 September 2009; pp. 152–157.
23. Jalal, A.; Uddin, M.Z.; Kim, T.S. Depth video-based human activity recognition system using translation and scaling invariant features for life logging at smart home. *IEEE Trans. Consum. Electron.* **2012**, *58*, 863–871. [CrossRef]
24. Jalal, A.; Mahmood, M.; Hasan, A.S. Multi-features descriptors for human activity tracking and recognition in Indoor-outdoor environments. In Proceedings of the 2019 16th International Bhurban Conference on Applied Sciences and Technology (IBCAST), Islamabad, Pakistan, 8–12 January 2019; pp. 371–376.
25. Albiol, A.; Silla, M.J.; Albiol, A.; Mossi, J.M. Video analysis using corner motion statistics. In Proceedings of the IEEE International Workshop on Performance Evaluation of Tracking and Surveillance, Miami, FL, USA, 25 June 2009; pp. 31–38.
26. Li, N.; Wu, X.; Guo, H.; Xu, D.; Ou, Y.; Chen, Y.L. Anomaly detection in video surveillance via gaussian process. *Int. J. Pattern Recognit. Artif. Intell.* **2015**, *29*, 1555011. [CrossRef]
27. Chan, A.B.; Morrow, M.; Vasconcelos, N. Analysis of crowded scenes using holistic properties. In Proceedings of the Performance Evaluation of Tracking and Surveillance Workshop at CVPR, Miami, FL, USA, 25 June 2009; pp. 101–108.
28. Sharif, M.H.; Djeraba, C. An entropy approach for abnormal activities detection in video streams. *Pattern Recognit.* **2012**, *45*, 2543–2561. [CrossRef]
29. Jalal, A.; Sarif, N.; Kim, J.T.; Kim, T.S. Human activity recognition via recognized body parts of human depth silhouettes for residents monitoring services at smart home. *Indoor Built Environ.* **2013**, *22*, 271–279. [CrossRef]
30. Jalal, A.; Kamal, S. Real-time life logging via a depth silhouette-based human activity recognition system for smart home services. In Proceedings of the 2014 11th IEEE International Conference on Advanced Video and Signal Based Surveillance (AVSS), Seoul, Korea, 26–29 August 2014; pp. 74–80.
31. Raghavendra, R.; Del Bue, A.; Cristani, M.; Murino, V. Abnormal crowd behavior detection by social force optimization. In Proceedings of the International Workshop on Human Behavior Understanding, Amsterdam, The Netherlands, 16 November 2011; Springer: Berlin/Heidelberg, Germany, 2011; pp. 134–145.
32. Javeed, M.; Gochoo, M.; Jalal, A.; Kim, K. HF-SPHR: Hybrid Features for Sustainable Physical Healthcare Pattern Recognition Using Deep Belief Networks. *Sustainability* **2021**, *13*, 1699 [CrossRef]
33. Chan, A.B.; Liang, Z.S.J.; Vasconcelos, N. Privacy preserving crowd monitoring: Counting people without people models or tracking. In Proceedings of the 2008 IEEE Conference on Computer Vision and Pattern Recognition, Anchorage, AK, USA, 23–28 June 2008; pp. 1–7.
34. Jalal, A.; Mahmood, M. Students' behavior mining in e-learning environment using cognitive processes with information technologies. *Educ. Inf. Technol.* **2019**, *24*, 2797–2821. [CrossRef]
35. Dey, P.; Roberts, D. A conceptual framework for modelling crowd behavior. In Proceedings of the 11th IEEE International Symposium on Distributed Simulation and Real-Time Applications, Chania, Greece, 22–26 October 2007; pp. 193–200.
36. Bellomo, N.; Dogbe, C. On the modelling crowd dynamics from scaling to hyperbolic macroscopic models. *Math. Models Methods Appl. Sci.* **2008**, *18* (Suppl. 1), 1317–1345. [CrossRef]
37. Roy, S.; Annunziato, M.; Borzì, A. A Fokker–Planck feedback control-constrained approach for modelling crowd motion. *J. Comput. Theor. Transp.* **2016**, *45*, 442–458. [CrossRef]
38. Ansar, H.; Jalal, A.; Gochoo, M.; Kim, K. Hand Gesture Recognition Based on Auto-Landmark Localization and Reweighted Genetic Algorithm for Healthcare Muscle Activities. *Sustainability* **2021**, *13*, 2961. [CrossRef]
39. Javeed, M.; Jalal, A.; Kim, K. Wearable Sensors based Exertion Recognition using Statistical Features and Random Forest for Physical Healthcare Monitoring. In Proceedings of the 2021 International Bhurban Conference on Applied Sciences and Technologies (IBCAST), Islamabad, Pakistan, 12–16 January 2021; pp. 512–517.
40. Kremer, M.; Haworth, B.; Kapadia, M.; Faloutsos, P. Modelling distracted agents in crowd simulations. *Vis. Comput.* **2021**, *37*, 107–118. [CrossRef]
41. Khalid, N.; Gochoo, M.; Jalal, A.; Kim, K. Modeling Two-Person Segmentation and Locomotion for Stereoscopic Action Identification: A Sustainable Video Surveillance System. *Sustainability* **2021**, *13*, 970. [CrossRef]
42. Jalal, A.; Kamal, S. Improved Behavior Monitoring and Classification Using Cues Parameters Extraction from Camera Array Images. *Int. J. Interact. Multimed. Artif. Intell.* **2019**, *5*, 71–78. [CrossRef]
43. Elaiw, A.; Al-Turki, Y.; Alghamdi, M. A critical analysis of behavioral crowd dynamics—From a modelling strategy to kinetic theory methods. *Symmetry* **2019**, *11*, 851. [CrossRef]
44. Nady, A.; Atia, A.; Abutabl, A. Real-time abnormal event detection in crowded scenes. *J. Theory Appl. Inf. Technol.* **2018**, *96*, 6064–6075.
45. He, F.; Xiang, Y.; Zhao, X.; Wang, H. Informative scene decomposition for crowd analysis, comparison and simulation guidance. *ACM Trans. Graph. (TOG)* **2020**, *39*, 50–51. [CrossRef]

46. Jalal, A.; Batool, M.; Kim, K. Sustainable Wearable System: Human Behavior Modeling for Life-Logging Activities Using K-Ary Tree Hashing Classifier. *Sustainability* **2020**, *12*, 10324. [CrossRef]
47. Pennisi, A.; Bloisi, D.D.; Iocchi, L. Online real-time crowd behavior detection in video sequences. *Comput. Vis. Image Underst.* **2016**, *144*, 166–176. [CrossRef]
48. Jalal, A.; Kamal, S.; Kim, D. A depth video sensor-based life-logging human activity recognition system for elderly care in smart indoor environments. *Sensors* **2014**, *14*, 11735–11759. [CrossRef] [PubMed]
49. Jalal, A.; Quaid, M.A.K.; Kim, K. A Study of Accelerometer and Gyroscope Measurements in Physical Life-Log Activities Detection Systems. *Sensors* **2020**, *20*, 6670. [CrossRef]
50. Bellomo, N.; Clarke, D.; Gibelli, L.; Townsend, P.; Vreugdenhil, B.J. Human behaviours in evacuation crowd dynamics: From modelling to "big data" toward crisis management. *Phys. Life Rev.* **2016**, *18*, 1–21. [CrossRef]
51. Nadeem, A.; Jalal, A.; Kim, K. Human actions tracking and recognition based on body parts detection via Artificial neural network. In Proceedings of the 2020 3rd International Conference on Advancements in Computational Sciences (ICACS), Lahore, Pakistan, 17–19 February 2020; pp. 1–6.
52. Fakhar, B.; Kanan, H.R.; Behrad, A. Event detection in soccer videos using unsupervised learning of Spatio-temporal features based on pooled spatial pyramid model. *Multimed. Tools Appl.* **2019**, *78*, 16995–17025. [CrossRef]
53. Shannon, C.E. A mathematical theory of communication. *Bell Syst. Tech. J.* **1948**, *27*, 379–423. [CrossRef]
54. Zhang, X.; Yu, Q.; Yu, H. Physics inspired methods for crowd video surveillance and analysis: A survey. *IEEE Access* **2018**, *6*, 66816–66830. [CrossRef]
55. Jalal, A.; Khalid, N.; Kim, K. Automatic recognition of human interaction via hybrid descriptors and maximum entropy markov model using depth sensors. *Entropy* **2020**, *22*, 817. [CrossRef] [PubMed]
56. Ryan, D.; Denman, S.; Fookes, C.; Sridharan, S. Crowd counting using group tracking and local features. In Proceedings of the 2010 7th IEEE International Conference on Advanced Video and Signal Based Surveillance, Boston, MA, USA, 29 August–1 September 2010; pp. 218–224.
57. Cong, Y.; Gong, H.; Zhu, S.C.; Tang, Y. Flow mosaicking: Real-time pedestrian counting without scene-specific learning. In Proceedings of the 2009 IEEE Conference on Computer Vision and Pattern Recognition, Miami, FL, USA, 20–25 June 2009; pp. 1093–1100.
58. Nadeem, A.; Jalal, A.; Kim, K. Automatic human posture estimation for sport activity recognition with robust body parts detection and entropy markov model. *Multimed. Tools Appl.* **2021**, *80*, 1–34.
59. Jalal, A.; Akhtar, I.; Kim, K. Human Posture Estimation and Sustainable Events Classification via Pseudo-2D Stick Model and K-ary Tree Hashing. *Sustainability* **2020**, *12*, 9814. [CrossRef]
60. Wu, T.; Toet, A. Color-to-grayscale conversion through weighted multiresolution channel fusion. *J. Electron. Imaging* **2014**, *23*, 043004. [CrossRef]
61. Jalal, A.; Kim, Y.; Kim, D. Ridge body parts features for human pose estimation and recognition from RGB-D video data. In Proceedings of the Fifth International Conference on Computing, Communications and Networking Technologies (ICCCNT), Hefei, China, 11–13 July 2014; pp. 1–6.
62. Wang, J.; Xu, Z. Spatio-temporal texture modelling for real-time crowd anomaly detection. *Comput. Vis. Image Underst.* **2016**, *144*, 177–187. [CrossRef]
63. Zitouni, M.S.; Bhaskar, H.; Dias, J.; Al-Mualla, M.E. Advances and trends in visual crowd analysis: A systematic survey and evaluation of crowd modelling techniques. *Neurocomputing* **2016**, *186*, 139–159. [CrossRef]
64. Pretorius, M.; Gwynne, S.; Galea, E.R. Large crowd modelling: An analysis of the Duisburg Love Parade disaster. *Fire Mater.* **2015**, *39*, 301–322. [CrossRef]
65. Quaid, M.A.K.; Jalal, A. Wearable sensors based human behavioral pattern recognition using statistical features and reweighted genetic algorithm. *Multimed. Tools Appl.* **2020**, *79*, 6061–6083. [CrossRef]
66. Jalal, A.; Quaid, M.A.K.; Kim, K. A wrist worn acceleration based human motion analysis and classification for ambient smart home system. *J. Electr. Eng. Technol.* **2019**, *14*, 1733–1739. [CrossRef]
67. Jalal, A.; Lee, S.; Kim, J.T.; Kim, T.S. Human activity recognition via the features of labeled depth body parts. In Proceedings of the International Conference on Smart Homes and Health Telematics, Artiminio, Italy, 12–15 June 202; Springer: Berlin/Heidelberg, Germany, 2012; pp. 246–249.
68. Chen, D.; Wang, L.; Wu, X.; Chen, J.; Khan, S.U.; Kołodziej, J.; Tian, M.; Huang, F.; Liu, W. Hybrid modelling and simulation of huge crowd over a hierarchical grid architecture. *Future Gener. Comput. Syst.* **2013**, *29*, 1309–1317. [CrossRef]
69. Jalal, A.; Mahmood, M.; Sidduqi, M.A. Robust spatio-temporal features for human interaction recognition via artificial neural network. In Proceedings of the IEEE International Conference on Frontiers of Information Technology, Islamabad, Pakistan, 17–19 December 2018.
70. Alguliyev, R.M.; Aliguliyev, R.M.; Sukhostat, L.V. Efficient algorithm for big data clustering on single machine. *CAAI Trans. Intell. Technol.* **2020**, *5*, 9–14. [CrossRef]
71. Jalal, A.; Batool, M.; Kim, K. Stochastic recognition of physical activity and healthcare using tri-axial inertial wearable sensors. *Appl. Sci.* **2020**, *10*, 7122. [CrossRef]
72. Jalal, A.; Kim, Y.H.; Kim, Y.J.; Kamal, S.; Kim, D. Robust human activity recognition from depth video using spatiotemporal multi-fused features. *Pattern Recognit.* **2017**, *61*, 295–308. [CrossRef]

73. Basavegowda, H.S.; Dagnew, G. Deep learning approach for microarray cancer data classification. *CAAI Trans. Intell. Technol.* **2020**, *5*, 22–33. [CrossRef]

74. Aylaj, B.; Bellomo, N.; Gibelli, L.; Knopoff, D. Crowd Dynamics by Kinetic Theory Modeling: Complexity, Modeling, Simulations, and Safety. *Synth. Lect. Math. Stat.* **2020**, *12*, 1–98.

75. Gochoo, M.; Akhter, I.; Jalal, A.; Kim, K. Stochastic Remote Sensing Event Classification over Adaptive Posture Estimation via Multifused Data and Deep Belief Network. *Remote Sens.* **2021**, *13*, 912. [CrossRef]

76. Rizwan, S.A.; Jalal, A.; Gochoo, M.; Kim, K. Robust Active Shape Model via Hierarchical Feature Extraction with SFS-Optimized Convolution Neural Network for Invariant Human Age Classification. *Electronics* **2021**, *10*, 465. [CrossRef]

77. Wu, S.; Wong, H.S.; Yu, Z. A Bayesian model for crowd escape behavior detection. *IEEE Trans. Circuits Syst. Video Technol.* **2013**, *24*, 85–98. [CrossRef]

78. Choudhary, S.; Ojha, N.; Singh, V. Real-time crowd behavior detection using SIFT feature extraction technique in video sequences. In Proceedings of the 2017 International Conference on Intelligent Computing and Control Systems (ICICCS), Madurai, India, 15–16 June 2017; pp. 936–940.

79. Direkoglu, C.; Sah, M.; O'Connor, N.E. Abnormal crowd behavior detection using novel optical flow-based features. In Proceedings of the 2017 14th IEEE International Conference on Advanced Video and Signal Based Surveillance (AVSS), Lecce, Italy, 29 August–1 September 2017; pp. 1–6.

80. Aguilar, W.G.; Luna, M.A.; Ruiz, H.; Moya, J.F.; Luna, M.P.; Abad, V.; Parra, H. Statistical abnormal crowd behavior detection and simulation for real-time applications. In Proceedings of the International Conference on Intelligent Robotics and Applications, Wuhan, China, 16–18 August 2017; Springer: Cham, Switzerland, 2017; pp. 671–682.

81. Shehzed, A.; Jalal, A.; Kim, K. Multi-person tracking in smart surveillance system for crowd counting and normal/abnormal events detection. In Proceedings of the 2019 International Conference on Applied and Engineering Mathematics (ICAEM), Taxila, Pakistan, 27–29 August 2017; pp. 163–168.

82. Ren, W.Y.; Li, G.H.; Chen, J.; Liang, H.Z. Abnormal crowd behavior detection using behavior entropy model. In Proceedings of the 2012 International Conference on Wavelet Analysis and Pattern Recognition, Xi'an, China, 15–17 July 2012; pp. 212–221.

83. Wang, G.; Fu, H.; Liu, Y. Real time abnormal crowd behavior detection based on adjacent flow location estimation. In Proceedings of the 2016 4th International Conference on Cloud Computing and Intelligence Systems (CCIS), Beijing, China, 17–19 August 2016; pp. 476–479.

84. Zhao, Z.; Fu, S.; Wang, Y. Eye Tracking Based on the Template Matching and the Pyramidal Lucas-Kanade Algorithm. In Proceedings of the 2012 International Conference on Computer Science and Service System, Nanjing, China, 11–13 August 2012; pp. 2277–2280.

85. Mehran, R.; Oyama, A.; Shah, M. Abnormal crowd behavior detection using social force model. In Proceedings of the 2009 IEEE Conference on Computer Vision and Pattern Recognition, Miami, FL, USA, 20–25 June 2009; pp. 935–942.

86. Bellomo, N.; Bellouquid, A. On multiscale models of pedestrian crowds from mesoscopic to macroscopic. *Commun. Math. Sci.* **2015**, *13*, 1649–1664. [CrossRef]

87. Colombo, R.M.; Rossi, E. Modelling crowd movements in domains with boundaries. *IMA J. Appl. Math.* **2019**, *84*, 833–853. [CrossRef]

88. Khan, S.D.; Basalamah, S. Multi-Scale Person Localization with Multi-Stage Deep Sequential Framework. *Int. J. Comput. Intell. Syst.* **2021**, *14*, 1217–1228. [CrossRef]

89. Zhang, A.; Jiang, X.; Zhang, B.; Cao, X. Multi-scale Supervised Attentive Encoder-Decoder Network for Crowd Counting. *ACM Trans. Multimed. Comput. Commun. Appl. (TOMM)* **2020**, *16*, 1–20. [CrossRef]

90. Jiang, R.; Mou, X.; Shi, S.; Zhou, Y.; Wang, Q.; Dong, M.; Chen, S. Object tracking on event cameras with offline–online learning. *CAAI Trans. Intell. Technol.* **2020**, *5*, 165–171. [CrossRef]

91. Jalal, A.; Kamal, S.; Kim, D. Shape and motion features approach for activity tracking and recognition from kinect video camera. In Proceedings of the 2015 IEEE 29th International Conference on Advanced Information Networking and Applications Workshops, Gwangju, Korea, 24–27 March 2015; pp. 445–450.

92. Keshtegar, B.; Nehdi, M.L. Machine learning model for dynamical response of nano-composite pipe conveying fluid under seismic loading. *Int. J. Hydromechatron.* **2020**, *3*, 38–50. [CrossRef]

93. Antonini, G.; Martinez, S.V.; Bierlaire, M.; Thiran, J.P. Behavioral priors for detection and tracking of pedestrians in video sequences. *Int. J. Comput. Vis.* **2006**, *69*, 159–180. [CrossRef]

94. Choudri, S.; Ferryman, J.M.; Badii, A. Robust background model for pixel based people counting using a single uncalibrated camera. In Proceedings of the 2009 Twelfth IEEE International Workshop on Performance Evaluation of Tracking and Surveillance, Snowbird, UT, USA, 7–9 December 2009; pp. 1–8.

95. Chen, H.W.; McGurr, M. Improved color and intensity patch segmentation for human full-body and body-parts detection and tracking. In Proceedings of the 2014 11th IEEE International Conference on Advanced Video and Signal Based Surveillance (AVSS), Seoul, Korea, 26–29 August 2014; pp. 361–368.

96. García, J.; Gardel, A.; Bravo, I.; Lázaro, J.L.; Martínez, M.; Rodríguez, D. Directional people counter based on head tracking. *IEEE Trans. Ind. Electron.* **2012**, *60*, 3991–4000. [CrossRef]

97. Vinod, M.; Sravanthi, T.; Reddy, B. An adaptive algorithm for object tracking and counting. *Int. J. Eng. Innov. Technol.* **2012**, *2*, 64–69.

98. Liu, G.; Liu, S.; Muhammad, K.; Sangaiah, A.K.; Doctor, F. Object tracking in vary lighting conditions for fog based intelligent surveillance of public spaces. *IEEE Access* **2018**, *6*, 29283–29296. [CrossRef]
99. Ristani, E.; Tomasi, C. Features for multi-target multi-camera tracking and re-identification. In Proceedings of the IEEE Conference on Computer Vision and Pattern Recognition, Salt Lake City, UT, USA, 18–22 June 2018; pp. 6036–6046.
100. Xu, H.; Lv, P.; Meng, L. A people counting system based on head-shoulder detection and tracking in surveillance video. In Proceedings of the 2010 International Conference on Computer Design and Applications, Qinhuangdao, China, 25–27 June 2010; Volume 1, pp. 394–398.
101. Kabbai, L.; Abdellaoui, M.; Douik, A. Image classification by combining local and global features. *Vis. Comput.* **2019**, *35*, 679–693. [CrossRef]
102. Lisin, D.A.; Mattar, M.A.; Blaschko, M.B.; Learned-Miller, E.G.; Benfield, M.C. Combining local and global image features for object class recognition. In Proceedings of the 2005 IEEE Computer Society Conference on Computer Vision and Pattern Recognition (CVPR'05)-Workshops, San Diego, CA, USA, 21–23 September 2005; p. 47.
103. Pirsiavash, H.; Ramanan, D.; Fowlkes, C.C. Globally-optimal greedy algorithms for tracking a variable number of objects. In Proceedings of the CVPR 2011, Colorado Springs, CO, USA, 20–25 June 2011; pp. 1201–1208.
104. Bay, H.; Tuytelaars, T.; Van Gool, L. Surf: Speeded up robust features. In Proceedings of the European Conference on Computer Vision, Graz, Austria, 7–13 May 2006; Springer: Berlin/Heidelberg, Germany, 2006; pp. 404–417.
105. Lowe, D.G. Distinctive image features from scale-invariant keypoints. *Int. J. Comput. Vis.* **2004**, *60*, 91–110. [CrossRef]
106. Führ, G.; Jung, C.R. Camera self-calibration based on nonlinear optimization and applications in surveillance systems. *IEEE Trans. Circuits Syst. Video Technol.* **2015**, *27*, 1132–1142. [CrossRef]
107. Gandomi, A.H.; Yang, X.S.; Alavi, A.H.; Talatahari, S. Bat algorithm for constrained optimization tasks. *Neural Comput. Appl.* **2013**, *22*, 1239–1255. [CrossRef]
108. Murlidhar, B.R.; Sinha, R.K.; Mohamad, E.T.; Sonkar, R.; Khorami, M. The effects of particle swarm optimisation and genetic algorithm on ANN results in predicting pile bearing capacity. *Int. J. Hydromechatron.* **2020**, *3*, 69–87. [CrossRef]
109. Shahgoli, A.F.; Zandi, Y.; Heirati, A.; Khorami, M.; Mehrabi, P.; Petkovic, D. Optimisation of propylene conversion response by neuro-fuzzy approach. *Int. J. Hydromechatron.* **2020**, *3*, 228–237. [CrossRef]
110. Yang, X.S. A new metaheuristic bat-inspired algorithm. In *Nature Inspired Cooperative Strategies for Optimization (NICSO 2010)*; Springer: Berlin/Heidelberg, Germany, 2010; pp. 65–74.
111. Alahi, A.; Jacques, L.; Boursier, Y.; Vandergheynst, P. Sparsity-driven people localization algorithm: Evaluation in crowded scenes environments. In Proceedings of the 2009 Twelfth IEEE International Workshop on Performance Evaluation of Tracking and Surveillance, Snowbird, UT, USA, 7–9 December 2009; pp. 1–8.
112. Berclaz, J.; Fleuret, F.; Fua, P. Multiple object tracking using flow linear programming. In Proceedings of the 2009 Twelfth IEEE international workshop on performance evaluation of tracking and surveillance, Snowbird, UT, USA, 7–9 December 2009; pp. 1–8.
113. Neiswanger, W.; Wood, F.; Xing, E. The dependent Dirichlet process mixture of objects for detection-free tracking and object modeling. In Proceedings of the Seventeenth International Conference on Artificial Intelligence and Statistics, PMLR, Reykjavik, Iceland, 22–25 April 2014; Volume 33, pp. 660–668.
114. Conte, D.; Foggia, P.; Percannella, G.; Vento, M. Performance evaluation of a people tracking system on pets2009 database. In Proceedings of the 2010 7th IEEE International Conference on Advanced Video and Signal Based Surveillance, Boston, MA, USA, 29 August–1 September 2010; pp. 119–126.

Article

Adding Prior Information in FWI through Relative Entropy

Danilo Santos Cruz [1,*,†], João M. de Araújo [1,2,*,†], Carlos A. N. da Costa [2,†] and Carlos C. N. da Silva [3,†]

1   Programa de Pós-Graduação em Ciência e Engenharia do Petróleo, Universidade Federal do Rio Grande do Norte, Natal 59064-741, Brazil
2   Departamento de Física Teórica e Experimental, Universidade Federal do Rio Grande do Norte, Natal 59064-741, Brazil; alexandrecosta17@gmail.com
3   Departamento de Geofísica, Universidade Federal do Rio Grande do Norte, Natal 59064-741, Brazil; carloscesar@geofisica.ufrn.br
*   Correspondence: danilosc@fisica.ufrn.br (D.S.C.); joaomedeiros@fisica.ufrn.br (J.M.d.A.)
†   These authors contributed equally to this work.

Citation: Cruz, D.S.; de Araújo, J.M.; da Costa, C.A.N.; da Silva, C.C.N. Adding Prior Information in FWI through Relative Entropy. *Entropy* **2021**, *23*, 599. https://doi.org/10.3390/e23050599

Academic Editor: Amelia Carolina Sparavigna

Received: 21 April 2021
Accepted: 8 May 2021
Published: 13 May 2021

Publisher's Note: MDPI stays neutral with regard to jurisdictional claims in published maps and institutional affiliations.

Abstract: Full waveform inversion is an advantageous technique for obtaining high-resolution subsurface information. In the petroleum industry, mainly in reservoir characterisation, it is common to use information from wells as previous information to decrease the ambiguity of the obtained results. For this, we propose adding a relative entropy term to the formalism of the full waveform inversion. In this context, entropy will be just a nomenclature for regularisation and will have the role of helping the converge to the global minimum. The application of entropy in inverse problems usually involves formulating the problem, so that it is possible to use statistical concepts. To avoid this step, we propose a deterministic application to the full waveform inversion. We will discuss some aspects of relative entropy and show three different ways of using them to add prior information through entropy in the inverse problem. We use a dynamic weighting scheme to add prior information through entropy. The idea is that the prior information can help to find the path of the global minimum at the beginning of the inversion process. In all cases, the prior information can be incorporated very quickly into the full waveform inversion and lead the inversion to the desired solution. When we include the logarithmic weighting that constitutes entropy to the inverse problem, we will suppress the low-intensity ripples and sharpen the point events. Thus, the addition of entropy relative to full waveform inversion can provide a result with better resolution. In regions where salt is present in the BP 2004 model, we obtained a significant improvement by adding prior information through the relative entropy for synthetic data. We will show that the prior information added through entropy in full-waveform inversion formalism will prove to be a way to avoid local minimums.

Keywords: prior information; entropy; fwi; regularization; inverse problems

1. Introduction

The subsurface image, more specifically, a detailed image of the oil reservoir, is essential in oil and gas exploration and production and requires appropriate data acquisition, processing to remove unwanted information, building a velocity model to use in an appropriate migration algorithm. The quality of the image obtained is, generally, controlled by the subsurface velocity model. When the geology of the area of interest is composed of salt bodies with complex geometrical shapes, the construction of a precise velocity model is more complicated. Full waveform inversion (FWI) is a tool that can provide us with a velocity model with greater precision and resolution. Wang and Rao [1] is, for the first time, applying FWI for the industrial standard reflection seismic data. Through the use of amplitude and travel time content of the acquired seismic data, this technique, theoretically, has the potential to be the most accurate method for the construction of subsurface velocity models [2,3]. Wang and Houseman [4] proposed the joint inversion that uses both the

95

amplitude and travel time data simultaneously, so as to mitigate the ambiguity of reflector geometry and the interval velocities between reflectors.

FWI is a nonlinear and ill-posed data fitting method that usually uses local optimisation methods and, thereforem its solution depends heavily on the initial model. In order to avoid the cycle skipping, the initial model should predict errors in the arrival times less than half the wavelength [3]. One can minimise the issue of non-linearity and the cycle skipping problem with the use of a multi-scale strategy, where we begin from the lowest to the highest frequencies, helping the convergence to the global minimum [5].

The effects of non-uniqueness of the ill-posed inverse problem are usually decreased by the use of regularisation techniques. These regularisation techniques help to stabilise the inversion scheme by incorporating a specific structure or characteristic of the model (e.g., smoothness, sparsity). The most used regularisation scheme is the one that was proposed by Tikhonov and Arsenin [6]. This method incorporated in the inversion scheme aims to find a smooth model that can justify the data. In FWI, in some cases, an $l1$ model penalty is used as a regularisation strategy to preserve edges and contrasts [7]. However, in some cases, prior information, such as sonic records, stratigraphic data, or geological restrictions, about the model is available. To mitigate the problems of non-uniqueness and stability of the solution, we can use a regularisation scheme that is composed of the model norm and first-order Tikhonov regularisation to act as a smoothing operator, as proposed in [8]. They also suggest that the weighting of the term that is responsible for adding the prior information be done dynamically. This strategy proved to be useful in avoiding local minimums in the FWI. Peters and Hermmann [9] showed a way of including constraints on spatial variations and values ranges of the inverted velocities in FWI.

One way to add prior information to the inversion scheme is through relative entropy. In Thermodynamics, we can introduce the notion of entropy to characterise the degree of disorder of a system. The notion of entropy has been the subject of many controversies and different formulations [10]. Here, we will use entropy just as nomenclature used to restrict the solutions of the inverse problem. In a minimisation problem, when we compare entropy with the model norm, the logarithmic operation will suppress the low-intensity ripples. Thus, the entropy method can deliver images with better resolution in some cases [11]. In other words, adding entropy to the FWI formalism, we introduce a smoothness characteristic in the objective function, which will lead to smoothed solutions that are consistent with the available data [12].

The principle of minimum relative entropy (PMRE) was introduced in [13], and it was first applied in the context of geophysicist by Shore [14] in spectral analysis. Other works applied the PMRE in the seismic deconvolution to make limited band extrapolation [15], diffraction seismic tomography [16], and different geophysical problems, such as inversion of interval velocities, removal of the alias effect, and seismic deconvolution [17]. In the context of the FWI, the entropy concept was applied by Chen and Peter [18], who proposed a misfit function based on entropy regularised optimal transport. da Silva et al. [19] proposed a misfit function for FWI based on Shannon entropy for deeper velocity model updates. None of them made use of prior information and, all of them, in some way, use statistical formalism.

Recently, many works have been developed formulating the FWI in terms of Bayesian formalism. In this sense, Zhu et al. [20] show a Bayesian approach to estimate uncertainty for full-waveform inversion using a priori information from depth migration. Singh et al. [21] propose a robust way to constrain the inversion workflow using per-facies rock-physics relationships derived from well logs. Carvalho et al. [22] show Full-waveform inversion with subsurface fractal information and variable model uncertainties. Zhang and Curtis [23] present the first application of variational full-waveform inversion (VFWI) to seismic reflection data where they imposed realistically weak prior information on seismic velocity.

Usually, when working with entropy, we use probability distribution information or Bayesian formalism, which requires some additional step in the formulation of the problem, such as representing initial and prior models as posterior and prior probability distributions [17]. Our proposal is to add prior information to the FWI using relative entropy without explicitly using the concept of a probability distribution or Bayesian formalism. We will do this in a deterministic and direct way. We present three distinct ways to add prior information to the FWI formalism through relative entropy. We will discuss some aspects of the relative entropy for obtaining velocity models and show how these ways of adding prior information will contribute to recovering the velocity model in areas that are affected by the presence of bodies of salt through a synthetic application on the BP model.

2. Theory

Full waveform inversion is an example of a nonlinear ill-posed problem. In general, the solution to this problem is to minimise the discrepancies between the observed and modelled data. From the mathematical point of view, this is a nonlinear problem that is usually being treated as a linearised least-squares problem. However, even the linearised problem still ill-posed and, consequently, several solutions can provide a satisfactory fit between the observed and modelled data.

One way to circumvent this ambiguity of solutions is by introducing prior information adding to the formalism of the inverse problem a terms of regularisation. Thus, we will briefly present the mathematical methodology of the inversion algorithm with the contribution of prior information.

Let F be a generic cost function that is given by:

$$F(\mathbf{m}) = \Phi_d(\mathbf{m}) + \alpha \Psi_m(\mathbf{m}).$$ (1)

For FWI, the term $\Phi_d(\mathbf{m})$ is usually constructed through the L_2 norm of the residue between the modeled and observed data, which is:

$$\Phi_d(\mathbf{m}) = \sum_{ns} \frac{1}{2} [(\mathbf{d}_{obs} - \mathbf{d}_{cal}(\mathbf{m}))^t (\mathbf{d}_{obs} - \mathbf{d}_{cal}(\mathbf{m}))],$$ (2)

where $\mathbf{d}_{obs}$ and $\mathbf{d}_{cal}(\mathbf{m})$ represent the observed and calculated data vectors, respectively. In our work, we use a time domain approach and each component of these vectors are samples of time domain seismograms recorded at each of the receivers for a seismic source. The misfit function is the result of the sum realized over all ns sources of the seismic acquisition. There is a non-linear dependence on the modeled data and the model parameters, as represented by $\mathbf{m}$. The model parameters are determined by an inverse process that aims to reduce the residue between the modeled and observed data.

In our case, the second term of the cost function will be responsible for adding the prior information (this a prior information can, for example, come from well profiles) to the inversion process. Here, the prior information will be denoted by $\mathbf{m^r}$. This prior information can be added in different ways to the inversion. In FWI, the model norm with this intention is used in [8]. We use this form for comparison purposes, so we can write Ψ as:

$$\Psi_m(\mathbf{m}) = \sum_{i=1}^{N} (m_i - m_i^r)^2.$$ (3)

The first form that we studied in this experiment was the relative entropy described by [24,25]:

$$\Psi = \sum_{i=1}^{N} m_i ln(\frac{m_i}{m_i^r}),$$ (4)

the Equation (4) is the *Kullback–Leibler's* distance from $\mathbf{m^r}$ to $\mathbf{m}$. Usually, this equation is used in association with probability distribution. This makes the relative entropy always positive. However, this is only true because the probabilities of the events fall between zero and one [26]. In order to avoid a reformulation of the problem, our idea is to use it in a deterministic way, which is, without using concepts of probability.

Formally, the Kullback–Leibler distance is a pseudo-distance, as it does not satisfy two properties of the metric definition; triangular inequality and symmetry [27]. The fact of non-symmetry led us to our second case study, which will be represented in the equation below:

$$\Psi = \sum_{i=1}^{N} m_i^r ln\left(\frac{m_i^r}{m_i}\right). \tag{5}$$

It becomes necessary to analyze the behavior of these functions. For this, we simulate a situation in which the prior information was constant ($\mathbf{m^r} = 10$, for example) and plot the graph of the function $y = xln(x/10)$ and $y = 10ln(10/x)$, as can be seen in the Figure 1. Analyzing Figure 1, we can infer that Ψ (in Equations (4) and (5)) can present positive and negative values. The graph of Equation (4) (red curve) shows that Ψ will always be negative when the values of the model parameters are less than the parameters of the reference model. The graph of Equation (5) (blue curve) shows that Ψ will always be negative when the values of the model parameters are greater than the parameters of the reference model. In addition, the graph of Equation (5) (red curve) shows that this function does not present a minimum.

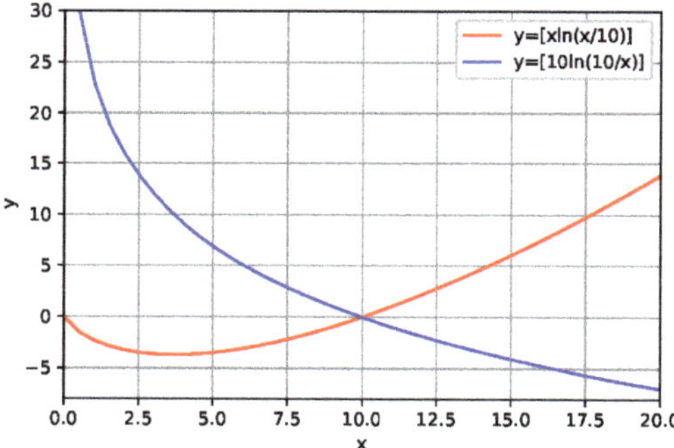

Figure 1. Graphic of the functions $y = xln(x/10)$ (red curve) and $y = 10ln(10/x)$ (blue curve). This function is not positive definite. The function $y = 10ln(10/x)$ does not present a minimum

This behavior (sometimes positive and negative) in both equations brings an inconvenience to inversion. At one point, we would be minimizing the function at another, maximizing the function.

The simplest way to transform Equation (4) into positive definite is to work with the quadratic form. Thus, we rewrite Equation (4), as follows:

$$\Psi = \sum_{i=1}^{N} \left[m_i ln\left(\frac{m_i}{m_i^r}\right) \right]^2. \tag{6}$$

Analogously, we can change Equation (5) into positive definite taking its quadratic form. Thus, we rewrite Equation (5), as follows:

$$\Psi = \sum_{i=1}^{N} \left[m_i^r ln\left(\frac{m_i^r}{m_i}\right) \right]^2. \tag{7}$$

As was done for Equations (4) and (5), we plotted a graph of the function described by Equations (6) and (7), as can be seen in Figure 2. Analyzing Figure 2, it can be seen that Equation (6) is positively defined. However, in the example that is illustrated in Figure 2, we can observe the presence of two minimums. This leads us to interpret that the use of Equation (6) in FWI can increase the problem of local minimums (this will be exemplified in numerical applications). Also analyzing Figure 2, we can expect that Equation (7) presents the characteristics that are favorable to the inversion process, which is, it is a definite positive function and that only presents a minimum.

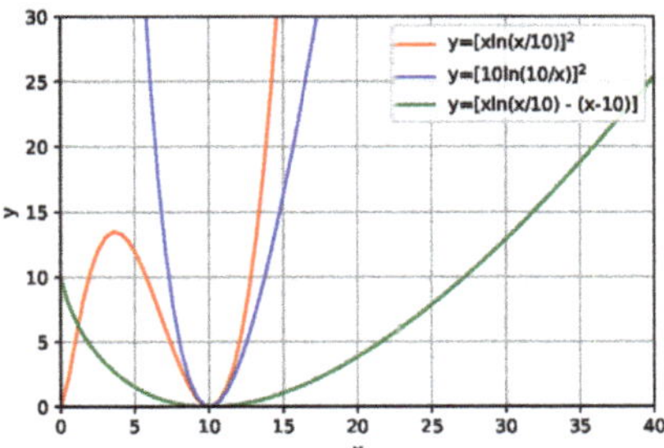

Figure 2. Graphic of the functions $y = [xln(x/10)]^2$ (red curve), $y = [10ln(10/x)]^2$ (blue curve) and $y = xln(x/10) - (x - 10)$ (green curve). The function $y = [xln(x/10)]^2$ is positive definite, but it has more than a minimum. The functions $y = [10ln(10/x)]^2$ and $y = xln(x/10) - (x - 10)$ are positive definite and present only a minimum.

Finally, we can add the prior information to FWI with the axiomatic form that is given by (8) [28,29]:

$$\Psi = \sum_{i=1}^{N} \left[m_i ln(\frac{m_i}{m_i^r}) - (m_i - m_i^r) \right], \tag{8}$$

the Equations (4) and (8) are similar. However, Equation (8) has a term referring to the difference of the models that leaves it with the characteristic that we expect (definite positive function), as can be seen in Figure 2 in a green curve.

If we minimize the objective function in the classical way, we obtain a system of equations that can be expressed as:

$$\mathcal{H}_F \Delta \mathbf{m} = -\mathcal{G}_F, \tag{9}$$

where $\mathcal{H}_F$ and $\mathcal{G}_F$ represent the Hessian and gradient of cost function, respectively. In this case, the gradient represents the sum of two terms. If the Ψ function is the model norm (Equation (3)), we have:

$$\mathcal{G}_F = \mathbf{J}^T(\mathbf{d}_{obs} - \mathbf{d}(\mathbf{m})) + 2\alpha(m_i - m_i^r). \tag{10}$$

For the Ψ function to be represented by Equation (6), the gradient will be represented, as follows:

$$\mathcal{G}_F = \mathbf{J}^T(\mathbf{d}_{obs} - \mathbf{d}(\mathbf{m})) + 2\alpha \left[\left(m_i ln(\frac{m_i}{m_i^r}) \right) \left(ln(\frac{m_i}{m_i^r}) + 1 \right) \right]. \tag{11}$$

When the Ψ function that is represented by Equation (7) is used, the gradient expression will be given by:

$$\mathcal{G}_F = \mathbf{J}^T(\mathbf{d}_{obs} - \mathbf{d}(\mathbf{m})) - 2\alpha \left[\left(m_i^r ln(\frac{m_i^r}{m_i}) \right) \left(\frac{m_i^r}{m_i} \right) \right]. \tag{12}$$

In case the Ψ function used is the one represented in Equation (8), the gradient will be described, as follows:

$$\mathcal{G}_F = \mathbf{J}^T(\mathbf{d}_{obs} - \mathbf{d}(\mathbf{m})) + \alpha \left[m_i ln(\frac{m_i}{m_i^r}) \right]. \tag{13}$$

The term $\mathbf{J}$ that is present in the gradient expressions is the sensitivity matrix. The sensitivity matrix is composed of the derivatives of the modeled data with respect to the model parameters $(\mathbf{J} = \partial \mathbf{d}(\mathbf{m})/\partial \mathbf{m})$. The elements of $\mathbf{J}$ are not explicitly calculated because they demand a high computational cost. For this reason, the adjoint formulation [30] is used for this purpose.

Asnaashari et al. [8] showed that we should work with a dynamic weighting of the term regularisation. The basic idea of this methodology is to help the inversion process to converge to the global minimum of the objective function by increasing the importance of prior information at the beginning of the process and gradually decreasing the penalty term weighting until, in the final iterations, the convergence driven by the term of the data. In this work, the $\alpha = \mu\gamma$ parameter is a dynamic weighting of the term regularisation and it has the role of decreasing the weight of the entropy term over the iterations. We built the dynamic term (γ) from the ratio of the gradients, as can be seen in Equation (14):

$$\gamma = \frac{\sum_{i=1}^{M}(\nabla\Phi_i)^2}{\sum_{i=1}^{M}(\nabla\Psi_i)^2}, \tag{14}$$

where Φ_i and Ψ_i are the elements of the gradients vectors of misfit and regularise, respectively. In our tests, as will be seen in numerical applications, the γ value proved to be inadequate (the initial value was large) and it needed to be adjusted. We made this adjustment using the μ parameter.

We calculated the gradients of the terms of those that are responsible for adding the prior information easily and directly added to the data gradient. The term of the Hessian matrices, which is composed of the second derivative of misfit function and relative entropy, is not explicitly resolved in this paper. We calculated the hessian using a limited quasi-newton method that is known in the literature as L-BFGS-B. The routine that was proposed by [31] considers that the inverse Hessian matrix is non-diagonal and roughly obtains its elements from the gradient vectors and previous models by performing a line search that satisfies Wolfe's conditions.

3. Numerical Tests

We only worked with the acoustic case (i.e., a P velocity model) and considered the homogeneous density distribution. We also considered a regular grid with 12.5 m spacing, which is used in both modelling and inversion. The data that were observed and modelled in time were obtained from the acoustic wave equation through finite-difference modelling, where an eighth order approximation for the Laplacian operator and a second-order approximation for the time derivative were considered. A CPML absorption boundary layer was employed to avoid boundary reflection [32,33]. The absorbing layer was applied to all sides of the model, using a width of 40 cells. The FWI worked here was performed in the time domain using all spectrum frequencies.

In this section, we will show the contribution of prior information added to the FWI through relative entropy. Therefore, we chose using the first and second part of the BP 2004 benchmark [34]. For the first part, the acquisition geometry consisted of 475 hydrophones distributed along a straight line 12.5 m deep, with 12.5 m spacing between each receiver. For the shots, 15 sources spaced 395 m arranged in line with 25 m deep. For all shots, a Ricker wavelet source with a central frequency of 10 Hz was used and the time record was 5.0 s. For the second part, the geometry acquisition is similar to that used in the first part, but the central frequency of Ricker was 12 Hz with a record time of 6.5 s.

Given that FWI is usually treated as an iterative process, an initial velocity model is required. For example, this model may be the result of a tomography that is based on the times of first arrivals and reflected events. For this work, we perform the smoothing of the real model (Figure 3), and we use it as an initial model in the FWI process.

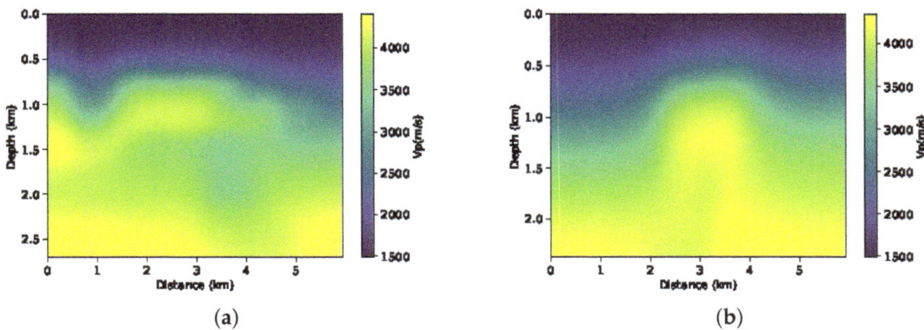

Figure 3. Illustration of the initial models used as an initial model to FWI. (a) Smoothing of the first part from the BP model. (b) Smoothing of the second part from the BP model.

For this study, we assume that we have information from exploration wells. The velocities profile are measurements that provide a good measure of the local depth velocity. Thus, we will use these sonic profiles to build our a priori information model. A linear interpolation was made between the two wells. In the other regions, we use an extrapolation of the well's velocity profile. We also apply a slight smoothing to this a priori model of velocity. We can see this interpolated model in Figure 4. Although not geologically significant, this model contains some travel time information, and it will be considered an a priori velocity model and incorporated into the FWI through the relative entropy of the model.

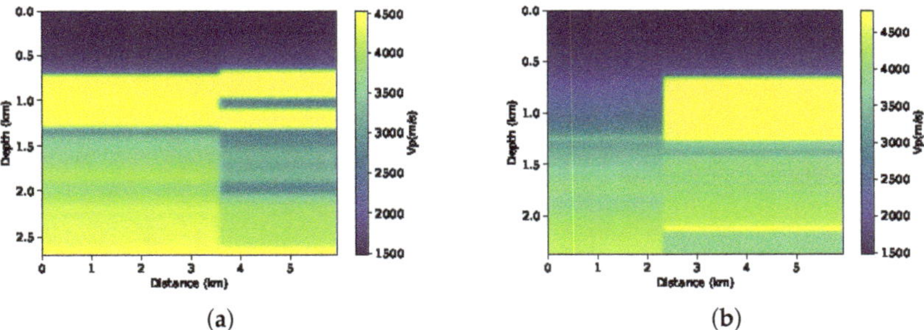

Figure 4. The prior models built by linear interpolation between the values and extrapolation outside from wells that will be added in the FWI. (a) Prior model of the first part from BP model, (b) Prior model of the second part from BP model.

First, we performed the inversion without adding prior information on both parts of the BP model that are illustrated in Figure 5a,b. This means that, in Equation (1), $\alpha = 0$. We used the initial models that are shown in Figure 3a,b. The results of the FWI for each of the cases are shown in Figure 5a,b. Clearly, in both cases, the conventional FWI (without any type of regularisation) converges to a local minimum. Asnaashari et al. [8] discussed some differences between the prior and initial models in the inversion procedure. In this case, the smoothed model that is shown in Figure 3 and the a priori model shown in Figure 4 have only part of the real model information. For the FWI result to converge to the desired result, both of the models must be used in a complementary way.

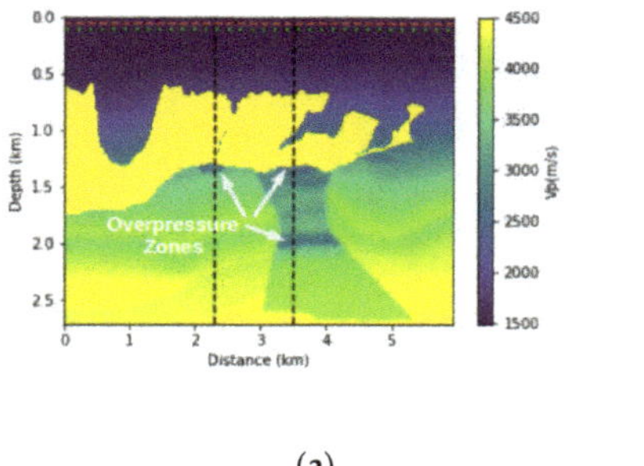

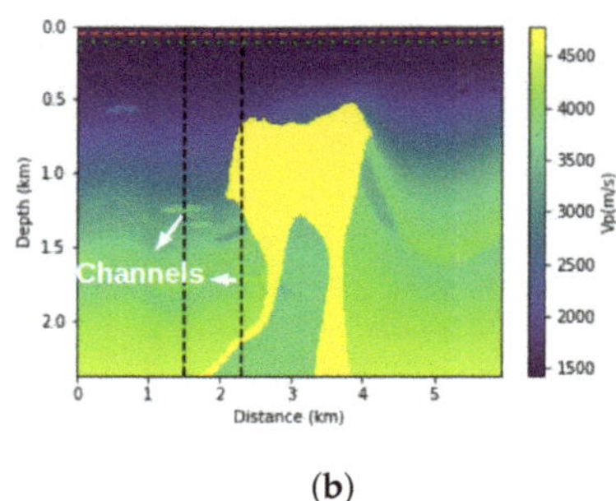

(a) (b)

Figure 5. True V_p velocity model, which are parts of the BP model and the scheme of acquisition. The red-dashed line represents the position of the receivers, while the green line represents the position of the sources. (a) First part from the BP model, the white arrows illustrate the target zones (overpressure zones) and the black-dashed lines represent the position of the two wells that cross overpressure zones. (b) Second part from the BP model, the white arrows illustrate the target zones (channels) and the black-dashed lines represent the position of the two wells that cross the channels.

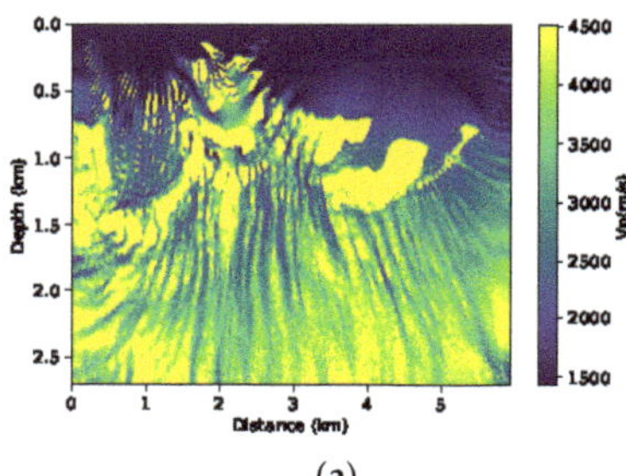

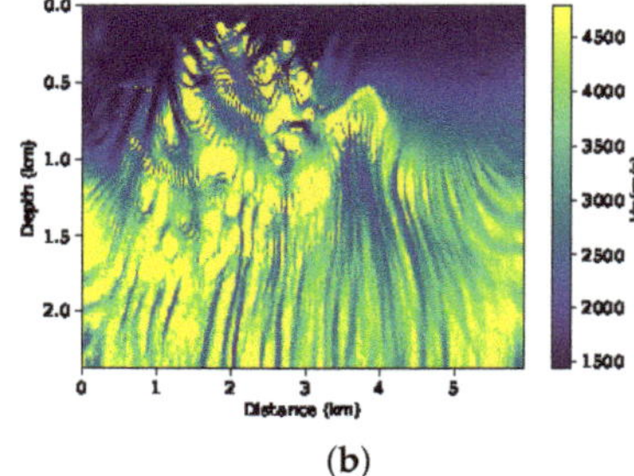

(a) (b)

Figure 6. (a) FWI results using the smoothed model and without prior information ($\alpha = 0$) (a) in first part of BP model and (b) in second part of BP model.

Given the obtained results, we will add the prior information to the FWI formalism in four different ways, as shown later on.

3.1. Model Norm

In this section, we add the prior information to the FWI using Equation (3). This method was used by [8] to incorporate the prior information in the FWI. Here, as previously mentioned, we will use the results obtained here to compare with the entropy methods that are the focus of this work. Therefore, we performed the FWI for both models (first and the second parts of the BP model), adding the prior information that is shown in Figure 4. The initial models used are those that are represented in Figure 3. Figures 7a and 8a show the results. When we compare these results with those that are obtained without adding prior information (Figure 6a,b), we observe an improvement in the quality of the FWI result. We observed that, in general, the body of salt was recovered in both models (although the first part of the BP model presents a small problem on the left side of the well positioned at $x = 2.3$ km). For a more detailed quality control, we can see the profiles in the positions of the wells in Figure 7b,c for the first part of the model, and Figure 8b,c for the second part of the model. By analysing the profiles, we can confirm that the addition of prior information in FWI through the model norm provides a good result.

A crucial point for the success of adding prior information in FWI scheme is the choice of the alpha parameter. In this work, as described in the theoretical section, the α parameter is the product of two terms. The first is a dynamic term (γ), as calculated from Equation (14). Using the model norm, the γ parameter initiated the inversion process equal to $6 \times 10^{+17}$ and $7.6 \times 10^{+17}$ for the first and second parts of the BP model, respectively. The initial value of the γ parameter proved to be inadequate and it needed to be adjusted. Consequently, we use a second term, the μ parameter to adjust the weight of the regularization term. After several tests, we found that the parameter should be 3.5×10^{-10} and 3×10^{-10} for the first and second part of the BP model, respectively. Once the values of the μ and γ parameters are found, determining the α parameter is straightforward. The evolution of the α parameter for each model can be seen in Figures 7d and 8d. Note that the α values are on the log (natural base) scale. We can see the expected behavior for the weight (α) given to the model norm term in Figures 7d and 8d. We observed that there was a sharp drop at the beginning of the inversion. As the model is updated, these terms will decrease and the seismic data will conduct the inversion.

The misfit data curve is shown in Figure 7e for the first part of the model and Figure 8e for the second part of the model. For all of the tests that we performed, we used a small stop criterion to ensure that the data adjustment was as large as possible, which resulted in a large number of iterations. For the stopping criterion, a tolerance limit of 9×10^{-9} was established in the total value of the model update (total gradient). The misfit data curves show a great difference between the convergence of conventional FWI (without any type of regularisation or prior information) and FWI with the addition of prior information through the model norm. Even with several iterations, the conventional FWI cannot reduce the misfit data, while the FWI with prior information shows a very sharp drop at the beginning of the iterations that continues until it reaches a relatively satisfactory result.

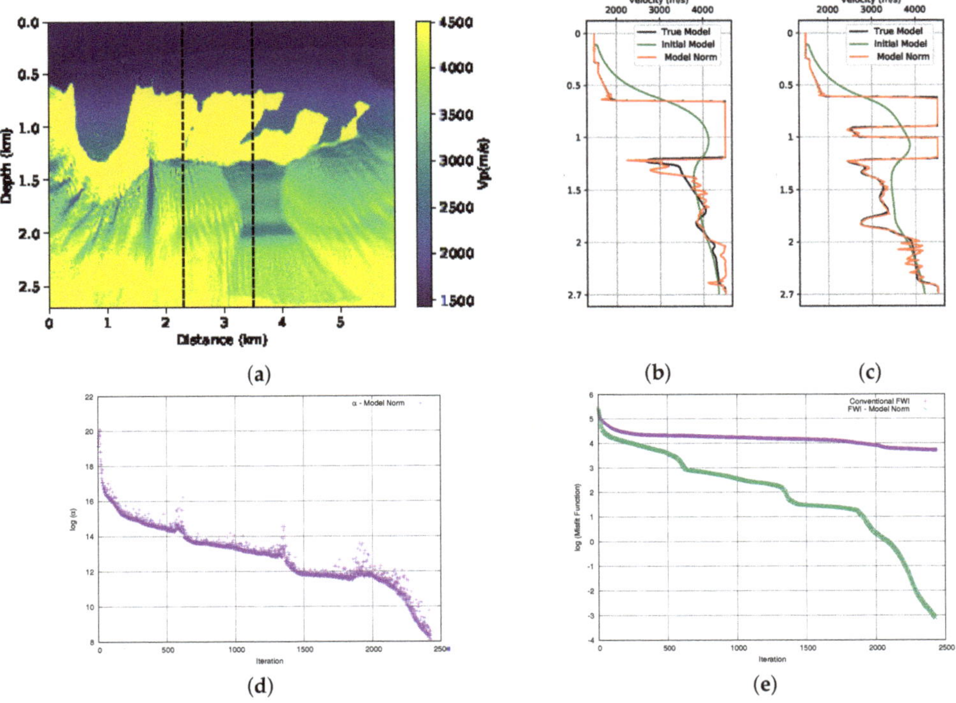

Figure 7. First part of BP model; (**a**) FWI result with adding prior information through the model norm; (**b**) profile in well at position $x = 2.3$ km, (**c**) profile in well at position $x = 3.5$ km and (**d**) dynamic term (α) progress, (**e**) misfit data function progress (logarithmic natural base scale).

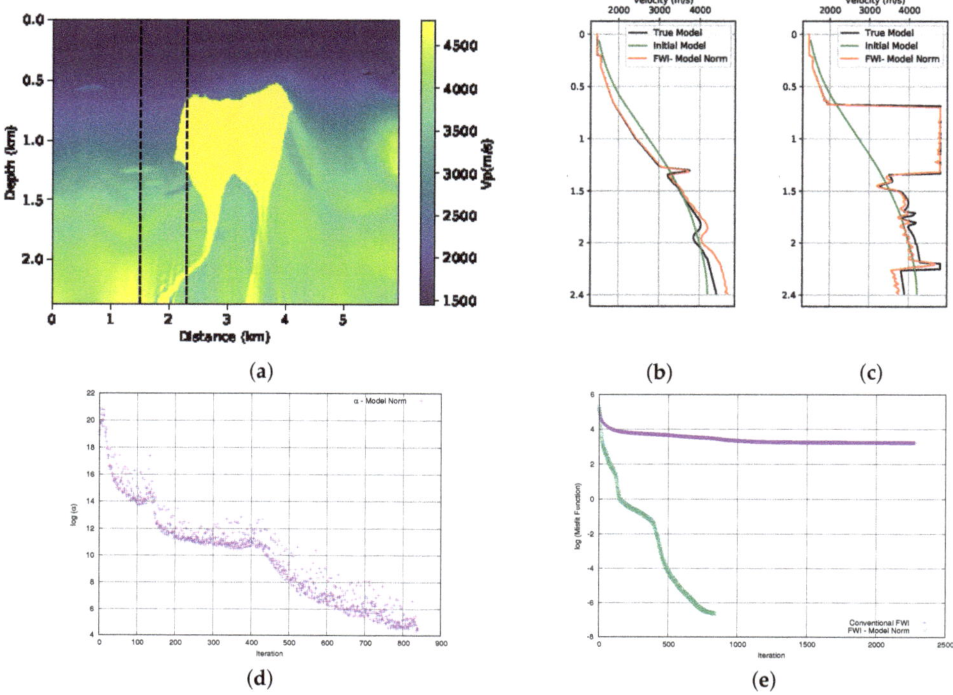

Figure 8. Second part of BP model; (**a**) FWI result with adding prior information through the model norm; (**b**) profile in well at position $x = 1.5$ km (**c**) profile in well at position $x = 2.3$ km, (**d**) dynamic term (α) progress (**e**) misfit data function progress (logarithmic natural base scale).

3.2. First Case: Kullback-Leibler's Distance

As mentioned earlier, our proposal is to add priori information to the FWI formalism through relative entropy. The first attempt to add prior information to FWI through entropy would be with the use of the relative entropy that is described by Equation (4). To use it, it would be necessary to represent entropy as a probability distribution function. This implies a normalization constraint, that is, $0 \leq p \leq 1$ [17]. For the addition of a priori information to be done in a simple and direct way, we will do this in a deterministic way. Thereby, the Equation (4) is not adequate as previously discussed. Equation (4) is not positively defined in any interval. The way to get around the problem that was brought by Equation (4) was to work with its quadratic form represented by Equation (6). Thus, we started the tests in the first part of the BP model using $\alpha = 2 \times 10^{+7}$. Even with the addition of prior information, the result converged to a local minimum and it is far from the expected result, as can be seen in Figure 9a. A natural idea would be that the initial weight given to the entropy term is inadequate. Consequently, we increased the value of this initial weight to $\alpha = 3 \times 10^{+7}$ and with few iterations we obtained the result that is illustrated in Figure 9b. In this result, which is still a local minimum, we note that the FWI is leading the solution for the prior model. Although we performed other tests with intermediate values for alpha, we did not achieve the desired success for this case.

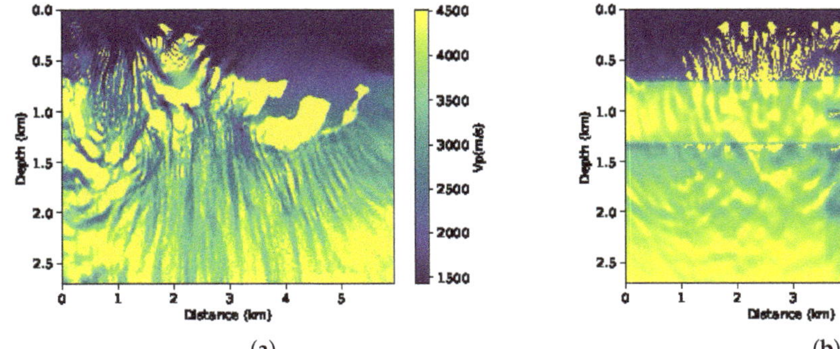

Figure 9. FWI results with adding prior information through the quadratic form of relative entropy for the first part of BP model. (**a**) initial $\alpha = 2 \times 10^7$ (**b**) initial $\alpha = 3 \times 10^7$.

Even with the failure in the first attempt to add the quadratic form of entropy relative to the formalities of the FWI, we performed the test in the second part of the BP model. Figure 10a shows the result. The analysis of the result that is illustrated in Figure 10a shows that in this case the addition of prior information through the quadratic form of the relative entropy enabled the FWI to converge on a satisfactory solution. When comparing the result that was obtained with the model norm (Figure 8a), a visible similarity is observed. A more detailed analysis of the position of the wells confirms this similarity when we compare the adjustment at position $x = 1.5$ km. The model norm (red curve) and the relative entropy (blue curve) both provide equivalent results, as can be seen in Figure 10b. However, when we see the position $x = 2.3$ km (Figure 10c), the adjustment that is provided by the relative entropy proved to be slightly better to the model norm, mainly in the deep part of the model. The initial alpha for this case was $\alpha \sim 8 \times 10^7$ (in this case, $\gamma \sim 2.7 \times 10^{17}$ and $\mu = 3 \times 10^{-10}$) and its evolution can be seen in Figure 10c. We observe that there is a marked decrease in the weight given to the end of the relative entropy, which allows us to avoid giving too much importance to the entropy term by ensuring that an adequate contribution of the prior information is maintained throughout the iterations.

Figure 10e illustrates the data misfit. As seen in the case of the model norm (green curve), the addition of prior information through relative entropy (light blue curve) causes the FWI to drastically reduce the misfit of the data, leading to a low misfit result. Even though the relative entropy leads to an inversion around iteration 470 through a path of misfit greater than that of the model norm, the values of the final misfit are close.

The failure in the first part and the success in the second part of the BP model led us to conclude that the initial reasoning made through the analysis of the graph that is shown in Figure 2 was correct. In other words, the quadratic form of the relative entropy can somehow increase the problem of local minimums, but, depending on the path that the inversion process takes, we can find the global minimum.

3.3. Second Case: Kullback-Leibler's Distance—Symmetric Form

There are few applications in the literature for the symmetric form of relative entropy that is shown in Equation (5). Probably the reason for this is the fact that this equation does not have a minimum region, as mentioned in the theoretical section. Therefore, our second proposal is to add the priori information in FWI is through the quadratic form that is represented in Equation (7). Analogously to the previous case, we will also add priori information to the problem directly, without using a probability distribution formalism. First, we performed the tests on the first and second parts of the BP model. The result can be seen in Figures 11a and 12a. We can observe, in both cases, that the addition of prior information through the use of the quadratic form of the symmetric relative entropy allows the FWI to provide satisfactory results.

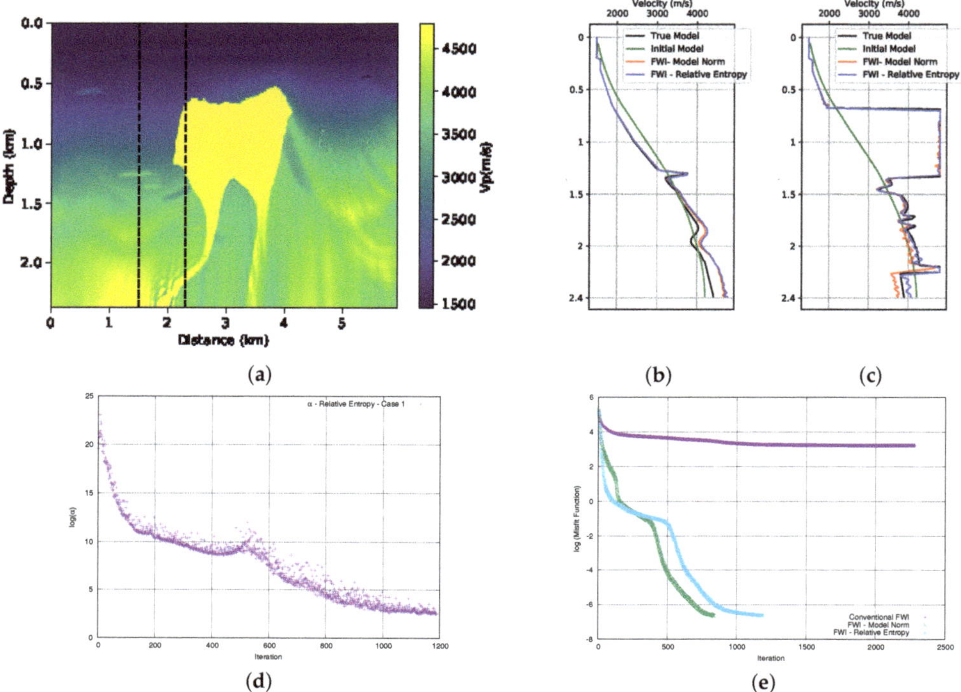

Figure 10. (**a**) FWI results of adding prior information through the quadratic form of the relative entropy in the second part of BP model; (**b**) velocity profile at position $x = 1.5$ km, (**c**) velocity profile at position $x = 2.3$ km and (**d**) α parameter evolution (note this curve is shown in logarithmic natural base scale), and (**e**) misfit data function progress (logarithmic natural base scale).

We compare the result of the FWI (for the first part of the model) using the symmetrical form of the relative entropy (Figure 11a) with that obtained with the model norm in Figure 7a. We observe that the addition of prior information through the symmetric relative entropy provides a better result on the left-hand side of the model. We can confirm the quality of the result by looking at the velocity profiles in the well positions, see Figure 11b,c. We can observe that, for the well at position $x = 2.3$ km (Figure 11b), the result of the FWI with the addition of prior information through the symmetric form of the relative entropy when compared to the use of the model norm provides a better approximation of the real value in the region below the salt. For the well at position $x = 3.5$ km (Figure 11c), the results of the FWI with the symmetric relative entropy and with the model norm provide equivalent adjustments. As for the second parts of the model, the result of the FWI with the symmetric relative entropy (Figures 12a) is visibly equivalent to the result that was obtained with the model norm (Figures 8a). However, when we observe the velocity profiles in the well positions that are shown in Figure 12b,c, we can see that adding the symmetrical case of the relative entropy in the FWI provides a slightly better fit in the deeper part of the model than the model norm.

After several tests, we concluded that the initial value of the α parameter for the first part of the BP model should be equal to $1.4 \times e^6$ (in this case, $\gamma \sim 2.8 \times 10^{17}$ and $\mu = 5 \times 10^{-10}$). Its progression can be seen in Figure 11d. For the second part of the BP model, the initial value of α parameter was equal to $1.3 \times e^6$ (in this case, $\gamma \sim 2.6 \times 10^{17}$

and $\mu = 5 \times 10^{-10}$, and we show its evolution in Figure 12d. As previously mentioned, we observed that the α parameter provides an adequate balance between the term of the data and the prior information added through the relative entropy. The misfit data curve for the first part of the model is illustrated in Figures 11e. The relative entropy (blue curve) that is added to the FWI provides a result with a better fit than with the model norm (green curve) and conventional FWI (purple curve). For the second part of the model, we can see the misfit data curve in Figure 12e. We observe that the addition of information through the relative entropy (light-blue curve) to the FWI also provides an extensive decay in the misfit data when compared to the classic FWI (purple curve). When comparing the misfit data of FWI with the model norm (green curve), we see that the final adjustment is close, although the misfit with relative entropy is a little better.

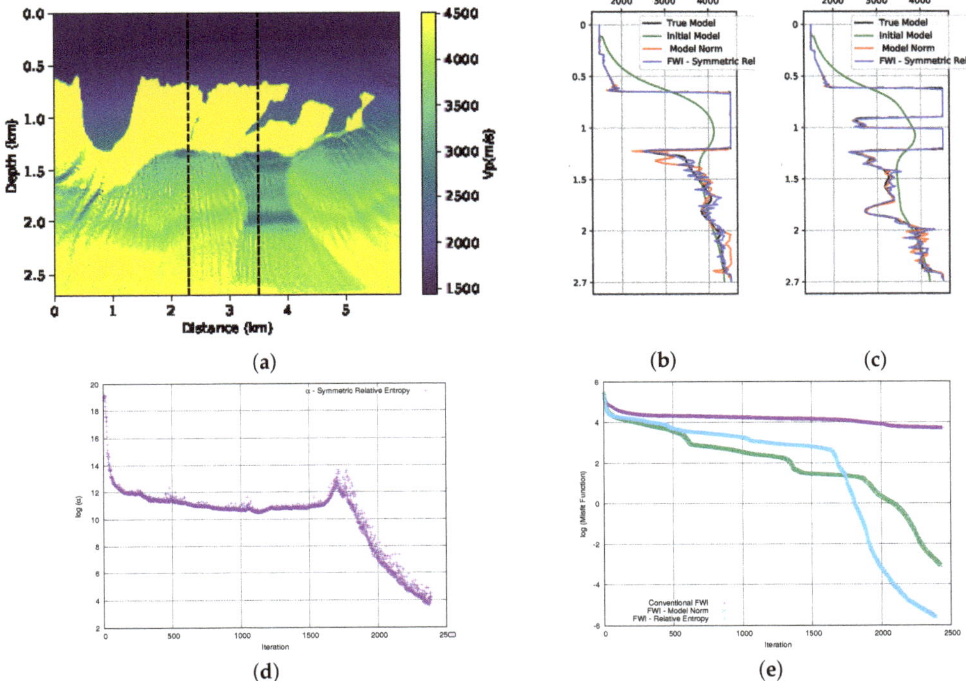

Figure 11. (**a**) FWI result with adding prior information through the quadratic form of the relative entropy in first part of BP model; (**b**) velocity log at position $x = 2.3$ km; (**c**) velocity log at position $x = 3.5$ km; (**d**) α parameter evolution (Note this curve is shown in logarithmic natural base scale); and, (**e**) misfit data function progress (logarithmic natural base scale).

3.4. Third Case: Axiomatic Form

Finally, our third proposal to add relative entropy in FWI is the Axiomatic form. We also use this form to add priori information to the FWI. Its form is described in the Equation (8). The advantage of the relative entropy described by Equation (8) is that it is positively defined, which makes the application straightforward without the need for any adjustments. Thus, we used Equation (8) and performed the FWI in the first of the BP model, and we show the result in Figure 13a. As in the previous cases, when we compare the result of the FWI with the use of the model norm (Figure 7a) to add the prior information, the relative entropy (8) added to the FWI provides a slightly better quality result on the left-hand side of the model. The analysis of the velocity profiles in the well positions show that the FWI with the addition of the relative entropy (blue curve) as compared to the FWI with the addition of the model norm (red curve) provides an adjustment that is closer to the desired in the well in the position $x = 2.3$ km (Figure 13b) and a similar adjustment at position $x = 3.5$ km (Figure 13c). We also performed FWI in the second part of BP model, and it can be seen in Figure 14a. In this case, we can see the similarity of the results when we compare this result with that obtained using the model norm (Figure 8a). The analysis of the velocity profiles in the well positions shows an equivalent result in the well position at $x = 1.5$ km (Figure 14b) and, in the position $x = 2.3$ km, we observe a slightly better result of the FWI with the relative entropy (Figure 14c).

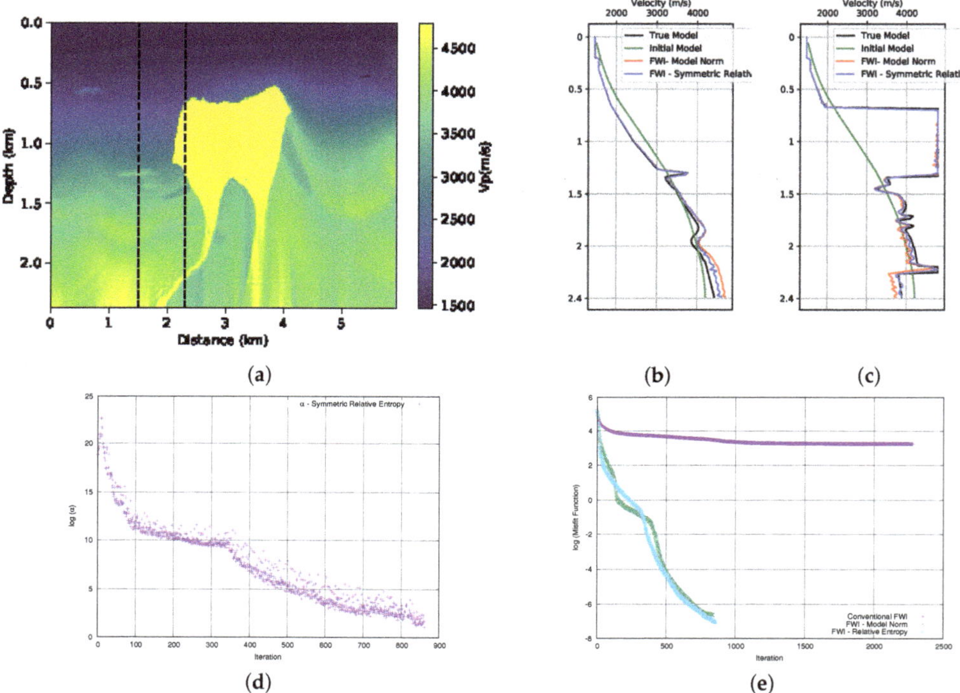

Figure 12. FWI result with adding prior information through the quadratic form of the relative entropy in second part of BP model; (**b**) velocity log at position $x = 1.5$ km; (**c**) velocity log at position $x = 2.3$ km; (**d**) α parameter evolution (Note this curve is shown in logarithmic natural base scale); and, (**e**) misfit data function progress (logarithmic natural base scale).

After several tests, we concluded that, for this case, the weight value of the entropy term should start at $\alpha = 25$ (in this case, $\gamma \sim 2.8 \times 10^4$ and $\mu = 9 \times 10^{-4}$) for the first part of the BP model. The evolution of the α parameter can be seen in Figure 13d. For the second part of the BP model, we conclude that the alpha parameter should be $\alpha = 50$ (in this case, $\gamma \sim 5.6 \times 10^4$ and $\mu = 9 \times 10^{-4}$) and its evolution can be seen in Figure 14d. In Figures 13e and 14e, we see the misfit data curve for the first and second parts of the model, respectively. We observe that in the case of the first part of the model, the FWI with the relative entropy (light-blue curve) starts the inversion in a path that is very close to the conventional FWI (purple curve). However, around the iteration 1350, the data adjustment begins to improve significantly, finishing the inversion at a lower misfit data value than the FWI with the model norm (green curve). For the second part of the model, we observe that the FWI with the relative entropy (light-blue curve) starts the inversion in a path that is very close to the FWI with the model norm (green curve). However, at iteration 200, the FWI with relative entropy takes a path of a larger misfit, so that more iterations are needed to obtain a misfit data result that is close to that obtained by the FWI with the model norm.

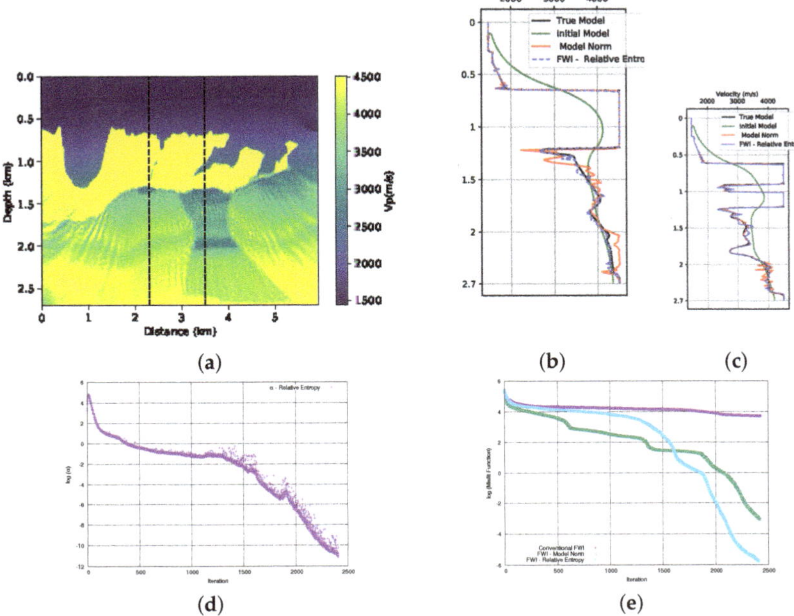

(a)

(b)

(c)

(d)

(e)

Figure 13. (**a**) FWI result of adding prior information through the relative entropy in the first part of the BP model: (**b**) velocity log at position $x = 2.3$ km; (**c**) velocity log at position $x = 3.5$ km; (**d**) α parameter evolution (Note this curve is shown in logarithmic natural base scale); and, (**e**) misfit data function progress (logarithmic natural base scale).

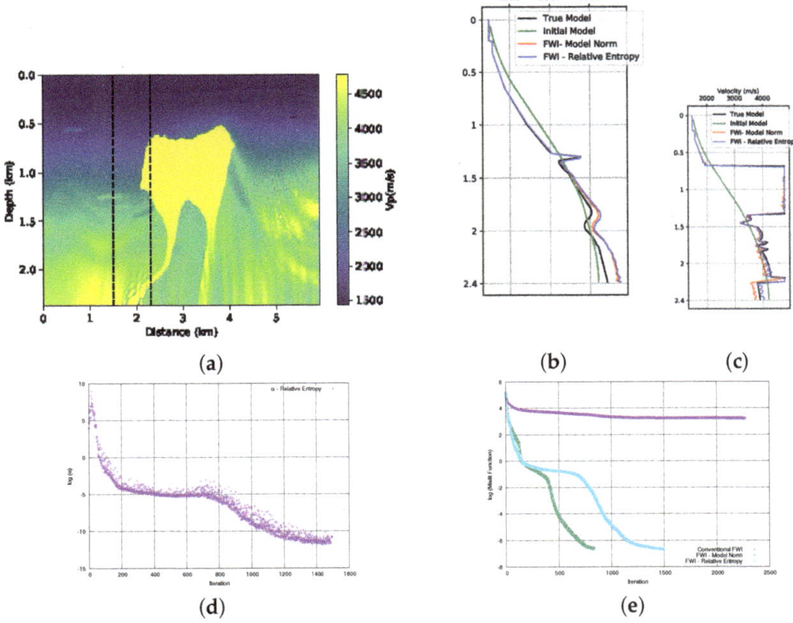

(a)

(b)

(c)

(d)

(e)

Figure 14. (**a**) FWI result with adding prior information through the relative entropy in second part of BP model: (**b**) velocity log at position $x = 1.5$ km; (**c**) velocity log at position $x = 2.3$ km; (**d**) α parameter evolution (Note this curve is shown in logarithmic natural base scale); and, (**e**) misfit data function progress (logarithmic natural base scale).

3.5. Discussion

We have shown that the addition of a priori information to the FWI scheme can be an effective strategy for driving the inversion process to converge toward the global minimum. Here, we have incorporated the priori information in different ways and, to quantify the accuracy of the inversion results for each one, we have computed the normalized model misfit using the following equation:

$$\epsilon = \frac{\left[\sum_{i=1}^{n} \left(m_i^{true} - m_i^{inv} \right)^2 \right]^{1/2}}{\left[\sum_{i=1}^{n} \left(m_i^{true} \right)^2 \right]^{1/2}}, \tag{15}$$

where m^{true} is the true model and model and m^{inv} is the inverted model using an FWI method [35,36]. An ϵ close to 0 means low error. The ϵ values are shown in the Table 1.

In the analysis of the results that are presented in Table 1, we observe that axiomatic form (third case) We note that the third case provides the model with the smallest error in both cases. Although it does not provide the smallest error in the model, our proposal to use the quadratic form of the symmetric form of the relative entropy (second case) is also robust. It is possible to see that the model error with this strategy is less than using the model norm. Finally, it is possible to observe that our proposal to use the quadratic form of relative entropy (first case), although its result depends on the inversion path, can also provide a good result. For the first part of the model, we were not successful in our tests, but we calculated the error for the results that are shown in Figure 9a,b, respectively. For the second part of the model, we can observe that the model error is less than the conventional case.

Table 1. Misfit model of the FWI results. ($*$) This is the model misfit for the result that is illustrated in Figure 9a, while ($**$) is the model misfit for the result illustrated in Figure 9b.

Strategy	First Part	Second Part
	ϵ	ϵ
Conventional FWI	0.6389	0.4415
FWI + Model Norm	0.0114	0.1132
Our proposal: FWI + First case	0.04963 * /0.0091 **	0.1415
Our proposal: FWI + Second case	0.0063	0.0968
Our proposal: FWI + Third case	0.0025	0.0876

4. Conclusions

In this work, we propose adding the relative entropy in the FWI formalism. We use relative entropy to include priori information in the FWI to reduce the difficulty of the uniqueness of the solution in this kind of inverse problem. The addition of the prior information was done in a deterministic way, which is, it was done in a direct way, avoiding the formulation in terms of a probability distribution. We have applied this scheme in two regions of the BP model, which presents a complex lateral velocity variation due to the halokinesis of salt layers. The numerical tests show the quality improvement in the result that was obtained when compared with the conventional FWI. In addition to the visual analysis, we calculated the misfit model to show the improvement that was brought by our proposal.

We present three different ways of introducing prior information through entropy to the FWI formalism: the literature as Kullback–Leibler's distance and its symmetrical form and an axiomatic form. In the first case, we show that in the deterministic approach, the Kullback–Leibler's distance is not positively defined in its entire domain. To avoid this misfortune, we suggest using its quadratic form. We have seen that this quadratic form will not always help to solve the local minimum problem. We graphically illustrate that the function described by the quadratic form of the Kullback–Leibler's distance has two regions of minimum. Therefore, the result will depend on the path taken by the FWI throughout the iterations: while it was not possible to obtain a satisfactory solution for the first part of the model after several tests, for the second part of the model, the result was satisfactory.

The second case was the symmetrical form of Kullback–Leibler's distance. We graphically illustrate that this function is also not positively defined, which makes it difficult to define the problem as a maximization or minimization. In addition, the symmetrical form has no minimum region. To avoid these inconveniences, we propose the use of its quadratic form. We have shown graphically that this quadratic shape has characteristics that can help the FWI to avoid local minimums. In addition to being positively defined, it presents a minimum region. The addition of previous information in the FWI through this quadratic form enables the FWI to deliver a satisfactory result for both cases.

The third case that we studied was the addition of prior information through an axiom of relative entropy. Graphically, we show that this shape is positively defined and it has a minimum region. These features allow this form to be used to add information prior to the FWI formalism in a straightforward manner. The results of the FWI with this regularization scheme were also satisfactory for both models.

The FWI results that were obtained using the relative entropy were compared to the result with the model norm. We observed that for the first part of the BP model, the FWI result with real entropy is slightly better (mainly on the left-hand side of the model). This fact is corroborated by the analysis of the velocity profiles in the position of the wells. We saw that the FWI result with the relative entropy provides an adjustment closer to the desired result when compared to the FWI result with the model norm. In addition, we saw that the misfit data are less when we add prior information through entropy. For the second part of the BP model, the results are visibly similar. We have seen that the adjustment of the data is close, although the first and third cases of relative entropy require a little more iteration. However, when comparing the well profiles, we observed that the adjustment of the FWI with the addition of prior information through entropy provides a better fit early in the deepest region of the model.

Author Contributions: Conceptualization, D.S.C., C.C.N.d.S. and J.M.d.A.; methodology, D.S.C., C.A.N.d.C., C.C.N.d.S. and J.M.d.A.; software, D.S.C. and J.M.d.A.; validation, D.S.C. and J.M.d.A.; formal analysis, D.S.C., C.A N.d.C., C.C.N.d.S. and J.M.d.A.; investigation, D.S.C., C.A.N.d.C., C.C.N.d.S. and J.M.d.A.; writing–original draft preparation, D.S.C., C.A.N.d.C., C.C.N.d.S. and J.M.d.A.; visualization, D.S.C., C.A.N.d.C., C.C.N.d.S. and J.M.d.A.; supervision, C.C.N.d.S. and J.M.d.A. All authors have read and agreed to the published version of the manuscript.

Funding: This work did not have external funding.

Data Availability Statement: The data presented in this study are available on request from the corresponding author.

Acknowledgments: The authors acknowledge the support of Petrobras. We thank the High Performance Computing Center at UFRN for providing the computational facilities to run the simulations. J.M.A. thanks CNPq for his productivity fellowship (grant No. 313431/2018-3). The authors also thank BP and Frederic Billette who provided the BP velocity model.

Conflicts of Interest: The authors declare no conflict of interest.

Abbreviations

The following abbreviations are used in this manuscript:

FWI   Full Waveform Inversion

References

1.   Wang, Y.; Rao, Y. Reflection seismic waveform tomography. *J. Geophys. Res. Solid Earth* **2009**, *114*, B03304. [CrossRef]

2.   Tarantola, A. Linearized inversion of seismic reflection data. *Geophys. Prospect.* **1984**, *32*, 998–1015. [CrossRef]

3.   Virieux, J.; Operto, S. An overview of full-waveform inversion in exploration geophysics. *Geophysics* **2009**, *74*, WCC1–WCC26. [CrossRef]

4.   Wang, Y.; Houseman, G.A. Tomographic inversion of reflection seismic amplitude data for velocity variation. *Geophys. J. Int.* **1995**, *123*, 355–372. [CrossRef]

5.   Bunks, C.; Saleck, F.M.; Zaleski, S.; Chavent, G. Multiscale seismic waveform inversion. *Geophysics* **1995**, *60*, 1457–1573. [CrossRef]

6.   Tikhonov, A.; Arsenin, V. *Solution of Ill-Posed Problems*; Winston: Washington, DC, USA; Halsted Press: New York, NY, USA, 1977; 258p, ISBN 0470991240.

7.   Guitton, A. A blocky regularization scheme for full waveform inversion. *SEG Tech. Program Expand. Abstr.* **2011**, 2418–2422. [CrossRef]

8.   Asnaashari, A.; Brossie, R.; Garambois, S.; Audebert, F.; Thore, P. Regularized seismic full waveform inversion with prior model information. *Geophysics* **2013**, *78*, R25–R36. [CrossRef]

9.   Peters, B.; Herrman, F.J. Constraints versus penalties for edge-preserving full-waveform inversion. *Lead. Edge* **2017**, *36*. [CrossRef]

10.   Gull, S.F.; Skilling, J. The maximum entropy method. In *Maximum-Entropy and Bayesian Methods in Inverse Problems*; Springer: Dordrecht, The Netherlands, 1984.

11.   Wang, Y. *Seismic Inversion: Theory and Applications*; John Wiley and Sons: Hoboken, NJ, USA, 2016.

12.   Muniz, W.B.; Ramos, F.M.; de Campos Velho, H.F. Entropy- and Tikhonov-Based Regularization Techniques Applied to the Backwards Heat Equation. *Comput. Math. Appl.* **2000**, *40*, 1071–1084. [CrossRef]

13.   Kullback, S.; Leibler, R.A. On Information and Sufficiency. *Ann. Math. Stat.* **1951**, *22*, 79–86. [CrossRef]

14.   Shore, J. Minimum cross-entropy spectral analysis. *IEEE Trans. Acoust. Speech Signal Process.* **1981**, *29*, 230–237. [CrossRef]

15.   Jacobs, F.J.; van der Geest, P.A.G. Spiking band-limited traces with a relative entropy algorithm. *SEG Tech. Program Expand. Abstr.* **1988**, *56*. [CrossRef]

16.   Lo, T.-W.; Duckworth, G.L.; Toksöz, M.N. Minimum cross entropy seismic diffraction tomography. *J. Acoust. Soc. Am.* **1990**, *87*, 026102. [CrossRef]

17.   Ulrych, T.; Bassrei, A.; Lane, M. Minimum Relative Entropy Inversion of 1D Data with Application. *Geophys. Prospect.* **1990**, *38*, 465–487. [CrossRef]

18.   Chen, F.; Peter, D. A misfit function based on entropy regularized optimal transport for full-waveform inversion. *SEG Tech. Program Expand. Abstr.* **2018**, 1314–1318. [CrossRef]

19.   da Silva, S.L.E.F.; Carvalho, P.T.C.; da Costa, C.A.N.; de Araujo, J.M.; Corso, G. Misfit function for full waveform inversion based on Shannon entropy for deeper velocity model updates. *SEG Tech. Program Expand. Abstr.* **2019**, 1556–1559. [CrossRef]

20.   Zhu, H.; Li, S.; Fomel, S.; Stadler, G.; Ghattas, O. A Bayesian approach to estimate uncertainty for full-waveform inversion using a priori information from depth migration. *Geophysics* **2016**, *81*, R307–R323. [CrossRef]

21.   Singh, S.; Tsvankin, I.; Naeini, E.Z. Bayesian approach to facies-constrained waveform inversion for VTI media. *SEG Tech. Program Expand. Abstr.* **2019**, 1370–1374. [CrossRef]

22.   Carvalho, P.T.C.; Corso, G.; da Silva, S.L.E.F.; de Araujo, J.M.; Lucena, L. Full-waveform inversion with subsurface fractal information and variable model uncertainties. *SEG Tech. Program Expand. Abstr.* **2019**. [CrossRef]

23.   Zhang, X.; Curtis, A. Bayesian Full-Waveform Inversion with Realistc Priors. *arXiv* **2021**, arXiv:2104.04775.

24.   Jaynes, E.T. Information Theory and Statistical Mechanics. *Phys. Rev.* **1957**, *106*, 620. [CrossRef]

25.   Kullback, S. Information Theory and Statistics. In *A Wiley Publication in Mathematical Statistics*; Wiley: Gloucester, UK, 1978.

26.   Marinescu, D.C.; Marinescu, G.M. *Classical and Quantum Information*; Elsevier: Amsterdam, The Netherlands, 2011.

27.   Cover, T.M.; Thomas, J. *Elements of Information Theory*; Wiley-Interscience: Hoboken, NJ, USA, 1991.

28.   Shore, J.E.; Johnson, R.W. Axiomatic Derivation of the Principle of Maximum Entropy and the Principle of Minimum Cross-Entropy. *IEEE Trans. Inf. Theory* **1980**, *26*, 26–39. [CrossRef]

29.   Tikochinsky, Y.; Tishby, N.Z.; Levine, R.D. Consistent Inference of Probabilities for Reproducible Experiments. *IEEE Trans. Inf. Theory* **1984**, *52*, 1357–1360. [CrossRef]

30.   Plessix, R.E. A review of the adjoint-state method for computing the gradient of a functional with geophysical applications. *Geophys. J. Int.* **2006**, *167*, 495–503. [CrossRef]

31.   Byrd, R.; Lu, P.; Nocedal, J. A limited memory algorithm for bound constrained optimization. *SIAM J. Sci. Stat. Comput.* **1995**, *16*, 1190–1208. [CrossRef]

32.   Komatitsch, D.; Martin, R. An unsplit convolutional perfectly matched layer improved at grazing incidence for the seismic wave equation. *Geophysics* **2007**, *72*. [CrossRef]

33.   Pasalic, D.; McGarry, R. Convolutional perfectly matched layer for isotropic and anisotropic acoustic wave equations. *SEG Tech. Program Expand. Abstr.* **2010**. [CrossRef]

34.   Billette, F.; Brandsberg-Dahl, S. The 2004 BP velocity benchmark. In Proceedings of the 67th Meeting, Madrid, Spain, 13–16 June 2005; EAGE, Extended Abstracts, B035.

35.   Ji, S.; Zhang, H.; Wang, Y.; Rong, L.; Shi, Y.; Chen, Y. Three-dimensional inversion of full magnetic gradient tensor data based on hybrid regularization method. *Geophys. Prospect.* **2019**, *67*, 226–261. [CrossRef]

36.   Song, X.; Xu, Y.; Dong, F. A hybrid regularization method combining Tikhonov with total variation for electrical resistance tomography. *Flow Meas. Instrum.* **2015**, *46*, 268–275. [CrossRef]

Article

Evaluation of Non-Uniform Image Quality Caused by Anode Heel Effect in Digital Radiography Using Mutual Information

Ming-Chung Chou [1,2,3]

1   Department of Medical Imaging and Radiological Sciences, Kaohsiung Medical University, Kaohsiung 80708, Taiwan; mcchou@kmu.edu.tw
2   Center for Big Data Research, Kaohsiung Medical University, Kaohsiung 80708, Taiwan
3   Department of Medical Research, Kaohsiung Medical University Hospital, Kaohsiung 80708, Taiwan

Abstract: Anode heel effects are known to cause non-uniform image quality, but no method has been proposed to evaluate the non-uniform image quality caused by the heel effect. Therefore, the purpose of this study was to evaluate non-uniform image quality in digital radiographs using a novel circular step-wedge (CSW) phantom and normalized mutual information (nMI). All X-ray images were acquired from a digital radiography system equipped with a CsI flat panel detector. A new acrylic CSW phantom was imaged ten times at various kVp and mAs to evaluate overall and non-uniform image quality with nMI metrics. For comparisons, a conventional contrast-detail resolution phantom was imaged ten times at identical exposure parameters to evaluate overall image quality with visible ratio (VR) metrics, and the phantom was placed in different orientations to assess non-uniform image quality. In addition, heel effect correction (HEC) was executed to elucidate the impact of its effect on image quality. The results showed that both nMI and VR metrics significantly changed with kVp and mAs, and had a significant positive correlation. The positive correlation is suggestive that the nMI metrics have a similar performance to the VR metrics in assessing the overall image quality of digital radiographs. The nMI metrics significantly changed with orientations and also significantly increased after HEC in the anode direction. However, the VR metrics did not change significantly with orientations or with HEC. The results indicate that the nMI metrics were more sensitive than the VR metrics with regards to non-uniform image quality caused by the anode heel effect. In conclusion, the proposed nMI metrics with a CSW phantom outperformed the conventional VR metrics in detecting non-uniform image quality caused by the heel effect, and thus are suitable for quantitatively evaluating non-uniform image quality in digital radiographs with and without HEC.

Keywords: circular-step wedge; contrast-detail; mutual information; visible ratio; anode heel effect

Citation: Chou, M.-C. Evaluation of Non-Uniform Image Quality Caused by Anode Heel Effect in Digital Radiography Using Mutual Information. *Entropy* **2021**, *23*, 525. https://doi.org/10.3390/e23050525

Academic Editor: Amelia Carolina Sparavigna

Received: 30 March 2021
Accepted: 24 April 2021
Published: 25 April 2021

1. Introduction

Image quality is an essential requirement in digital X-ray imaging and is closely associated with the accuracy of disease diagnosis. The fundamental metrics of static image quality are contrast, spatial resolution, and noise, which can be evaluated through the measurements of modulation transfer function (MTF), point-spread function, and noise power spectrum (NPS) [1–3]. Although these metrics can be measured from an X-ray imaging system, the individual metrics cannot correctly reflect the overall image quality. Detective quantum efficiency (DQE), which is a function of MTF, NPS, and system gain, is the most commonly used metric to quantify the overall performance of X-ray imaging systems [4–6]; however, DQE cannot reflect entire imaging chains, such as image processing and correction [7]. In contrast, a more practical approach to quantifying overall image quality of a radiograph is to use contrast-detail phantoms [8–12]. Previously, an emerging metric, termed as mutual information (MI), was shown to successfully quantify the overall image quality of a digital radiograph with the use of a linear step-wedge phantom [13,14]. Although these metrics were shown to be capable of quantifying overall image quality, none are suitable for evaluating the non-uniform image quality of an image caused by the anode heel effect.

In radiography, the "heel effect" causes less X-ray fluence and higher mean radiation energy in the anode direction due to the absorption of low-energy photons by the anode heel [15]. The non-uniform distribution of X-ray fluence may result in non-uniform image quality, especially in the anode-cathode direction. However, there were limited previous works quantifying the influence of anode heel effect on image quality in digital radiographs [16]. Previous studies demonstrated that the heel effect significantly impacted the signal-to-noise ratio (SNR) using an anthropomorphic phantom, but the image quality was not significantly different between pelvic radiographs with the head towards the anode and cathode directions [17,18]. Moreover, some previous studies performed post-processing heel effect correction (HEC) to minimize the inhomogeneous intensity in radiographs [19–21]. However, no suitable method has been presented that can objectively quantify the non-uniform image quality in radiographs. Moreover, no methods can elucidate how much the image quality can be improved in the radiographs with HEC. Therefore, the purposes of this study were three-fold: (1) to design a circular step-wedge (CSW) phantom for evaluating overall and non-uniform image quality, (2) to compare other image quality metrics measured from a contrast-detail phantom, and (3) to understand how much HEC can improve the image quality.

2. Materials and Methods

2.1. Circular Step-Wedge Phantom

In information theory, MI is a measure of mutual dependence between two random variables, and is calculated from their individual entropy and joint entropy, defined as

$$MI = H(X) + H(Y) - H(X, Y),$$

where $H(X)$ and $H(Y)$ are individual entropy of random variables (X and Y), and $H(X, Y)$ is their joint entropy [22]. As MI reflects the amount of information of one random variable that is observed from the other random variable, it is possible to utilize the MI metrics to reflect the image quality using a linear step-wedge phantom [13,14]. However, the original design can only measure MI in one direction parallel to the long axis of the phantom, so it is unable to evaluate non-uniform image quality in radiographs caused by anode heel effect. Therefore, the present study designed a CSW phantom with acrylic material to estimate the MI metrics in different directions from a single image. The phantom was fabricated using 14 pieces of circular acrylic board with radii from 4 cm to 30 cm, which were precisely (±0.1 mm) laser cut from a 2 mm thick acrylic plastic sheet. After a 1 mm hole (diameter) was drilled in the center, 14 circular acrylic boards were piled up sequentially from large to small and were aligned and glued together at the center. The CSW phantom consisted of 14 steps with thickness from 2 mm to 28 mm and with radii from 4 cm to 30 cm, as shown in Figure 1.

2.2. Contrast-Detail Resolution Phantom

A commercial contrast-detail resolution (CDR) phantom was also used to evaluate the overall image quality of radiographic images. The phantom consists of 144 circular details with 12 sizes × 12 contrasts (TO16, Leeds Test Objects LTD, North Yorkshire, UK; https://www.leedstestobjects.com (accessed on 30 March 2021)) [9]. Of the 144 details, 72 larger details were arranged circularly in the outer region, and the remaining 72 smaller details were arranged linearly in the central region, as shown in Figure 2.

2.3. Image Data Acquisition

Image quality was evaluated using both CSW and CDR phantoms in a digital radiographic system (Toshiba/DRX-3724HD) that was equipped with a CsI flat panel detector (a-Si, TFT, CXD-70C wireless). The X-ray images were acquired from the two phantoms with matrix size = 2800 × 3408, pixel size = 0.13 × 0.13 mm^2, dynamic range = 4096, and source-to-detector distance = 100 cm. For statistical analysis, image acquisition was repeated ten times at 40, 45, 50, 55, and 60 kV (5 mAs), and at 5, 10, 20, 25, and 40 mAs (40 kVp), respectively. A posterior-anterior right-hand radiograph was acquired with 52 kVp and 10 mAs to show the impact of anode heel effect on image quality. The human study was approved by the local institutional review board (KMUHIRB-E(I)-20200274).

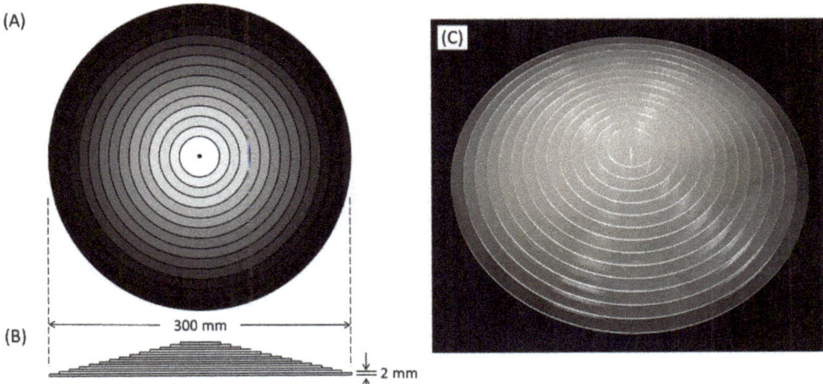

Figure 1. The top view (**A**), lateral view (**B**), and actual image (**C**) of the CSW phantom consisting of 14 steps with 2 mm incremental thickness (from 2 to 28 mm), and 20 mm incremental diameter (from 40 to 300 mm).

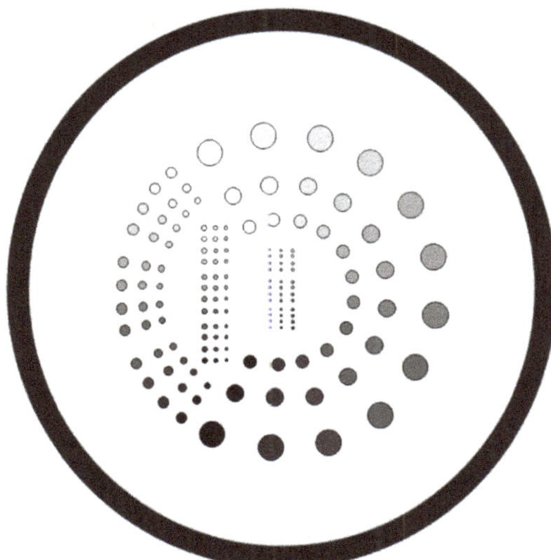

Figure 2. The arrangement of 144 disc details within the TO16 CDR phantom. In the phantom, 72 larger disc details are arranged circularly in the outer region, and 72 smaller ones are arranged linearly in the central region.

2.4. Mutual Information with a CSW Phantom

This study estimated MI from an X-ray image of the CSW phantom using a home-made script on a MATLAB software. First, the center of the CSW phantom in the image was detected by the center of gravity. Second, 14 circular regions-of-interest (ROIs), each containing 1941 pixels, were automatically placed on the center of 14 steps, respectively, in one direction, as shown in Figure 3. Subsequently, the 14 ROIs were rotated counterclockwise around the center every 10 degrees, from which 36 MI metrics were calculated. For each direction, the MI metrics were calculated according to the method reported by previous studies [13,14]. However, since a larger number of steps of the phantom would give rise to larger MI values (bits), the present study calculated a normalized MI (nMI) [23,24], defined as MI/log2(N) $\times$ 100 %. N is the number of steps in the CSW phantom. The resultant nMI ranges from 0 to 100%, and a larger nMI value indicates better image quality.

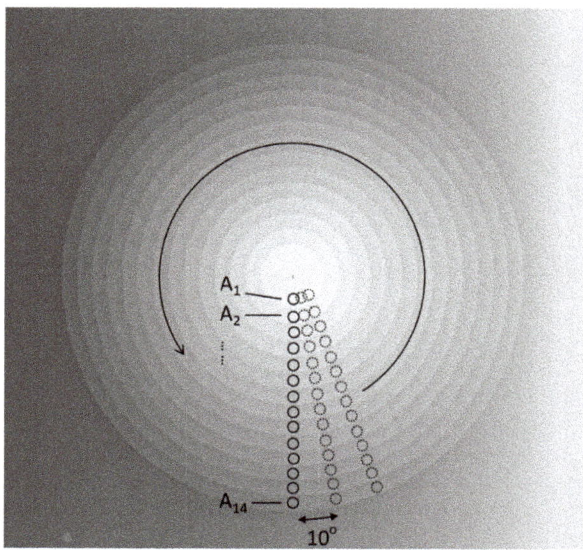

Figure 3. The estimation of the nMI metrics in 36 orientations separated by 10 degrees. 14 equal-sized circular ROIs (A1 to A14) are placed respectively on the step centers to calculate the nMI metrics. Afterwards, the 14 ROIs are rotated counterclockwise by multiples of 10 degrees to estimate the corresponding nMI metrics in other orientations.

2.5. Visible Ratio with a CDR Phantom

This study measured visible ratio (VR) metrics using a TO16 CDR phantom with a commercial AutoPIA tool (Leeds Test Objects LTD, North Yorkshire, UK). The phantom was rotated counterclockwise every 30 degrees from 0 to 180 degrees to understand whether the CDR phantom can adequately reflect the anode heel effect on image quality. For each orientation, ten repeated X-ray images of CDR phantoms were acquired for comparisons and were analyzed automatically to detect all possible details. In this step, the software calculated the contrast-to-noise ratio (CNR) for each of 144 details, defined as |(target signal − background signal)|/(background noise), and then those details with CNR higher than a predefined threshold were considered as visible details [9]. Finally, the VR metrics, defined as (number of successfully detected details)/(total number of details) $\times$ 100 %, were calculated to give a value between 0 to 100%. Similarly, a larger VR metrics indicates better image quality and higher performance in detecting details.

2.6. Heel Effect Correction

This study performed a retrospective correction method that minimizes the intensity inhomogeneity in the X-ray images by fitting the background signals to a 2nd order polynomial function in the anode-cathode direction to understand how the HEC impacts the image quality. Subsequently, the phantom image was subtracted by the fitted curve and added by a minimum value of the curve to keep similar image brightness, as shown in Figure 4. Finally, nMI and VR metrics were estimated from the phantom images with and without HEC.

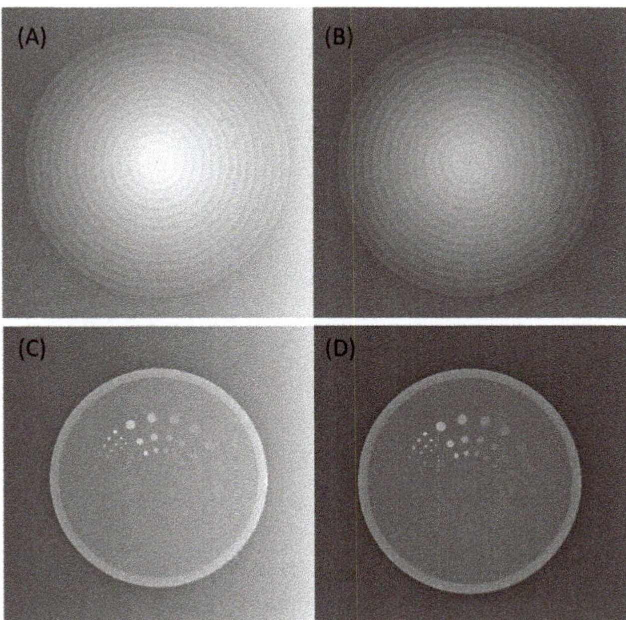

Figure 4. The CSW (**A**) and CDR (**C**) images acquired with 40 kVp and 5 mAs exhibited inhomogeneous signal intensity in the anode-cathode (horizontal) direction due to the heel effect. The inhomogeneity was successfully removed in the corrected CSW (**B**) and CDR (**D**) images after HEC.

2.7. Statistical Analysis

A one-way analysis of variance (ANOVA) was performed to understand whether the image quality metrics significantly changed with kVp, mAs, and orientations before and after HEC, respectively. A post-hoc Mann–Whitney U test was used to compare the differences between two exposure parameters and between two orientations. The Wilcoxon signed rank test was conducted to show the difference in nMI and VR metrics before and after HEC [25]. Moreover, Pearson's correlation analysis was carried out to reveal the relationship between the two metrics before and after HEC, respectively [26]. Statistical significance (P) was deemed if $P < 0.05$.

3. Results

By varying kVp, one-way ANOVA analysis showed that both nMI and VR metrics significantly changed with kVp between 40 and 60 kVp at a constant 5 mAs. It was also found that nMI changed more prominently than VR in X-ray images with and without HEC, as shown in Figure 5. The Mann–Whitney U test highlighted a significant difference in nMI metrics between any two kVps; however, the VR metrics were not significantly different between 45 to 50 kVp, 45 to 60 kVp, or 50 to 60 kVp in images with and without HEC. Moreover, the nMI metrics were significantly increased after HEC; however, no significant change was noted in the VR metrics at different kVps after HEC.

By varying mAs, one-way ANOVA analysis showed that both nMI and VR metrics also significantly changed with mAs between 5 and 40 mAs at a constant 40 kVp, as shown in Figure 6.

The post-hoc Mann–Whitney U test showed that both nMI and VR were significantly different between any two mAs in the images with and without HEC. The nMI metrics were significantly increased after HEC; however, no significant change was noted in the VR metrics at any of the mAs after HEC. Moreover, the averaged nMI and VR metrics significantly correlated in the images without HEC, as shown in Figure 7.

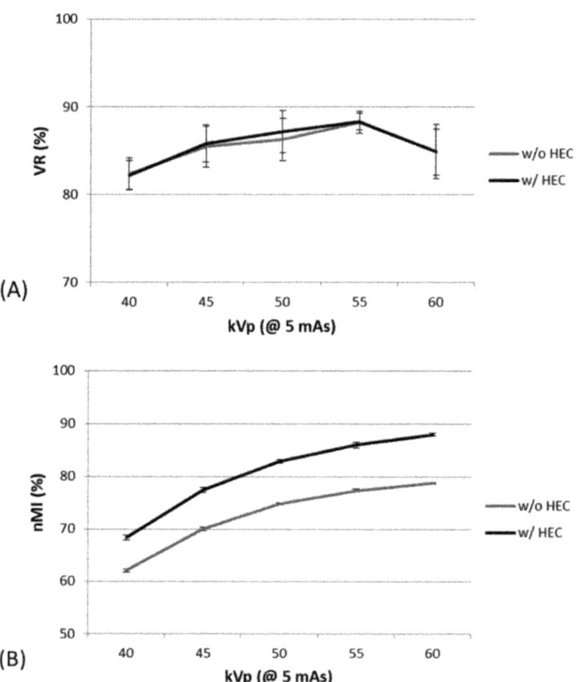

Figure 5. The VR (**A**) and nMI (**B**) metrics changed significantly with kVp, at 5 mAs, before and after HEC.

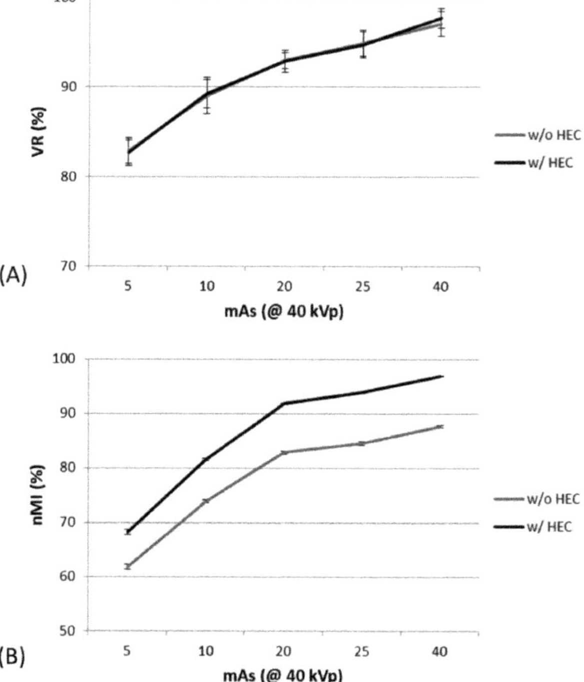

Figure 6. The VR (**A**) and nMI (**B**) metrics changed significantly with mAs, at 40 kVp, before and after HEC.

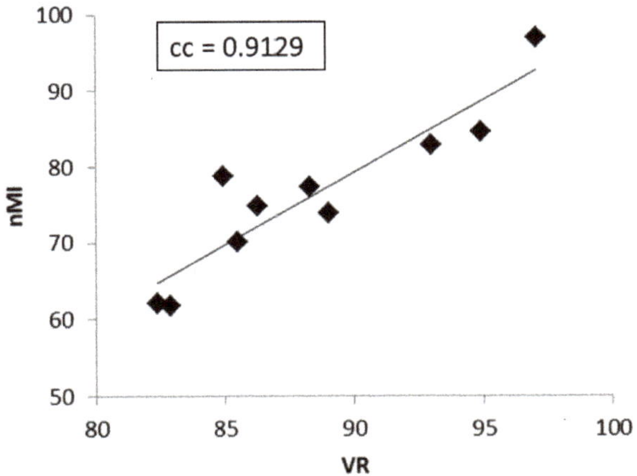

Figure 7. The significant correlation (cc = 0.9129, P < 0.05) between the VR and nMI metrics measured from all exposure parameters in images without HEC.

By varying orientation in the measurement, without HEC, there were significant changes in nMI with orientations between 0 and 180 degrees (the results were symmetric around 180 degrees). However, without HEC, there were no significant changes in VR with orientations between 0 and 180 degrees, as shown in Figure 8. The post-hoc Mann–Whitney U test showed that the nMI metrics were significantly different between two orientations in images with and without HEC. Although the nMI metrics came to be more uniform across different orientations, there remains slight difference in nMI metrics between 30 and 150 degrees.

A posterior-anterior right-hand X-ray image (Figure 9) demonstrated inhomogeneous signal intensity in the ancde-cathode direction due to the heel effect, where lower signal intensity (higher X-ray exposure) was noted in the finger than the wrist direction (Figure 9A,C,E). By applying the HEC, the inhomogeneity issue was minimized across the entire image, and small bony structures were more conspicuous in the corrected image than the raw image displayed with an identical window level and width (Figure 9B,D,F). Although the bony structures of the wrist in the raw image can be visualized by adjusting the window level and width, the bony structures of the fingers will be too dark to be visualized. This inhomogeneous issue can be reflected by the inconsistent nMI metrics in radial direction, as shown in Figure 8.

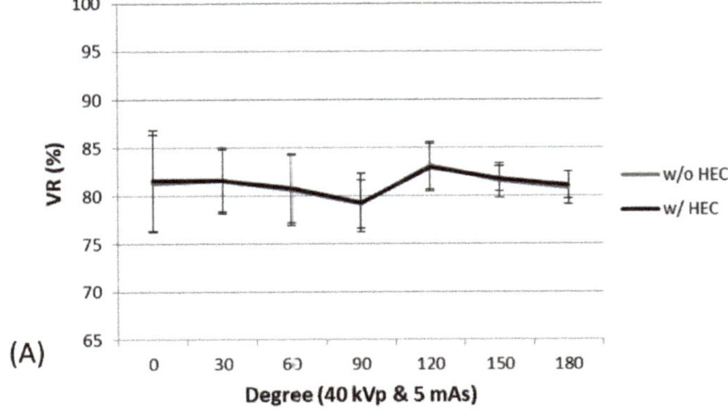

Figure 8. *Cont.*

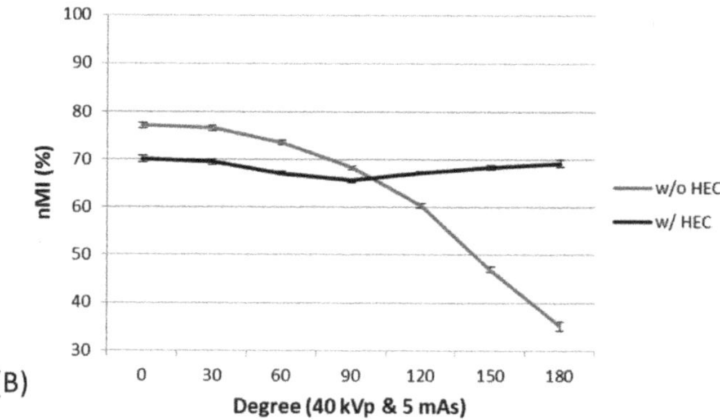

Figure 8. The VR (**A**) and nMI (**B**) metrics measured as a function of orientation at 40 kVp and 5 mAs. The VR metrics did not change significantly with orientations, whereas the nMI metrics changed significantly with orientations in the images without HEC.

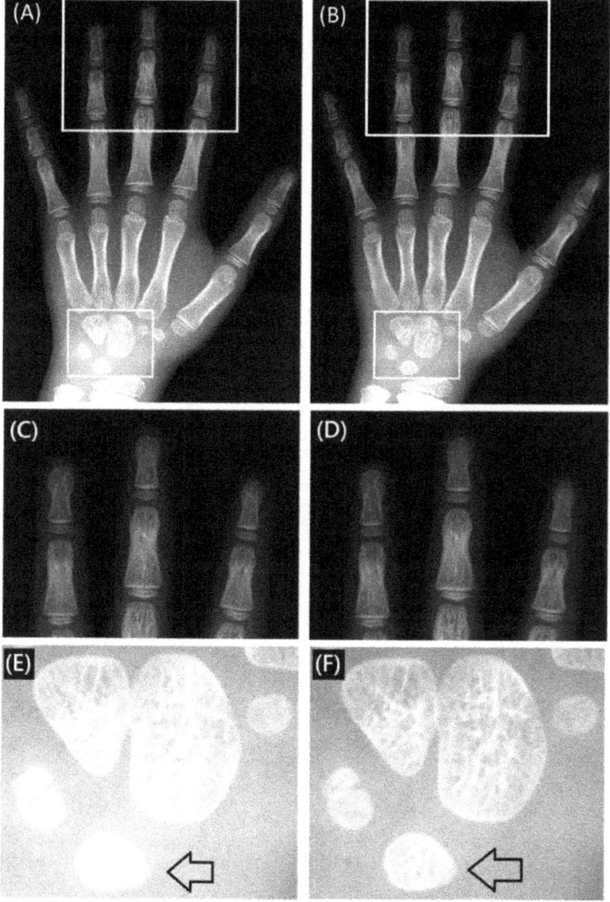

Figure 9. A posterior-anterior right-hand image acquired with 52 kVp and 10 mAs before (**A,C,E**) and after (**B,D,F**) HEC. The arrows indicate the bony structures of the lunate that were more conspicuous in the image with (**F**) than without (**E**) HEC.

4. Discussion

In radiography, the heel effect causes less X-ray fluence and higher mean radiation energy in the anode direction, and results in non-uniform image quality. Although there have been some methods proposed to reduce the heel effect [19–21], no suitable method has been presented that can objectively quantify the overall and non-uniform image quality caused by the heel effect. This study designed a CSW phantom for quantification of overall and non-uniform image quality in X-ray radiographs using nMI metrics based on information theory. The nMI metrics were demonstrated to be associated with imaging SNR, contrast, and resolution [13,14]. In the present study, the evaluated image quality was compared between the nMI (CSW phantom) and conventional VR (CDR phantom) metrics in digital X-ray images acquired at various exposure parameters and orientations, and with and without HEC. The results highlight that both metrics significantly changed with kVp (from 40 to 60 kVp at 5 mAs) and mAs (from 5 to 40 mAs at 40 kVp). The overall image quality assessed by nMI and VR metrics exhibited a similar trend with high correlation, suggesting that both metrics are capable of reflecting image quality in digital X-ray images. In addition, the nMI metrics were found to be more sensitive to changes in exposure parameters (kVp and mAs) than the VR metrics. It is postulated that the increased sensitivity is due to the fact the CSW phantom was made of acrylic material and had a small difference in thickness.

It is known that the anode heel effect may lead to heterogeneous X-ray exposure that can deteriorate overall image quality. The results of the present study demonstrated that the heel effect significantly deteriorated the overall image quality. Furthermore, the image quality reflected by the nMI metrics can be significantly improved with HEC in the anode direction; this correction resulted in improved homogeneity of image quality and higher conspicuity of bony structures in the hand X-ray images. However, the conventional VR metrics were not significantly changed with orientations before and after HEC, suggesting that the nMI metrics were more sensitive than the VR metrics to non-uniform image quality.

The insensitivity of VR metrics to detect non-uniformity of image quality was likely attributable to the fact that the disk details were embedded in the central area of the CDR phantom, as shown in Figure 2. Although the centralized disk details in the CDR phantom were suitable for measuring the image quality in the central field of view, the design itself rendered it less sensitive to inhomogeneous image quality that occurred in the outer region. On the contrary, the nMI metrics were calculated from the image of CSW phantom made of acrylic material and with a suitable size that fits the flat panel detector. A previous study showed that the image quality reflected by the correctly identified holes (%) of the CDRAD phantom was more sensitive to changes in exposure parameters than the number of detected details in a CDR phantom [12], suggesting that the acrylic material of the CDRAD phantom was sensitive to changes in signal intensity. Similarly, our results demonstrated that the nMI metrics (CSW phantom) were more sensitive to changes in exposure parameters and orientations than the VR metrics (CDR phantom). The results indicated that the nMI with the CSW phantom could potentially be a quantifiable metric for non-uniform image quality in digital X-ray images.

Some limitations, however, warrant discussion. First, a small range of exposure parameters was utilized in this study. A study using a broader range of exposure parameters may provide more comprehensive comparisons between the two metrics. Second, the nMI metrics with the CSW phantom have an intrinsic disadvantage of less sensitivity to changes in spatial resolution [13]. However, the circular nature of CSW phantom can be used to estimate radial MTF, as proposed by a previous study [27], so in addition to nMI, the CSW phantom can be utilized to evaluate the radial MTF in X-ray images. Third, the CSW phantom was designed with acrylic material, so it may not be suitable to measure the image quality at high kVp and high mAs. A CSW phantom with a combination of aluminum and acrylic materials may be helpful to reflect image quality of X-ray images acquired using clinical parameter settings. Further investigations will be needed to compare the results between phantoms made of different materials.

5. Conclusions

In conclusion, the nMI with the CSW phantom performs as well as VR does with the CDR phantom in evaluating overall image quality in digital X-ray images. Moreover, both metrics had a significantly high correlation at various exposure parameters. The nMI metrics further outperformed the VR metrics in detecting heel effects associated with non-uniform image quality. The nMI metrics also had higher sensitivity to changes in image quality after HEC. Therefore, we concluded that the proposed nMI metrics with the CSW phantom are suitable for evaluating overall and non-uniform image quality in digital X-ray images.

Funding: This study was supported by a grant MOST107-2314-B-037-050-MY2 from Ministry of Science and Technology of Taiwan.

Institutional Review Board Statement: The study was conducted according to the guidelines of the Declaration of Helsinki and approved by the Institutional Review Board of Kaohsiung Medical University Hospital (KMUHIRB-E(I)-20200274).

Informed Consent Statement: Informed consent was waived due to the nature of retrospective study.

Data Availability Statement: The data presented in this study are available on request from the corresponding author.

Conflicts of Interest: The author declares no conflict of interest.

Abbreviations

CSW	Circular Step-Wedge
CDR	Contrast-Detail Resolution
HEC	Heel Effect Correction
MTF	Modulation Transfer Function
nMI	normalized Mutual Information
NPS	Noise Power Spectrum
ROI	Region of Interest
VR	Visible Ratio

References

1. Fujita, H.; Doi, K.; Giger, M.L. Investigation of Basic Imaging Properties in Digital Radiography. 6. Mtfs of Ii-Tv Digital Imag-ing-Systems. *Med. Phys.* **1985**, *12*, 713–720. [CrossRef]
2. Giger, M.L.; Doi, K. Investigation of Basic Imaging Properties in Digital Radiography. 3. Effect of Pixel Size on Snr and Thresh-old Contrast. *Med. Phys.* **1985**, *12*, 201–208. [CrossRef] [PubMed]
3. Avakyan, A.K.; Dergacheva, I.L.; Elanchik, A.A.; Korovkin, D.Y.; Krylova, T.A.; Lobzhanidze, T.K.; Polikhov, S.A.; Smirnov, V.P. Method for Determining the Point Spread Function for a Digital Radiography System. *At. Energy* **2020**, *127*, 310–315. [CrossRef]
4. Samei, E.; Ranger, N.T.; MacKenzie, A.; Honey, I.D.; Dobbins, J.T.; Ravin, C.E. Detector or System? Extending the Concept of Detec-tive Quantum Efficiency to Characterize the Performance of Digital Radiographic Imaging Systems. *Radiology* **2008**, *249*, 926–937. [CrossRef]
5. Drangova, M.; Rowlands, J.A. Optical factors affecting the detective quantum efficiency of radiographic screens. *Med. Phys.* **1986**, *13*, 150–157. [CrossRef] [PubMed]
6. Bunch, P.C.; Huff, K.E.; Van Metter, R. Analysis of the detective quantum efficiency of a radiographic screen–film combination. *J. Opt. Soc. Am. A* **1987**, *4*, 902–909. [CrossRef] [PubMed]
7. Sund, P.; Bath, M.; Kheddache, S.; Mansson, L.G. Comparison of visual grading analysis and determination of detective quantum efficiency for evaluating system performance in digital chest radiography. *Eur. Radiol.* **2004**, *14*, 48–58. [PubMed]
8. Uffmann, M.; Schaefer-Prokop, C.; Neitzel, U.; Weber, M.; Herold, C.J.; Prokop, M. Skeletal applications for flat-panel versus stor-age-phosphor radiography: Effect of exposure on detection of low-contrast details. *Radiology* **2004**, *231*, 506–514. [CrossRef] [PubMed]
9. Lu, Z.; Nickoloff, E.L.; So, J.C.; Dutta, A.K. Comparison of computed radiography and film/screen combination using a con-trast-detail phantom. *J. Appl. Clin. Med. Phys.* **2001**, *4*, 91–98. [CrossRef]
10. Konst, B.; Weedon-Fekjaer, H.; Båth, M. Image quality and radiation dose in planar imaging—Image quality figure of merits from the CDRAD phantom. *J. Appl. Clin. Med. Phys.* **2019**, *20*, 151–159. [CrossRef] [PubMed]

11. De Crop, A.; Bacher, K.; Van Hoof, T.; Smeets, P.V.; Smet, B.S.; Vergauwen, M.; Kiendys, U.; Duyck, P.; Verstraete, K.; D'Herde, K.; et al. Correlation of Contrast-Detail Analysis and Clinical Image Quality Assessment in Chest Radiography with a Human Ca-daver Study. *Radiology* **2012**, *262*, 298–304. [CrossRef]
12. Weir, A.; Salo, E.-N.; Janeczko, A.J.; Douglas, J.; Weir, N.W. Evaluation of CDRAD and TO20 test objects and associated software in digital radiography. *Biomed. Phys. Eng. Express* **2019**, *5*, 065001. [CrossRef]
13. Matsuyama, E.; Tsai, D.Y.; Lee, Y. Mutual information-based evaluation of image quality with its preliminary application to as-sessment of medical imaging systems. *J. Electron. Imaging* **2009**, *18*, 033011. [CrossRef]
14. Tsai, D.-Y.; Lee, Y.; Matsuyama, E. Information Entropy Measure for Evaluation of Image Quality. *J. Digit. Imaging* **2007**, *21*, 338–347. [CrossRef] [PubMed]
15. Mesbahi, A.; Zakariaee, S.S. Effect of anode angle on photon beam spectra and depth dose characteristics for X-RAD320 or-thovoltage unit. *Rep. Pract. Oncol. Radiother.* **2013**, *18*, 148–152. [CrossRef]
16. Kusk, M.W.; Jensen, J.M.; Gram, E.H.; Nielsen, J.; Precht, H. Anode heel effect: Does it impact image quality in digital radiography? A systematic literature review. *Radiography* **2021**. [CrossRef]
17. Mraity, H.; Walton, L.; England, A.; Thompson, J.; Lanca, L.; Hogg, P. Can the anode heel effect be used to optimise radiation dose and image quality for AP pelvis radiography? *Radiography* **2020**, *26*, e103–e108. [CrossRef] [PubMed]
18. Buissink, C.; Bowdler, M.; Abdullah, A.; Al-Murshedi, S.; Custódio, S.; Huhn, A.; Jorge, J.; Ali, M.; Peters, A.L.; Rey, Y.; et al. *Impact of the Anode Heel Effect on Image Quality and Effective Dose for AP Pelvis: A Pilot Study*; University of Salford: Salford, UK, 2017.
19. Behiels, G.; Maes, F.; Vandermeulen, D.; Suetens, P. Retrospective correction of the heel effect in hand radiographs. *Med. Image Anal.* **2002**, *6*, 183–190. [CrossRef]
20. Yu, Y.; Wang, J. Heel effect adaptive flat field correction of digital X-ray detectors. *Med. Phys.* **2013**, *40*, 081913. [CrossRef]
21. do Nascimento, M.Z.; Frere, A.F.; Germano, F. An automatic correction method for the heel effect in digitized mammography images. *J. Digit. Imaging* **2008**, *21*, 177–187. [CrossRef]
22. Cover, T.M.; Thomas, J.A. *Elements of Information Theory*; A Wiley-Interscience Publication: New York, NY, USA, 1991.
23. Szczepanski, J.; Arnold, M.; Wajnryb, E.; Amigo, J.M.; Sanchez-Vives, M.V. Mutual information and redundancy in spontaneous communication between cortical neurons. *Biol. Cybern.* **2011**, *104*, 161–174. [CrossRef]
24. Pregowska, A.; Szczepanski, J.; Wajnryb, E. Mutual information against correlations in binary communication channels. *BMC Neurosci.* **2015**, *16*, 32. [CrossRef] [PubMed]
25. Kim, H.Y. Statistical notes for clinical researchers: Post-hoc multiple comparisons. *Restor. Dent. Endod.* **2015**, *40*, 172–176. [CrossRef] [PubMed]
26. Saccenti, E.; Hendriks, M.; Smilde, A.K. Corruption of the Pearson correlation coefficient by measurement error and its estima-tion, bias, and correction under different error models. *Sci. Rep.* **2020**, *10*, 438. [CrossRef] [PubMed]
27. Friedman, S.N.; Fung, G.S.K.; Siewerdsen, J.H.; Tsui, B.M.W. A simple approach to measure computed tomography (CT) modula-tion transfer function (MTF) and noise-power spectrum (NPS) using the American College of Radiology (ACR) accreditation phantom. *Med. Phys.* **2013**, *40*, 051907. [CrossRef] [PubMed]

Article

Hyperchaotic Image Encryption Based on Multiple Bit Permutation and Diffusion

Taiyong Li * and Duzhong Zhang

School of Economic Information Engineering, Southwestern University of Finance and Economics, Chengdu 611130, China; zhangduzhong@swufe.edu.cn
* Correspondence: litaiyong@gmail.com

Abstract: Image security is a hot topic in the era of Internet and big data. Hyperchaotic image encryption, which can effectively prevent unauthorized users from accessing image content, has become more and more popular in the community of image security. In general, such approaches conduct encryption on pixel-level, bit-level, DNA-level data or their combinations, lacking diversity of processed data levels and limiting security. This paper proposes a novel hyperchaotic image encryption scheme via multiple bit permutation and diffusion, namely MBPD, to cope with this issue. Specifically, a four-dimensional hyperchaotic system with three positive Lyapunov exponents is firstly proposed. Second, a hyperchaotic sequence is generated from the proposed hyperchaotic system for consequent encryption operations. Third, multiple bit permutation and diffusion (permutation and/or diffusion can be conducted with 1–8 or more bits) determined by the hyperchaotic sequence is designed. Finally, the proposed MBPD is applied to image encryption. We conduct extensive experiments on a couple of public test images to validate the proposed MBPD. The results verify that the MBPD can effectively resist different types of attacks and has better performance than the compared popular encryption methods.

Keywords: hyperchaotic; image encryption; permutation; diffusion; multiple bit operation

Citation: Li, T.; Zhang, D. Hyperchaotic Image Encryption Based on Multiple Bit Permutation and Diffusion. *Entropy* **2021**, *23*, 510. https://dx.doi.org/10.3390/e23050510

Academic Editor: Amelia Carolina Sparavigna

Received: 19 March 2021
Accepted: 22 April 2021
Published: 23 April 2021

Publisher's Note: MDPI stays neutral with regard to jurisdictional claims in published maps and institutional affiliations.

1. Introduction

In the current era of Internet and big data, billions of images are produced, stored and transmitted every day. How to protect image content from illegal acquisition, especially for military, medical, and privacy purposes, has become a hot topic in recent years. Because of some attributes of images, such as high redundancy, strong correlation, and bulky data, traditional encryption methods for common text and data are usually not the best choice for image encryption. In recent years, various chaos-based image encryption approaches have emerged and they have been demonstrated very effective in improving image security. The reason why chaotic image encryption has become so popular is that chaotic systems have some characteristics that are very suitable for image encryption, such as extreme sensitivity to initial values, unpredictability, pseudorandomness, and ergodicity [1–3].

In chaotic image encryption, chaotic sequences are generated from the chaotic systems and they usually are applied to change the positions and/or values of image data. Early schemes usually used single low-dimensional chaotic systems, such as Logistic map, Tent map, Baker map, Cat map, etc., to encrypt images [4–8]. For example, Chen et al. extended 2D Cat map to a 3D one and designed a fast symmetric encryption approach, and the experiments demonstrated the approach was superior to the compared methods in terms of security and speed [4]. Fisarchik et al. proposed a pixel-by-pixel image encryption with Logistic maps [7]. Although these schemes achieved satisfactory encryption results at that time, the relatively simple structure of low-dimensional chaotic systems made them have a certain risk of being cracked. To solve this issue, possible directions are to use a more complex chaotic system or to combine two or more simple chaotic systems. In recent years, a variety of researchers have attempted to improve the performance of image encryption in

125

these two directions. According to the theory of chaos, Lyapunov exponent (LE) can be used to characterize the variable of a chaotic/hyperchaotic system. A dynamic system is chaotic if it has one positive LE, while it is hyperchaotic if it has two or more positive LEs. In general, image encryption schemes based on hyperchaotic systems are more secure than those with chaotic systems. In real-world encryption, Lorenz system and its extensions are among the most popular chaotic/hyperchaotic systems [9–12]. Wang and Zhang applied a 4D Lorenz-like hyperchaotic system with two positive LEs and genetic recombination to image encryption [10]. Li et al. used 5D and 7D hyperchaotic systems, dynamic filtering, DNA permutation and bit cuboid operations for image encryption, and the experimental results prove the effectiveness [13,14]. Unlike most schemes that carry out encryption in spatial domain, Wu et al. used 2D discrete wavelet transform (2D DWT) and a 6D hyperchaotic system to encrypt images in both spatial domain and frequency domain [15]. In addition to integer-order hyperchaotic systems, fractional-order hyperchaotic systems are becoming more and more popular with image encryption [16–19]. Zhu and Sun proposed a Logistic-Tent map for image encryption [20]. Luo et al. cooperated a piecewise linear chaotic map and a 4D hyperchaotic map for parallel image encryption [21]. Other combinations include Henon-Sine map [22], Logistic map and Lu system [23], Logistic-Sine map [24–29], Logistic-Tent-Sine map [30,31], Rossler-Sine map [32], etc. These combinations have been proven effective in improving the security of encryption.

The aim of image encryption is to prevent unauthorized users from discovering any meaningful content in the image. In other words, encrypted images are entirely random-like for them. There are many operations to convert informative images (plain images) to random-like ones (cipher images), among which permutation and diffusion are two major ones. Permutation changes the positions of image content, while diffusion changes the values of images. Most existing image encryption schemes adopted both the operations, separately or jointly, to achieve good security [33–37]. Among them, pixel-level (8 bits) data or bit-level (1 bit) data are the most widely used encryption units. In recent years, DNA computing has been introduced into image encryption; hence, DNA-level (2 bits) data has also be used to encrypt images [38–42]. Most studies focus on one or two bit levels in image encryption and the bit levels of encrypted data need to be enhanced to improve the effectiveness of image encryption. In fact, besides the mentioned 1 bit, 2 bit and/or 8 bit encryption operations, other multiple bit data, such as 3–7 bit data, can also be used to permutate and/or diffuse images. More bit-level data enhances the diversity of encrypted units and may have the potential to improve encryption performance. However, few existing studies have paid attention to this point.

Motivated by the above analysis, this paper proposes multiple bit permutation and diffusion, namely MBPD, for hyperchaotic image encryption. We first extend a modified 3D Lorenz chaotic system to a 4D hyperchaotic system with 3 positive LEs, and its characteristics are analyzed. Second, the hyperchaotic system is used to generate a hyperchaotic sequence for the consequent encryption operations. The initial values of the hyperchaotic system are considered as keys for the purpose of encryption. Then, the operations of multiple bit permutation and multiple bit diffusion are presented to encrypt images. The MBPD treats several bits (e.g., 3 bits) as a processing unit for permutation and diffusion, and different lengths of bits can be chosen for encryption. For the permutation, the order of the hyperchaotic sequence is used to scramble the multiple bit data, while the sequence will be converted into an integer mask for the diffusion. Finally, the proposed MBPD with different lengths of bits is applied to image encryption to improve security.

The contributions of this paper are the following:

(1) A new 4D hyperchaotic system with 3 positive LEs is presented, and some related hyperchaotic characteristics are analyzed.
(2) Multiple bit permutation and diffusion is proposed for image encryption, which is very different from most existing image encryption schemes that encrypt images only with 1 bit, 2 bit and/or 8 bit data. To the best of our knowledge, it is the first time that multiple bit operations are proposed for image encryption.

(3) Extensive experiments demonstrate that the proposed MBPD significantly outperforms the state-of-the-art compared image encryption schemes in terms of the evaluation indicators.

The rest of this paper is organized as follows: Section 2 presents a new 4D hyperchaotic system with 3 positive LEs. Section 3 proposes MBPD and details the encryption steps. In Section 4, experimental results are reported and analyzed. Finally, we conclude the paper in Section 5.

2. Presented 4D Hyperchaotic System

2.1. Lorenz System

Since the chaotic attractor was first found by Lorenz in 1963, chaos theory has attracted researchers from many fields, such as economics, mathematics, physics, and communications [9]. The initial Lorenz system has been extended to many versions. One modified generalized Lorenz system is formulated as Equation (1) [43].

$$\begin{cases} \dot{x} = -ax + by \\ \dot{y} = cx + dy - xz \ , \\ \dot{z} = -ez + x^2 \end{cases} \tag{1}$$

where a, b, and e are positive real constants, and c and d are real parameters meeting $d > -\frac{bc}{a}$ [44]. By introducing a 1D linear system to Equation (1), a new 4D system can be obtained, as Equation (2).

$$\begin{cases} \dot{x} = -ax + ay \\ \dot{y} = bx + cy - xz \\ \dot{z} = -dz + x^2 \\ \dot{w} = ey + fw \end{cases} \tag{2}$$

In this system, a, b, c. d, and f are real constant parameters, while f is a coupling parameter. When the parameters $(a,b,c,d,e,f) = (35,7,35,5,1.5,1)$, the system has the following LEs: $LE_1 = 1.284559$, $LE_2 = 0.937533$, $LE_3 = 0.007986$, and $LE_4 = -38.230078$. Since three LEs are positive, the system is hyperchaotic [44].

2.2. 4D Hyperchaotic System

Although Equation (2) is hyperchaotic, the introduced component w will increase exponentially after a certain number of iterations, and then its value will become positive infinity and its applications will be limited. To cope with this issue, we modify the fourth item of Equation (2) and add the component w to the first equation. A new 4D system can be obtained, as shown in Equation (3).

$$\begin{cases} \dot{x} = -ax + ay + w \\ \dot{y} = bx + cy - xz \\ \dot{z} = -dz + x^2 \\ \dot{w} = ey + fwsin(w) \end{cases} \tag{3}$$

where parameters $a - f$ are the same as Equation (2).

We use the 4th-order Runge-Kutta method to plot the attractors of the presented 4D hyperchaotic system with parameters $(a,b,c,d,e,f) = (35,7,35,5,1.5,1)$ and initial values $(x_0,y_0,z_0,w_0) = (0.12,0.23,0.34,0.45)$ in 2D space and 3D space, as shown in Figure 1. From this figure, we can see that the component w falls within an appropriate range.

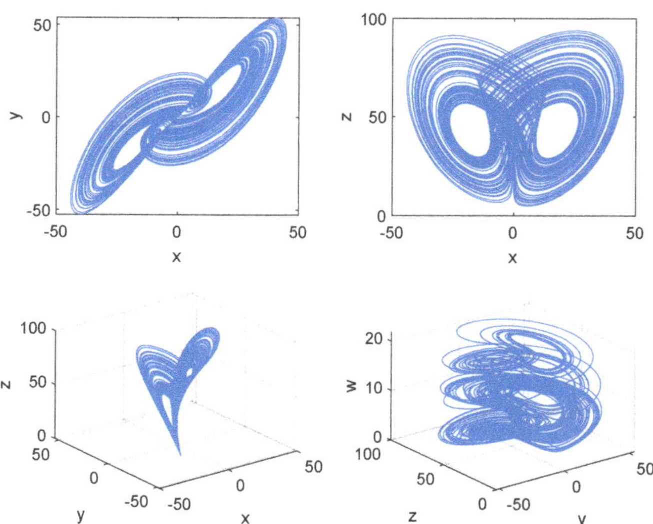

Figure 1. Attractors of the presented 4D hyperchaotic system with the parameters $(a, b, c, d, e, f) = (35, 7, 35, 5, 1.5, 1)$ and initial values $(x_0, y_0, z_0, w_0) = (0.12, 0.23, 0.34, 0.45)$.

By using Wolf's method [45], we fix $(a, b, c, d, e) = (35, 7, 35, 5, 1.5)$ and let f vary from 0 to 2 to plot the dynamics of LEs, as shown in Figure 2. We can see that the new system has three positive LEs in many ranges. For example, when $f = 1$, the LEs of the system are $LE_1 = 2.253019$, $LE_2 = 1.406374$, $LE_3 = 0.054342$, and $LE_4 = -38.339706$ and the three positive LEs (LE_1, LE_2, and LE_3) are much larger than the corresponding positive LEs of Reference [43]. Therefore, the new system is also hyperchaotic, and it is better than Equation (2).

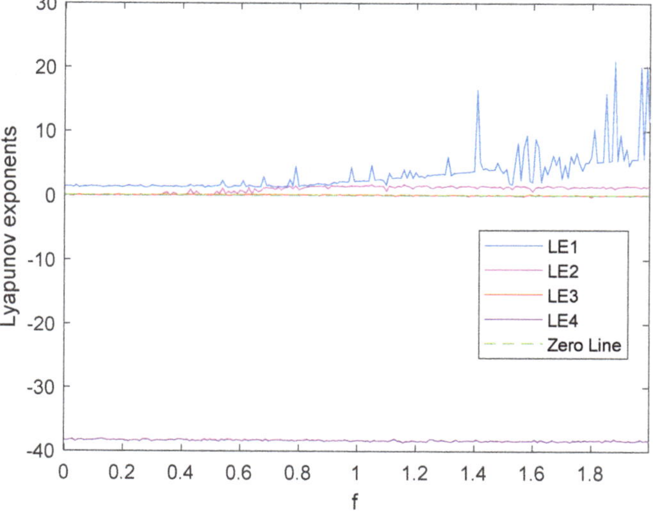

Figure 2. Dynamics of Lyapunov exponents of the proposed 4D hyperchaotic system with the parameters $(a, b, c, d, e) = (35, 7, 35, 5, 1.5)$, variable f from 0 to 2, and initial values $(x_0, y_0, z_0, w_0) = (0.12, 0.23, 0.34, 0.45)$.

In this paper, we will use the new 4D hyperchaotic system for image encryption. The reasons lie in: (1) Although it has only 4 dimensions in total, it has 3 positive LEs. The hyperchaotic characteristics make it very suitable for image encryption. (2) It has a simpler mathematical form when compared with some hyperchaotic systems of higher dimensions. (3) All the components fall within appropriate ranges, making it easy to sort for permutation and convert hyperchaotic sequences into integers for diffusion.

3. MBPD: Multiple Bit Permutation and Diffusion

This section will detail the steps of multiple bit permutation and diffusion for image encryption, including how to generate the hyperchaotic sequence, operations of multiple bit permutation and multiple bit diffusion, and the encryption algorithm.

3.1. Hyperchaotic Sequence Generation

In chaotic image encryption, a chaotic sequence is required to generate index for permutation and a mask for diffusion. Given the parameters and the initial values, we use the Fourth-order Runge-Kutta method and an interval of 0.001 to solve the presented 4D hyperchaotic system in Section 2.2 and then construct the hyperchaotic sequence for encryption. The detailed steps are as follows:

Step 1: Given the initial values $IV = \{x_0, y_0, z_0, w_0\}$, we solve the 4D hyperchaotic system to obtain long enough state values. The state values in the $i-$th iteration can be denoted as $s_i = \{x_i, y_i, z_i, w_i\}$.

Step 2: To remove the adverse effects, the state values obtained by the first it_0 iterations are discarded.

Step 3: When the iteration terminates, we can get a hyperchaotic sequence H by concatenating all the $s_j (j = 1, 2, \cdots, N)$ as Equation (4):

$$H = \{s_1, s_2, \cdots, s_N\} = \{x_1, y_1, z_1, w_1, \cdots, x_N, y_N, z_N, w_N\}$$
$$= \{h_1, h_2, h_3, h_4, \cdots, h_{4N-3}, h_{4N-2}, h_{4N-1}, h_{4N}\}, \tag{4}$$

where N is the iteration times excluding it_0.

Step 4: Since the elements in H come from different equations in Equation (3) and, hence, have different ranges, we use the following formulation to further map each element in H to a uniform interval $[0, 1)$.

$$h_i = \left| h_i \times 10^8 \right| - \left\lfloor \left| h_i \times 10^8 \right| \right\rfloor, \tag{5}$$

where $|\cdot|$ and $\lfloor \cdot \rfloor$ are the mathematical computation of absolute value and flooring, respectively.

It can be seen that each element in H is a real value in $[0, 1)$. With an element h_i, we can use the following formula to map it to an integer I in the range of $[0, N]$:

$$I = \lfloor ((|h_i| - \lfloor |h_i| \rfloor) \times 10^{14}) \rfloor \% N, \tag{6}$$

where % is the modulo operation.

3.2. Multiple Bit Permutation

Permutation is to rearrange the image content on a certain basis. For the permutation in chaotic image encryption, the positions of the image data to be permuted are usually determined by an index vector that can be obtained by sorting a hyperchaotic sequence. Typical permutation is conducted on pixel-level, DNA-level, and/or bit-level data [46]. The pixel-level data and DNA-level data in the current encryption technique actually refer to 8-bit data and 2-bit data, respectively. Few studies have focused on other numbers of bit data for encryption, such as 3–7 bits. In this paper, multiple bit permutation means conducting permutation on different numbers of bit data. n-bit permutation refers to using n bits as a minimum permutation unit.

All multiple bit operations require a bit stream of an image. Without loss of generality, given a bit stream B of length L, a hyperchaotic sequence H, and the number of bits to be permutated n, the first step is to calculate the number of permutation units PU and the remaining bits RB by $PU = \lfloor L/n \rfloor$ and $RB = L\%n$, respectively. It is clear that $RB < n$. Then, the PU units need to rearrange according to the index of sorting PU values in H and the RB bits can be embedding into the rearranged bit stream at a position decided by a value in H.

The n-bit permutation can be described as Algorithm 1:

Algorithm 1 n-bit permutation.

Input: a bit stream B, a hyperchaotic sequence H, and the number of bits to be permuted in a unit n
Output: a permutated bit stream PB, the number of used elements PU in H
1: **function** BitPermute(B, H, n)
2:     $L \leftarrow$ length(B); //length of B
3:     $PU \leftarrow \lfloor L/n \rfloor$;
4:     $RB \leftarrow L\%n$;
5:     $PB \leftarrow reshape(B(1 : PU * n), [PU, n])$; //reshape the first $PU * n$ bits in B into a vector PB having PU n-bit units
6:     $[v, idx] \leftarrow sort(H(1 : PU))$; // ascending sort to get the index vector idx
7:     $PB(1 : idx) \leftarrow PB$; // permute the PU units
8:     $PB \leftarrow reshape(PB, [1, PU \times n])$; // reshape the matrix PB to a bit stream
9:     **if** $RB <> 0$ **then**
10:         $PU \leftarrow PU + 1$;
11:         Generate a random position pos in the range of $[1, PU]$ from $H(PU)$ via Equation (6);
12:         Insert the remaining RB bits $B(L - RB + 1 : L)$ into PB at pos;
13:     **end if**
14:     **return** PB, PU;
15: **end function**

When n equals 1 or 8, Algorithm 1 degenerates to bit-level permutation or pixel-level permutation. Hence, the common bit-level permutation and pixel-level permutation are the special cases of Algorithm 1.

Here, we take 2-bit permutation and 3-bit permutation as an example to illustrate the detailed permutation procedure, as shown in Figure 3.

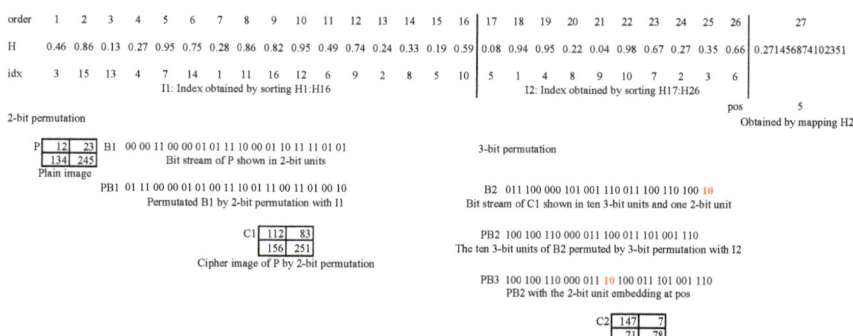

Figure 3. Illustration of 2-bit and 3-bit permutation.

A 2×2 plain image F is firstly converted into a bit stream $B1$ and the bits are grouped into 16 2-bit units. We get an index vector idx from the first 16 elements of the given hyperchaotic sequence H. It can be further grouped into two small vectors: $I1$ and $I2$. Then, we use $I1$ to rearrange the 16 units to get the permutated $PB1$. Since the RB is equal to 0 for this 2-bit permutation, there are no remaining bits needed to be embedded. Up to now, the 2-bit permutation completes. The obtained $PB1$ by the 2-bit permutation is actually the cipher image $C1$, which is clearly different from P. Then, it starts to conduct 3-bit permutation on the $PB1$. The 32 bits can be grouped into 10 complete 3-bit units and 2 remaining bits, as shown by $B2$. For the 10 3-bit units, we can rearrange them by $I2$, and then obtain the $PE2$ before embedding. H_{27} can be mapped to an integer 5 using Equation (6), and the remaining 2 bits can be inserted after the 5-th 3-bit unit in $PB2$, shown as $PB3$ after embedding in the figure. The final $PB3$ is actually the cipher image $C2$, a totally different image from P. From this illustration, we can see that the proposed n-bit permutation can also cause the change of the pixel values in plain images.

3.3. Multiple Bit Diffusion

The purpose of diffusion is to change the values of image data. The existing image encryption schemes mainly conduct diffusion on pixel-level data and/or DNA-level data (two bits). Similar to n-bit permutation, we propose n-bit diffusion that can be conducted on n-bit data per unit.

With the B, L, H, and n given for n-bit permutation, the number of n-bit units to be diffused DU is equal to $\lfloor L/n \rfloor$ and the length of the last unit LL is $L\%n$. If LL equals 0, the last unit is null; otherwise, its length is less than n (we call it non-n-bit unit) and it needs special handling. In this paper, we use a ciphertext diffusion in crisscross pattern (CDCP)-like idea to conduct n-bit diffusion [47]. Specifically, the bit stream B is transformed into a vector P of n-bit unit and then divided into two parts, and the two parts are diffused in crisscross pattern with two rounds. A mask vector M and an initial n-bit integer V can be mapped from H. When DU is an even, the first n-bit unit of each part can be initialized by Equation (7).

$$\begin{cases} C_1 = P_1 \otimes ((V - M_1)\%2^n) \\ C_{DU/2+1} = P_{DU/2+1} \otimes ((C_1 - M_{DU/2+1})\%2^n) \end{cases} \quad (7)$$

where $\otimes$ is the bitwise XOR (exclusive or) operation, and C is the vector of an cipher image. After that, the other n-bit units of each part can be updated as Equation (8):

$$\begin{cases} C_i = P_i \otimes ((C_{DU/2+i-1} - M_i)\%2^n) \\ C_{DU/2+i} = P_{DU/2+i} \otimes ((C_i - M_{DU/2+i})\%2^n) \end{cases} , i = 2,3,\cdots,DU/2. \quad (8)$$

There are two cases that need to be handled specially. When DU is an odd, we use the following formulation to encrypt the $(DU+1)/2$-th unit.

$$C_{(DU+1)/2} = P_{(DU+1)/2} \otimes ((C_{DU} - M_{(DU+1)/2})\%2^n). \quad (9)$$

Another case is about the non-n-bit unit. When it exists, we use the following formulation which is similar to Equation (9) to handle it.

$$C_{DU+1} = P_{DU+1} \otimes ((C_{DU} - M_{DU+1})\%2^{LL}). \quad (10)$$

The second round diffusion is the same as the first round, except that C_{DU} is used as the initial value to replace V in Equation (7).

The n-bit diffusion can be described as Algorithm 2.

Algorithm 2 n-bit diffusion.

Input: a bit stream B, a hyperchaotic sequence H, and the number of bits to be diffused in a unit n

Output: a diffused bit stream DB, the number of used elements PU in H

 1: **function** BITDIFFUSE(B, H, n)
 2:     $L \leftarrow \text{length}(B)$; //length of B
 3:     $DU \leftarrow \lfloor L/n \rfloor$;
 4:     $LL \leftarrow L\%n$;
 5:     $P \leftarrow reshape(B(1 : DU * n), [DU, n])$; //reshape the first $DU * n$ bits in B into a vector P having DU n-bit units
 6:     $P_{DU+1} = B(DU * n + 1 : end)$ //Use P_{DU+1} to denote the remaining $L\%n$ bits in B if they exist;
 7:     $PU \leftarrow L/8$;
 8:     Map $H(1 : PU)$ to a vector of 8-bit unsigned integers M;
 9:     Map $H(PU + 1)$ to a n-bit unsigned integer V;
 10:     $PU \leftarrow PU + 1$;
 11:     Conduct the first round diffusion with P, V and M by Equations (7)–(10);
 12:     $P \leftarrow C, V \leftarrow C_{DU}$;
 13:     Conduct the second round diffusion with P, V and M by Equations (7)–(10);
 14:     $DB = reshape(P, [1, L])$;
 15:     **return** DB, PU;
 16: **end function**

An illustration on 2-bit diffusion and 3-bit diffusion is shown in Figure 4.

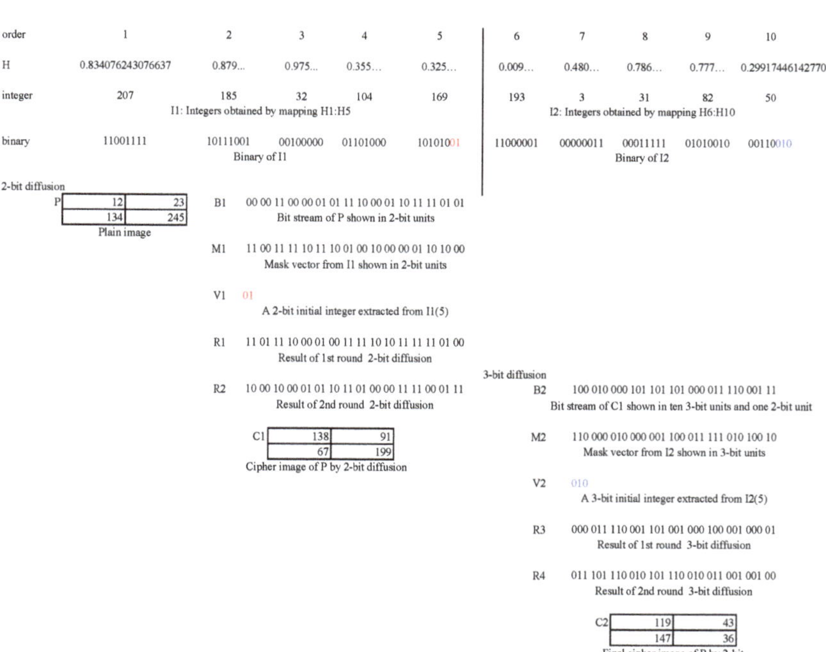

Figure 4. Illustration of 2-bit and 3-bit diffusion.

As done in Figure 3, the same plain image P is transformed into a binary sequence $B1$. A hyperchaotic sequence H having 10 elements are mapped into 8-bit integers and a further binary sequence ("binary" in the figure) by Equation (6). The sequence can also be split into $I1$ and $I2$. Note that some elements in H only show their first four digits to save spaces of the figure. The $I1$ can be further shown in a two-bit format as $M1$. The initial values $V1$ is extracted from the last 2 bits from $I1(5)$, as shown in red. When the 2-bit diffusion completes using Equation (7)/Equation (8) for the first/second round, we can obtain $R1$ and $R2$, respectively. $R2$ is actually the cipher image $C1$, which is totally different from the plain image P. Similarly, $C1$ can be encrypted by performing 3-bit diffusion. Note that since it has a 2-bit unit, when all the 3-bit units are encrypted, the remaining 2-bit unit needs to be encrypted by Equation (10). After the first and the second round 3-bit diffusion, we can obtain $R3$ and $R4$, respectively. $R4$ represents the final cipher image $C2$, where we can not find any visually information of the plain image P.

3.4. MBPD: Multiple Bit Permutation and Diffusion for Image Encryption

The main characteristic of the proposed MBPD lies in the permutation and diffusion can be conducted on multiple bit level data, which is very different from the common 1-bit, 2-bit (DNA) and/or 8-bit (pixel) permutation and 8-bit diffusion operations in most existing image encryption schemes.

With the aforementioned analysis, the detailed steps of the proposed MBPD are described as Algorithm 3.

Algorithm 3 MBPD: Multiple bit permutation and diffusion.

Input: a plain image P, initial values $IV = (x_0, y_0, z_0, w_0)$ for the hyperchaotic system, and iteration numbers of the discarded sequence it_0
Output: a cipher image C

1: **function** MBPD(P, IV, it_0)
2:     Generate a hyperchaotic sequence H with IV and it_0 as described in Section 3.1;
3:     Get the height h and the width w of P;
4:     Convert P to a bit stream B of length $L = h * w * 8$;
5:     $i \leftarrow 0$;
6:     **for** $n = 1 \rightarrow 8$ **do**
7:         $B, ul \leftarrow BITPERMUTE(B, H(i+1 : end), n)$;   //$n$-bit permutation
8:         $i \leftarrow i + ul$;
9:         $B, ul \leftarrow BITDIFFUSE(B, H(i+1 : end), n)$;   //$n$-bit diffusion
10:        $i \leftarrow i + ul$;
11:     **end for**
12:     Convert B to an image C;
13:     **return** C;
14: **end function**

The key steps of Algorithm 3 consist of a hyperchaotic sequence generation (Line 2), conversion the plain image to a bit stream (Line 3–4), conducting multiple bit permutation and diffusion on the bit stream (Line 6–11), and converting the bit stream back to an image (Line 12). Note that Algorithm 3 is proposed for gray images, but it can be easily extended for color images. The easiest way is to consider an RGB color image as three gray images and encrypt each gray image independently. The current proposed algorithm considers 8-bit permutation and diffusion at most, and it might be extended for 9-bit, 10-bit and even more bit permutation and diffusion. In addition, the proposed MBPD can be performed more than one round to enhance the effect of encryption. On the other hand, in real-world applications, it is not necessary to conduct all n-bit ($n = 1, 2, \cdots, 8$) operations to save time. The proposed MBPD can also be considered as a typical application of the strategy of "divide and conquer" [48,49].

To obtain a decrypted image, it only needs to execute the steps in Algorithm 3 reversely.

4. Experimental Results

4.1. Experimental Settings

We select the initial values $IV = (x_0, y_0, z_0, w_0)$ for the presented 4D hyperchaotic system as the security keys of the MBPD. Instead of conducting all bit levels permutation and diffusion, we only perform 6 types of n-bit permutation ($n = 1, 2, 3, 5, 6, 7$) and 2 types of n-bit diffusion ($n = 4, 8$). Specifically, we list all the parameters in Table 1. Although we use fixed security keys for all test images, they can also be optimized by evolutionary algorithms for each image [50–52].

Table 1. Experiment parameters.

Parameter Description	Value
Hyperchaotic system's parameters	$(a, b, c, d, e, f) = (35, 7, 35, 5, 1.5, 1)$
Security keys	$(x_0, y_0, z_0, w_0) = (0.12, 0.23, 0.34, 0.45)$
Iteration number to generate discarded sequence	$it_0 = 500$
Bit levels of permutation	$n = 1, 2, 3, 5, 6, 7$
Bit levels of diffusion	$n = 4, 8$
Rounds of encryption	1

We use 16 publicly accessible 256-level gray images as test images in most experiments. The size of each image is 256×256 or 512×512. We name each image by the format of "name+width". For example, "Lena512" represents gray Lena image of size 512×512. To demonstrate the performance of the proposed MBPD, we compare it with three popular gray image encryption schemes in most experiments: DFDLC [13], HCDNA [38], and CDCP [47].

All the experiments are conducted with MATLAB R2020b on a PC with 64-bit Windows 10 OS, an i5-9500 CPU @3.00 GHz, and 32 GB RAM.

4.2. Security Key Analysis

The security key is very important in cryptography, regardless of whether the encryption object is text, ordinary data or multimedia information. Key space and key sensitivity are two important indicators for evaluating security keys in image encryption.

4.2.1. Key Space

A good encryption scheme should have an enough large key space. An image encryption scheme with a key space larger than 2^{100} is able to resist brute-force attacks from modern computers. As far as the proposed MBPD is concerned, the initial value of the 4D hyperchaotic system can be considered as the security key. According to the IEEE standard, the precision of each element of the initial values is 10^{-15}; hence, the total key space is $(10^{15})^4 \approx 2^{199}$, which is far larger than 2^{100}. In addition, the parameters of the hyperchaotic system, the iteration number to generate discarded sequence, and the combination of permutation and/or diffusion at n bits can be thought of as parts of security key to further enlarge the key space. Therefore, the key space of the proposed MBPD is so large that it can resist brute-force attacks.

4.2.2. Key Sensitivity

A hyperchaotic system is extremely sensitive to the key. A tiny change in the key will produce a different hyperchaotic sequence and, hence, result in completely different decrypted images. To demonstrate the sensitivity of the proposed MBPD, we use the corrected security $K_1 = (x_0, y_0, z_0, w_0) = (0.12, 0.23, 0.34, 0.45)$ and a slightly different key $K_2 = (x_0 + 10^{-15}, y_0, z_0, w_0)$ to decrypt some cipher images. The decrypted images with K_1 and K_2 are shown in the first row and the second row of Figure 5, respectively.

Figure 5. Results of key sensitivity. The first row and the second row show the decrypted images by K_1 and K_2, respectively. From left to right: Clock256, Cameraman256, Finger512, Lena512, Baboon512, Bw512, Couple512, and Peppers512.

From this figure, we can find that the MBPD can decrypt the cipher images correctly with K_1 and even a tiny change (10^{-15}) that occurs in one element of K_1 will result in random-like images. It reveals that the MBPD is very sensitive to the security key.

To quantitatively demonstrate the sensitivity, we further use the SSIM to measure the structural similarity between the two decrypted images with K_1 and K_2 [53]. The lower the SSIM value, the higher the sensitivity. If the SSIM value is very close to 0, it reveals that the two images are almost completely different. Therefore, if a tiny change in the security key produces an SSIM value close to 0, we can say that the security key is very sensitive, from the review of decrypted images. The SSIM values of the decrypted images in Figure 5 are listed in Table 2. From this table, we can observe that all the SSIM values are very close to 0, showing the sensitivity of security keys.

Table 2. The SSIM values of decrypted images with K_1 and K_2.

Image Name	SSIM Value	Image Name	SSIM Value
Clock256	0.0083	Cameraman256	0.0087
Finger512	0.0081	Lena512	0.0093
Baboon512	0.0107	Bw512	0.0047
Couple512	0.0110	Peppers512	0.0098

We also use the SSIM to verify the structural similarity between the two cipher images by K_1 and K_2. The results are shown in Table 3. Again, we can find that the SSIM values are far below 0.01 and very close to 0, indicating that the security keys are very sensitive to cipher images.

Table 3. The SSIM values of cipher images with K_1 and K_2.

Image Name	SSIM Value	Image Name	SSIM Value
Clock256	0.0011	Cameraman256	0.0021
Finger512	0.0052	Lena512	0.0081
Baboon512	0.0087	Bw512	0.0060
Couple512	0.0052	Peppers512	0.0054

In summary, both the visual decrypted images and the quantitative analysis for decrypted images and cipher images show that the proposed MBPD has sensitive security keys for image encryption.

4.3. Statistical Analysis

In this subsection, we will analyze the MBPD via information entropies, histograms and correlations, which are all among the typical statistical analysis indicators in the area of image encryption.

4.3.1. Information Entropy

Entropy is an important concept in physics, communication, information theory, and others. It is often used to measure the uncertainty or randomness of a specific complex system. Given an L-level gray image I and the probability p_i of each gray level i occurs in the image, the information entropy of I, denoted by $E(I)$, can be calculated by:

$$E(I) = -\sum_{i=0}^{L-1} p_i log_2(p_i). \tag{11}$$

For a 256-level test image in the experiment, if it has only one level, for example, all-white image, its information entropy will equal the minimal value, 0. If each level appears with an identical probability, $\frac{1}{256}$, the corresponding information entropy is equal to the maximal value, 8. Therefore, the closer the information entropy to 8, the better the encrypted image. We list the information entropies of all plain images and their corresponding cipher images by the MBPD and the other compared schemes in Table 4, where the highest entropy of each image is shown in bold.

Table 4. Information entropies of the testing images.

Image Name	Plain Image	Cipher Image			
		MBPD	**DFDLC** [13]	**HCDNA** [38]	**CDCP** [47]
Airplane256	6.4523	7.9970	7.9974	7.9961	**7.9975**
Clock256	6.7057	7.9970	**7.9974**	7.9957	7.9972
Cameraman256	7.0492	**7.9976**	7.9972	7.9961	7.9975
Cameraman512	7.0480	**7.9994**	7.9992	7.9982	7.9993
Finger512	6.7279	**7.9994**	7.9993	7.9991	7.9993
Gray512	4.3923	7.9992	**7.9993**	7.9920	**7.9993**
Lena512	7.4460	7.9993	**7.9994**	7.9989	7.9993
Baboon512	7.1391	**7.9994**	**7.9994**	7.9993	7.9993
Barbara512	7.6321	7.9993	**7.9994**	7.9993	7.9993
Boat512	7.1914	7.9992	**7.9994**	7.9990	7.9993
Bw512	1.0000	7.9992	7.9992	7.9154	**7.9993**
Couple512	7.0572	7.9993	7.9993	7.9992	**7.9994**
Houses512	7.6548	**7.9993**	**7.9993**	**7.9993**	7.9992
Peppers512	7.5925	**7.9994**	**7.9994**	7.9992	**7.9994**
Pirate512	7.2367	**7.9993**	**7.9993**	7.9990	**7.9993**
Truck512	6.0274	**7.9994**	**7.9994**	7.9991	7.9993

We can see that all plain images' information entropies are much less than those of their cipher images. Specifically, the entropies of plain images fall in the range of $[1.0000, 7.6548]$, where the lower bound and the upper bound are achieved by Bw512 and House512, respectively. However, those values of cipher images of size 256×256 are greater than 7.9957. For cipher images of size 512×512, except for HCDNA's 7.9154 for Bw512, the lowest information entropy is 7.9920, which is very close to the maximal value, 8. The proposed MBPD, DFDLC, HCDNA, and CDCP achieve the highest information entropies in 8, 10, 1, and 6 out of 16 cases, respectively. In terms of information entropy, the MBPD significantly outperforms HCDNA and achieves comparable results with CDCP and DFDLC, revealing that the MBPD is able to resist entropy attacks effectively.

4.3.2. Histogram

The histogram of an image reflects the distribution of pixel levels. A natural image often has a histogram with certain irregular shapes, such as mountain peaks and valleys. A well-designed encryption scheme should break the original distribution of gray-levels and make the new distribution as even as possible. The histograms obtained by the MBPD are shown in Figure 6, where the test images' orders are the same as in Table 4.

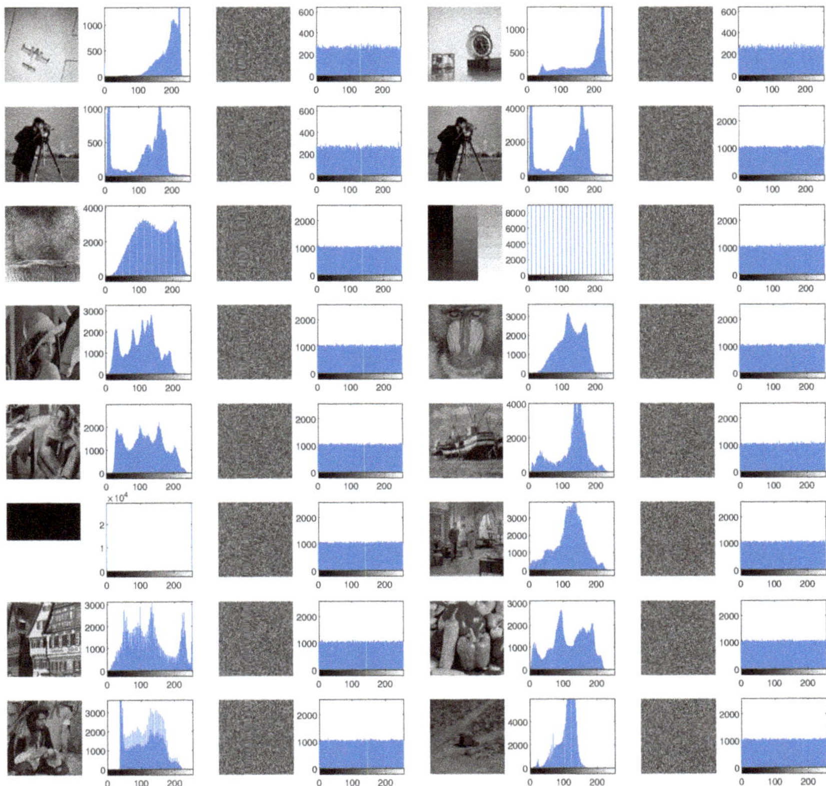

Figure 6. Histograms of plain images and their corresponding cipher images. Each plain image is followed by its histogram, the corresponding cipher image, and its histogram.

From this figure, we can find that all the natural images (those except for Gray512 Bw512) appear irregular histograms. Since plain Gray512 and Bw512 have evenly 21 and 2 gray-levels, respectively, they only have 21 and 2 bars in the histograms. However, the distributions of the pixel values of the cipher images are so uniform that the tops of the bars in the histograms appear as horizontal lines, even for Gray512 and Bw512. The results reveal that the proposed MBPD can effectively break the distributions of cipher images and produce sufficiently uniform histograms.

4.3.3. Correlation

Strong correlation among neighboring pixels is a key attribute of plain images. A practical image encryption scheme should reduce such correlation significantly. The lower the correlation in cipher images, the better an encryption scheme. Given two sequences s_1 and s_2, the correlation (γ) between them can be computed by:

$$\gamma = \frac{\rho(s_1, s_2)}{\sqrt{D(s_1)D(s_2)}},\tag{12}$$

where ρ denotes the covariance of two sequences, and D is the standard deviation of a sequence. According to this equation, the highest value of correlation will be 1 if s_1 and s_2 are identical, while it will be 0 if they are independent.

Given an image, there are many ways to construct s_1 and s_2. Typically, when a pixel is put into s_1, its horizontal, vertical, or diagonal adjacent pixel can be placed in s_2. In this way, we can use Equation (12) to calculate the correlations at the horizontal (γ_h), vertical

(γ_v), and diagonal (γ_d) directions. We use all the pixels in an image to construct s_1, and then construct corresponding s_2 to compute γ_h, γ_v, and γ_d. The correlations of plain images and cipher images are shown in Table 5, where the best results are in bold.

Table 5. The correlation coefficients γ of the testing images.

Image Name	γ	Plain Image	Cipher Image			
			MBPD	**DFDLC** [13]	**HCDNA** [38]	**CDCP** [47]
Airplane256	γ_h	0.9562	-0.0062	0.0004	-0.0049	$-$**0.0003**
	γ_v	0.8742	**0.0006**	-0.0042	-0.0045	-0.0006
	γ_d	0.8995	0.0019	**0.0001**	0.0038	-0.0022
Clock256	γ_h	0.9540	-0.0024	0.0034	**0.0017**	0.0020
	γ_v	0.9734	-0.0107	0.0022	-0.0060	$-$**0.0017**
	γ_d	0.9376	**0.0013**	0.0026	-0.0015	-0.0031
Cameraman256	γ_h	0.9554	-0.0059	0.0015	$-$**0.0006**	-0.0013
	γ_v	0.9710	0.0007	0.0023	-0.0012	**0.0001**
	γ_d	0.9377	0.0052	-0.0053	**0.0012**	-0.0030
Cameraman512	γ_h	0.9830	-0.0013	-0.0016	-0.0014	**0.0011**
	γ_v	0.9887	**0.0014**	0.0015	0.0029	-0.0026
	γ_d	0.9746	-0.0017	**0.0002**	-0.0022	0.0010
Finger512	γ_h	0.9343	0.0022	-0.0010	-0.0012	$-$**0.0002**
	γ_v	0.9168	0.0007	-0.0012	$-$**0.0005**	$-$**0.0005**
	γ_d	0.8664	0.0017	**0.0001**	-0.0003	-0.0003
Gray512	γ_h	0.9913	0.0028	-0.0006	0.0017	**0.0001**
	γ_v	0.9989	**0.0009**	0.0021	-0.0010	0.0017
	γ_d	0.9964	**0.0005**	-0.0006	-0.0007	0.0007
Lena512	γ_h	0.9705	**0.0005**	0.0014	0.0022	-0.0028
	γ_v	0.9856	**0.0002**	-0.0004	-0.0004	0.0038
	γ_d	0.9649	**0.0000**	0.0021	-0.0008	0.0023
Baboon512	γ_h	0.8652	$-$**0.0018**	0.0024	0.0020	-0.0024
	γ_v	0.7524	-0.0017	$-$**0.0000**	0.0027	0.0024
	γ_d	0.7210	0.0022	0.0026	**0.0011**	0.0012
Barbara512	γ_h	0.8940	**0.0000**	0.0001	0.0007	-0.0001
	γ_v	0.9572	0.0002	0.0034	-0.0018	$-$**0.0001**
	γ_d	0.8942	0.0004	-0.0006	$-$**0.0001**	-0.0014
Boat512	γ_h	0.9368	0.0022	$-$**0.0015**	-0.0022	-0.0052
	γ_v	0.9709	0.0007	0.0008	$-$**0.0004**	0.0025
	γ_d	0.9240	$-$**0.0007**	0.0012	0.0015	0.0021
Bw512	γ_h	1.0000	-0.0013	-0.0009	$-$**0.0001**	-0.0022
	γ_v	0.9922	0.0031	0.0050	$-$**0.0016**	-0.0019
	γ_d	0.9961	-0.0011	-0.0019	**0.0002**	-0.0008
Couple512	γ_h	0.9451	0.0013	0.0020	**0.0007**	0.0019
	γ_v	0.9514	0.0011	**0.0002**	0.0020	0.0032
	γ_d	0.9116	-0.0015	-0.0011	-0.0006	**0.0001**
Houses512	γ_h	0.9077	$-$**0.0013**	0.0028	0.0014	-0.0030
	γ_v	0.9173	-0.0002	**0.0000**	-0.0026	0.0016
	γ_d	0.8439	0.0014	**0.0005**	0.0010	-0.0011
Peppers512	γ_h	0.9733	-0.0009	0.0006	-0.0004	**0.0001**
	γ_v	0.9763	-0.0021	-0.0024	0.0007	$-$**0.0006**
	γ_d	0.9650	**0.0005**	0.0007	0.0011	-0.0008
Pirate512	γ_h	0.9593	$-$**0.0006**	0.0014	-0.0020	-0.0020
	γ_v	0.9675	-0.0022	**0.0006**	-0.0008	**0.0006**
	γ_d	0.9432	**0.0002**	-0.0010	-0.0003	-0.0023
Truck512	γ_h	0.9610	0.0005	**0.0000**	0.0012	0.0035
	γ_v	0.9164	0.0018	**0.0001**	0.0003	-0.0004
	γ_d	0.9048	-0.0028	$-$**0.0003**	-0.0016	-0.0005

From this table, we can observe that all the plain images have high correlations. In particular, the γ_h of plain Bw512 is equal to the highest value, i.e., 1. However, these high correlations are reduced to a very low level by the encryption schemes. More specifically, the correlations by the encryption schemes are very close to or even equal to 0, showing that all the schemes can break the high correlations in plain images. As far as the four schemes, MBPD achieves the lowest correlations in 15 out of 48 times, followed by CDCP's 13 times, DFDLC's 12 times, and HCDNA's 11 times, indicating that MBPD performs better than the compared encryption schemes.

To further analyze the correlations, we randomly pick up 4000 pairs of horizontally adjacent pixels from plain images and cipher images by the proposed MBPD and then plot their gray levels as x-values and y-values in a 2D plane, as shown in Figure 7. We can observe that the plots of all the plain images except for Bw512 appear near the main diagonals, showing that there exist strong correlations in the cipher images. Since Bw512 has only two gray levels: 0 and 255, most points are piled up at $(0, 0)$ and $(255, 255)$, which are also on the main diagonal. In contrast, the plots of all the cipher images fill with the whole planes, suggesting low correlations in cipher images.

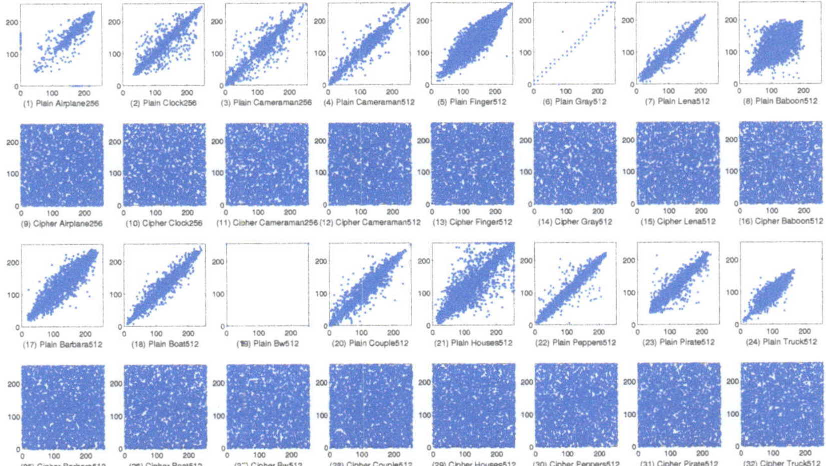

Figure 7. Horizontal correlations of plain images and their corresponding cipher images.

4.4. Differential Attack Analysis

Differential attacks compare the variations in a plain image with variations in the cipher image to find the plain image and/or desired security key. To resist differential attacks, a well-designed image encryption scheme must produce a completely different cipher image even for a tiny change in the corresponding plain image.

There are two popular indicators in the community of image security to measure image encryption schemes' capability of resisting differential attacks. One is the number of pixels change rate (NPCR), which can be defined as Equation (13). And the other is the unified average changing intensity (UACI) defined by Equation (14).

$$NPCR = \frac{\sum_{h=1}^{H} \sum_{w=1}^{W} d(h, w)}{H \cdot W} \times 100\%,$$

$$UACI = \frac{\sum_{h=1}^{H} \sum_{w=1}^{W} |C_1(h, w) - C_2(h, w)|}{255 \cdot H \cdot W} \times 100\%,$$

where H and W denote the height and the width of the cipher images C_1 and C_2, and $d(h, w)$ is used to judge whether the gray levels of C_1 and C_2 at the position (h, w) are different, as formulated by Equation (15).

$$d(w, h) = \begin{cases} 0, & C_1(h, w) = C_2(h, w) \\ 1, & C_1(h, w) \neq C_2(h, w) \end{cases}.$$

Given two 8-bit gray images, if they are identical, their both NPCR and UACI obtain the minimal value, 0. If one is all-white and the other is all-black, their NPCR and UACI values will be the maximal value, 1. Since the cipher images are all random-like, the NPCR and UACI values of a pair of cipher images usually fall into a certain range. The study by Wu et al. reveals that, given a significance level $\alpha = 0.05$ and a 256×256 8-bit gray levels image, if the NPCR is greater than $\mathcal{N}_{0.05}^1 = 99.5693\%$ and the UACI falls into the range of $(\mathcal{U}_{0.05}^{11}, \mathcal{U}_{0.05}^{1u}) = (33.2824\%, 33.6447\%)$, the encryption scheme is said to pass NPCR test and UACI test separately at $\alpha = 0.05$ [54]. Similarly, for a 512×512 image, the corresponding NPCR threshold and UACI range are $\mathcal{N}_{0.05}^2 = 99.5893\%$ and $(\mathcal{U}_{0.05}^{21}, \mathcal{U}_{0.05}^{2u}) = (33.3703\%, 33.5541\%)$, respectively.

We compute NPCR and UACI values from the cipher image by the exact plain image and a cipher image by a slightly changed plain image generated by adding one to the least significant bit of a random pixel. The computation procedure is repeated 20 times, and the average NPCR and UACI are reported in Tables 6 and 7, respectively, where the values that pass the tests are shown in bold. Moreover, the times of passing the test, the standard deviation, and the average value of the 16 test images by each scheme are shown in the last three lines of the tables.

From Table 6, we can find that the MBPD passes the NPCR test on all images, following by DFDLC and CDCP's in 15 out of 16 cases. The HCDNA fails to the test because it has no operations to expand a tiny change in the plain images to the whole cipher images. Although CDCP achieves the highest average NPCR value (99.6773%) for the 16 test images, but its standard deviation (0.0723%) is not as low as that of MBPD (0.0037%), indicating that the MBPD achieves the stablest NPCR values. Regarding UACI, again, MBPD passes the test on all test images and achieved the lowest standard deviation, and CDCP and DFDLC fails one image, i.e., Bw512 and Pirate512, respectively. HCDNA performs the worst and fails all the test images. To summarize, the proposed MBPD outperforms the other compared schemes in terms of NPCR and UACI and can effectively resist differential attacks.

Table 6. The average NPCR (%) of running the schemes 20 times.

Image	MBPD	DFDLC [13]	HCDNA [38]	CDCP [47]
Airplane256	**99.6014 %**	**99.6125%**	76.4828%	**99.6374%**
Clock256	**99.6114%**	**99.6085%**	65.7269%	**99.7081%**
Cameraman256	**99.6099%**	**99.6196%**	73.4785%	**99.7564%**
Cameraman512	**99.6112%**	**99.6078%**	67.1009%	**99.6590%**
Finger512	**99.6083%**	**99.6108%**	76.2949%	**99.6928%**
Gray512	**99.6088%**	**99.6131%**	61.1288%	**99.6767%**
Lena512	**99.6062%**	**99.6084%**	66.5552%	**99.6849%**
Baboon512	**99.6104%**	**99.6077%**	64.3461%	**99.6372%**
Barbara512	**99.6113%**	**99.6114%**	73.5446%	**99.5927%**
Boat512	**99.6093%**	**99.6089%**	75.0493%	99.4786%
Bw512	**99.5997%**	89.6501%	64.8879%	**99.7000%**
Couple512	**99.6033%**	**99.6045%**	63.5847%	**99.7910%**
Houses512	**99.6037%**	**99.6092%**	75.8256%	**99.7578%**
Peppers512	**99.6090%**	**99.6100%**	73.8790%	**99.6849%**
Pirate512	**99.6082%**	**99.6087%**	73.9838%	**99.6765%**
Truck512	**99.6063%**	**99.6070%**	66.5778%	**99.7033%**
Pass/Fail/All	16/0/16	15/1/16	0/16/16	15/1/16
Std.	0.0037%	2.4899%	5.3211%	0.0723%
Mean	99.6074%	98.9874%	69.9029%	99.6773%

Table 7. The average UACI (%) of running the schemes 20 times.

Image	MBPD	DFDLC [13]	HCDNA [38]	CDCP [47]
Airplane256	**33.4400** %	**33.4256**%	30.6926%	**33.4682**%
Clock256	**33.4610**%	**33.4992**%	28.3912%	**33.5090**%
Cameraman256	**33.4312**%	**33.4529**%	31.3096%	**33.4766**%
Cameraman512	**33.4618**%	**33.4547**%	27.7148%	**33.4765**%
Finger512	**33.4552**%	**33.4766**%	33.6617%	**33.4796**%
Gray512	**33.4628**%	**33.4638**%	25.1829%	**33.4842**%
Lena512	**33.4545**%	**33.4581**%	27.2038%	**33.4484**%
Baboon512	**33.4590**%	**33.4528**%	26.1169%	**33.4996**%
Barbara512	**33.4905**%	**33.4746**%	28.2405%	**33.5072**%
Boat512	**33.4684**%	**33.4781**%	31.6422%	**33.4881**%
Bw512	**33.4899**%	30.1296%	22.3338%	**33.4655**%
Couple512	**33.4661**%	**33.4853**%	25.9647%	**33.4975**%
Houses512	**33.4682**%	**33.4631**%	31.4138%	**33.4587**%
Peppers512	**33.4409**%	**33.4255**%	29.2497%	**33.4637**%
Pirate512	**33.4808**%	**33.4251**%	30.3032%	33.5917%
Truck512	**33.4612**%	**33.4609**%	28.0393%	**33.4589**%
Pass/Fail/All	16/0/16	15/1/16	0/16/16	15/1/16
Std.	0.0164%	0.8328%	2.8880%	0.0335%
Mean	33.4520%	33.2516%	28.5913%	33.4858%

4.5. Robustness

From the above analysis, we know that a tiny change in a plain image will result in a completely different cipher image for a well-designed image encryption scheme. However, contamination in cipher images is unavoidable during transmission and storage. Therefore, a good encryption scheme should recover a contaminated cipher image to some extent. Noise and cropping are two typical types of contamination.

To validate the robustness to noise and cropping, we first add 0.5%, 1%, 2%, 4%, and 10% salt-and-pepper noise to the cipher images, and decrypt them with the proposed MBPD. The results are shown in Figure 8, where we can find that when the noise level is less than 4%, the MBPD can recover the cipher images very well and even for 10% noise level, the profile of Lena can be clearly recognized. Then, we crop the images with 1%, 2.78%, 6.75%, 11.11%, and 25% data loss, the cropped cipher images and the corresponding decrypted images are shown in Figure 9. We can see that Lena can be easily recognized when the data loss levels are less than 11.11%. When the level equals to 25%, it is hard to recognize the profile of Lena. Another finding is that, even if the data loss is concentrated in the center of an encrypted image, the contaminated locations in the decrypted image are evenly distributed throughout the image.

To summarize, the MBPD can effectively resist noise and cropping attacks to some extent.

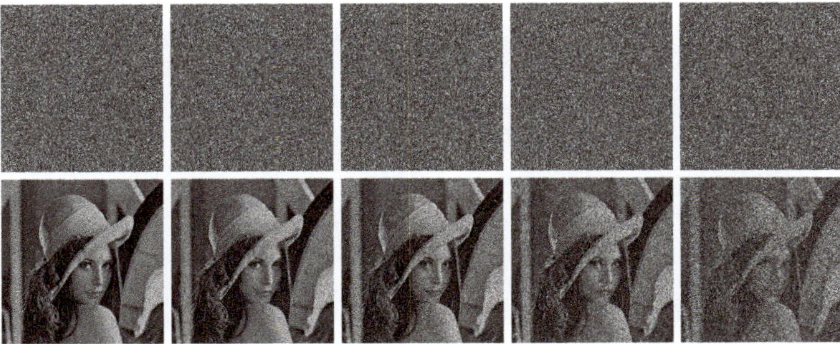

Figure 8. Noise test. The first row, from left to right: cipher images with 0.5%, 1%, 2%, 4%, and 10% salt-and-pepper noise added. The second row: the decrypted images from the corresponding cipher images in the first row.

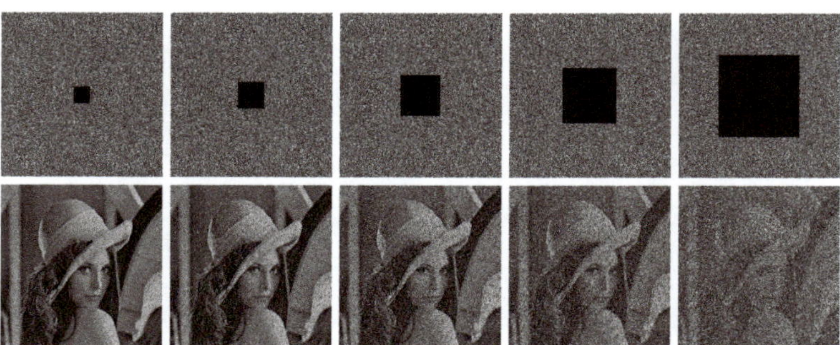

Figure 9. Cropping test. The first row, from left to right: cipher images with 1%, 2.78%, 6.75%, 11.11%, and 25% data loss. The second row: the decrypted images from the corresponding cipher images in the first row.

4.6. Running Time

Running time is used to measure the efficiency of the encrypted algorithms. Table 8 lists the running time of encryption and decryption operations on images with sizes 256 × 256 and 512 × 512. We can find that CDCP takes the least time among the four schemes, while the HCDNA takes the most time. The running time of HCDNA is about 30 times that of CDCP. The results of MBPD and DFDLC are somewhere in between and are very close but the former is slightly less than the latter. The major reasons why MBPD is somewhat time-consuming are that it conducts encryption at multiple bit levels and the operations with most multi-bit levels involve string operations. Two possible directions for decreasing running time are: using parallel computing and reducing the number of bit levels for multi-bit operations, e.g., encrypting images only with 1-bit permutation and 4-bit diffusion.

Table 8. Running time of encryption and decryption (in seconds).

Operation	Size	MBPD	DFDLC [13]	HCDNA [38]	CDCP [47]
Encryption	256 × 256	0.8158	0.8479	3.4635	0.1268
	512 × 512	3.3833	3.4261	14.2808	0.5240
Decryption	256 × 256	0.8136	0.8366	4.7181	0.1305
	512 × 512	3.3624	3.4116	19.5680	0.5126

4.7. Discussion

From the above experimental results and the corresponding analysis, we can see that the proposed MBPD is a promising scheme for image encryption.

In addition to the proposed 4D hyperchaotic system and the extensive experiments, the major contribution of the paper lies in proposing a novel multiple bit permutation and diffusion scheme for image encryption. The MBPD can encrypt images not only with 1-bit, 2-bit, and 8-bit (one pixel) data that are widely processed by existing image encryption schemes but also with 3–7 bit data that few studies have focused on.

The proposed MBPD's main advantage over the existing image encryption schemes is that it can perform permutation and diffusion with multiple different bits. The diversity of each encrypted unit's length is enhanced, and the proposed MBPD finally achieves promising results in terms of the evaluation metrics when compared with four state-of-the-art image encryption schemes, as demonstrated by the experiments.

Sixteen publicly accessible 256-level gray images of two sizes are used to evaluate the proposed MBPD. They include 14 natural images in different scenes, as well as two handcrafted images, which are very popular in the evaluation of image encryption schemes. The MBPD performs quite well with all the test images. Although the MBPD is proposed to encrypt gray images only in this paper, it can be easily extended for color image encryption. The simplest way is to treat each channel of a color image as a gray image, and each channel can be separately encrypted by the MBPD. Here, we use miscellaneous images of different sizes, different scenes and different channels (a 3-channel image means a color image) from the SIPI image database (http://sipi.usc.edu/database/database.php?volume=misc, accessed on 19 April 2021) to verify the generality of the proposed MBPD. Note that the data set has 39 images in total, consisting of 24 gray images and 15 color ones. Six of them have been tested in the above experiments; hence, they are excluded in this experiment. The results of entropy, γ_h, γ_v, γ_d, NPCR, and UACI of the rest 33 images obtained by the proposed MBPD are reported in Table 9, where the test images are sorted by size and image name. Note that the table reports the average of the three channels for color images.

Table 9. Results obtained by the proposed MBPD on miscellaneous images from the SIPI image database.

Image	Size	Entropy	γ_h	γ_v	γ_d	NPCR	UACI
5.1.10	256×256	7.9973	0.0010	-0.0024	-0.0003	99.6053%	33.4448%
5.1.13	256×256	7.9976	0.0011	0.0000	-0.0016	99.6127%	33.4409%
5.1.14	256×256	7.9974	-0.0022	0.0000	-0.0010	99.6114%	33.5188%
Moonsurface256	256×256	7.9973	0.0023	-0.0042	-0.0012	99.6093%	33.4342%
5.2.10	512×512	7.9992	0.0010	-0.0040	0.0005	99.6116%	33.4516%
7.1.02	512×512	7.9993	-0.0007	0.0018	-0.0004	99.6083%	33.4698%
7.1.03	512×512	7.9993	-0.0040	0.0017	-0.0004	99.6070%	33.4621%
7.1.04	512×512	7.9993	0.0001	-0.0006	-0.0027	99.6060%	33.4578%
7.1.05	512×512	7.9993	-0.0017	-0.0027	0.0013	99.6114%	33.4616%
7.1.06	512×512	7.9994	0.0028	0.0015	-0.0027	99.6074%	33.4786%
7.1.07	512×512	7.9993	0.0018	-0.0006	-0.0004	99.6066%	33.4736%
7.1.08	512×512	7.9994	0.0007	-0.0012	-0.0000	99.6093%	33.4505%
7.1.09	512×512	7.9993	0.0003	-0.0035	0.0011	99.6142%	33.4582%
7.1.10	512×512	7.9992	-0.0026	-0.0007	-0.0000	99.6109%	33.4435%
Aerial512	512×512	7.9993	0.0002	-0.0016	0.0004	99.6075%	33.4789%
ruler.512	512×512	7.9993	-0.0045	-0.0001	-0.0010	99.6134%	33.4642%
5.3.01	1024×1024	7.9998	-0.0006	-0.0017	0.0002	99.6117%	33.4545%
5.3.02	1024×1024	7.9998	0.0014	0.0000	0.0004	99.6090%	33.4637%
7.2.01	1024×1024	7.9998	0.0005	-0.0001	-0.0006	99.6127%	33.4609%
4.1.01	$256 \times 256 \times 3$	7.9969	0.0016	0.0031	0.0027	99.6155%	33.4652%
4.1.02	$256 \times 256 \times 3$	7.9975	-0.0068	-0.0037	0.0032	99.6149%	33.4376%
4.1.03	$256 \times 256 \times 3$	7.9971	0.0029	-0.0029	-0.0001	99.6168%	33.4728%
4.1.04	$256 \times 256 \times 3$	7.9972	0.0013	0.0024	0.0001	99.5991%	33.4596%
4.1.05	$256 \times 256 \times 3$	7.9974	0.0030	0.0031	-0.0008	99.6046%	33.4304%
4.1.06	$256 \times 256 \times 3$	7.9971	-0.0030	0.0009	0.0020	99.6051%	33.3983%
4.1.07	$256 \times 256 \times 3$	7.9972	0.0002	-0.0003	-0.0049	99.6139%	33.4285%
4.1.08	$256 \times 256 \times 3$	7.9970	0.0024	-0.0008	0.0003	99.6086%	33.4542%
4.2.01	$512 \times 512 \times 3$	7.9993	0.0014	-0.0005	-0.0006	99.6051%	33.4530%
4.2.03	$512 \times 512 \times 3$	7.9992	-0.0011	0.0032	0.0017	99.6102%	33.4833%
4.2.05	$512 \times 512 \times 3$	7.9994	-0.0004	-0.0002	-0.0004	99.6111%	33.4447%
4.2.06	$512 \times 512 \times 3$	7.9993	0.0004	-0.0016	-0.0006	99.6106%	33.4843%
4.2.07	$512 \times 512 \times 3$	7.9993	-0.0013	-0.0001	-0.0000	99.6087%	33.4446%
house	$512 \times 512 \times 3$	7.9992	-0.0012	0.0005	-0.0010	99.6079%	33.4962%

From this table, we can find that the experimental results are very ideal in terms of all the evaluation indicators, regardless of the image content, size, and the number of channels. Specifically, the entropies are very close to the theoretical best value, 8, and all the correlations in all directions are close to 0. All the images pass the NPCR and UACI tests. Therefore, the extensive test images demonstrate that the proposed MBPD has good generality.

5. Conclusions

Most existing image encryption schemes involve 1-bit level, 2-bit level (DNA computing), and/or 8-bit level (pixel) data. Few studies focus on other bit-level data, which limits the diversity of encrypted data units and ultimately negatively affects the encryption effect. To this end, this paper proposes a novel multi-bit permutation and diffusion scheme

(MBPD) for image encryption. The key characteristic of MBPD is that it can perform permutation and diffusion at different bit-level data, such as 1-bit permutation, 3-bit diffusion, and 6-bit permutation, to encrypt images. The results of extensive experiments demonstrate that the proposed MBPD can resist different types of attacks and has high security. One limitation of the MBPD is that it is somewhat time-consuming. In the future, we will study how to speed it up and apply it to color image encryption.

Author Contributions: Conceptualization, T.L.; Formal analysis, T.L.; Funding acquisition, T.L.; Investigation, T.L.; Methodology, T.L.; Software, T.L.; Supervision, T.L.; Validation, T.L.; Visualization, D.Z.; Writing—original draft, T.L.; Writing—review & editing, T.L. and D.Z. Both authors have read and agreed to the published version of the manuscript.

Funding: This work was supported by the Ministry of Education of Humanities and Social Science Project (Grant No. 19YJAZH047) and the Scientific Research Fund of Sichuan Provincial Education Department (Grant No. 17ZB0433).

Institutional Review Board Statement: Not applicable.

Informed Consent Statement: Not applicable.

Data Availability Statement: The used test images are all included in the paper.

Conflicts of Interest: The authors declare no conflict of interest.

References

1. Zhou, S.; Zhang, Q.; Wei, X.; Zhou, C. A Summarization on Image Encryption. *IETE Tech. Rev.* **2010**, *27*, 503–510. [CrossRef]
2. Cheng, G.; Wang, C.; Chen, H. A Novel Color Image Encryption Algorithm Based on Hyperchaotic System and Permutation-Diffusion Architecture. *Int. J. Bifurc. Chaos* **2019**, *29*. [CrossRef]
3. Lu, Q.; Zhu, C.; Deng, X. An Efficient Image Encryption Scheme Based on the LSS Chaotic Map and Single S-Box. *IEEE Access* **2020**, *8*, 25664–25678. [CrossRef]
4. Chen, G.; Mao, Y.; Chui, C.K. A symmetric image encryption scheme based on 3D chaotic cat maps. *Chaos Solitons Fractals* **2004**, *21*, 749–761. [CrossRef]
5. Mao, Y.; Chen, G.; Lian, S. A novel fast image encryption scheme based on 3D chaotic baker maps. *Int. J. Bifurc. Chaos* **2004**, *14*, 3613–3624. [CrossRef]
6. Pareek, N.K.; Patidar, V.; Sud, K.K. Image encryption using chaotic logistic map. *Image Vis. Comput.* **2006**, *24*, 926–934. [CrossRef]
7. Pisarchik, A.N.; Flores-Carmona, N.J.; Carpio-Valadez, M. Encryption and decryption of images with chaotic map lattices. *Chaos* **2006**, *16*. [CrossRef]
8. Yoon, J.W.; Kim, H. An image encryption scheme with a pseudorandom permutation based on chaotic maps. *Commun. Nonlinear Sci. Numer. Simul.* **2010**, *15*, 3998–4006. [CrossRef]
9. Lorenz, E.N. Deterministic nonperiodic flow. *J. Atmos. Sci.* **1963**, *20*, 130–141. [CrossRef]
10. Wang, X.; Zhang, H.l. A novel image encryption algorithm based on genetic recombination and hyper-chaotic systems. *Nonlinear Dyn.* **2016**, *83*, 333–346. [CrossRef]
11. Younas, I.; Khan, M. A New Efficient Digital Image Encryption Based on Inverse Left Almost Semi Group and Lorenz Chaotic System. *Entropy* **2018**, *20*, 913. [CrossRef]
12. Hu, T.; Liu, Y.; Gong, L.H.; Ouyang, C.J. An image encryption scheme combining chaos with cycle operation for DNA sequences. *Nonlinear Dyn.* **2017**, *87*, 51–66. [CrossRef]
13. Li, T.; Shi, J.; Li, X.; Wu, J.; Pan, F. Image encryption based on pixel-level diffusion with dynamic filtering and DNA-level permutation with 3D Latin cubes. *Entropy* **2019**, *21*, 319. [CrossRef] [PubMed]
14. Li, X.; Xie, Z.; Wu, J.; Li, T. Image encryption based on dynamic filtering and bit cuboid operations. *Complexity* **2019**, *2019*, 7485621. [CrossRef]
15. Wu, X.; Wang, D.; Kurths, J.; Kan, H. A novel lossless color image encryption scheme using 2D DWT and 6D hyperchaotic system. *Inf. Sci.* **2016**, *349*, 137–153. [CrossRef]
16. Li, T.; Yang, M.; Wu, J.; Jing, X. A novel image encryption algorithm based on a fractional-order hyperchaotic system and DNA computing. *Complexity* **2017**, *2017*, 9010251. [CrossRef]
17. Yang, Y.G.; Guan, B.W.; Zhou, Y.H.; Shi, W.M. Double image compression-encryption algorithm based on fractional order hyper chaotic system and DNA approach. *Multimed. Tools Appl.* **2021**, *80*, 691–710. [CrossRef]
18. Yang, F.; Mou, J.; Liu, J.; Ma, C.; Yan, H. Characteristic analysis of the fractional-order hyperchaotic complex system and its image encryption application. *Signal Process.* **2020**, *169*. [CrossRef]
19. Wang, Y.; Yang, F. A fractional-order CNN hyperchaotic system for image encryption algorithm. *Phys. Scr.* **2021**, *96*. [CrossRef]
20. Zhu, C.; Sun, K. Cryptanalyzing and Improving a Novel Color Image Encryption Algorithm Using RT-Enhanced Chaotic Tent Maps. *IEEE Access* **2018**, *6*, 18759–18770. [CrossRef]

21. Luo, Y.; Zhou, R.; Liu, J.; Cao, Y.; Ding, X. A parallel image encryption algorithm based on the piecewise linear chaotic map and hyper-chaotic map. *Nonlinear Dyn.* **2018**, *93*, 1165–1181. [CrossRef]
22. Wu, J.; Liao, X.; Yang, B. Image encryption using 2D Henon-Sine map and DNA approach. *Signal Process.* **2018**, *153*, 11–23. [CrossRef]
23. Fu, C.; Zhang, G.Y.; Zhu, M.; Chen, Z.; Lei, W.M. A New Chaos-Based Color Image Encryption Scheme with an Efficient Substitution Keystream Generation Strategy. *Secur. Commun. Networks* **2018**. [CrossRef]
24. Hua, Z.; Zhou, Y.; Pun, C.M.; Chen, C.L.P. 2D Sine Logistic modulation map for image encryption. *Inf. Sci.* **2015**, *297*, 80–94. [CrossRef]
25. Yu, C.; Li, J.; Li, X.; Ren, X.; Gupta, B.B. Four-image encryption scheme based on quaternion Fresnel transform, chaos and computer generated hologram. *Multimed. Tools Appl.* **2018**, *77*, 4585–4608. [CrossRef]
26. Ghazvini, M.; Mirzadi, M.; Parvar, N. A modified method for image encryption based on chaotic map and genetic algorithm. *Multimed. Tools Appl.* **2020**, *79*, 26927–26950. [CrossRef]
27. Si, Z.H.; Wei, W.; Li, B.S.; Feng, W.J. Analysis of DNA Image Encryption Effect by Logistic-Sine System Combined with Fractional Chaos Stability Theory. *J. Imaging Sci. Technol.* **2020**, *64*. [CrossRef]
28. Faragallah, O.S. Optical double color image encryption scheme in the Fresnel-based Hartley domain using Arnold transform and chaotic logistic adjusted sine phase masks. *Opt. Quantum Electron.* **2018**, *50*. [CrossRef]
29. Hua, Z.; Jin, F.; Xu, B.; Huang, H. 2D Logistic-Sine-coupling map for image encryption. *Signal Process.* **2018**, *149*, 148–161. [CrossRef]
30. Li, H.; Yu, C.; Wang, X. A novel 1D chaotic system for image encryption, authentication and compression in cloud. *Multimed. Tools Appl.* **2021**, *80*, 8721–8758 [CrossRef]
31. Pourjabbar Kari, A.; Habibizad Navin, A.; Bidgoli, A.M.; Mirnia, M. A new image encryption scheme based on hybrid chaotic maps. *Multimed. Tools Appl.* **2021**, *80*, 2753–2772. [CrossRef]
32. Sangavi, V.; Thangavel, P. An exotic multi-dimensional conceptualization for medical image encryption exerting Rossler systemand Sine map. *J. Inf. Secur. Appl.* **2020**, *55*. [CrossRef]
33. Wang, X.; Wang, Q. A fast image encryption algorithm based on only blocks in cipher text. *Chin. Phys. B* **2014**, *23*, 030503. [CrossRef]
34. Zhang, X.; Fan, X.; Wang, J.; Zhao, Z. A chaos-based image encryption scheme using 2D rectangular transform and dependent substitution. *Multimed. Tools Appl.* **2016**, *75*, 1745–1763. [CrossRef]
35. Zhou, N.; Hu, Y.; Gong, L.; Li, G. Quantum image encryption scheme with iterative generalized Arnold transforms and quantum image cycle shift operations. *Quantum Inf. Process.* **2017**, *16*. [CrossRef]
36. Cao, W.; Mao, Y.; Zhou, Y. Designing a 2D infinite collapse map for image encryption. *Signal Process.* **2020**, *171*. [CrossRef]
37. Li, T.; Shi, J.; Zhang, D. Color image encryption based on joint permutation and diffusion. *J. Electron. Imaging* **2021**, *30*, 013008. [CrossRef]
38. Zhan, K.; Wei, D.; Shi, J.; Yu, J. Cross-utilizing hyperchaotic and DNA sequences for image encryption. *J. Electron. Imaging* **2017**, *26*. [CrossRef]
39. Chai, X.; Fu, X.; Gan, Z.; Lu, Y.; Chen, Y. A color image cryptosystem based on dynamic DNA encryption and chaos. *Signal Process.* **2019**, *155*, 44–62. [CrossRef]
40. Zhu, X.; Liu, H.; Liang, Y.; Wu, J. Image encryption based on Kronecker product over finite fields and DNA operation. *Optik* **2020**, *224*. [CrossRef]
41. Ben Farah, M.A.; Guesmi, R.; Kachouri, A.; Samet, M. A novel chaos based optical image encryption using fractional Fourier transform and DNA sequence operation. *Opt. Laser Technol.* **2020**, *121*. [CrossRef]
42. Wu, J.; Shi, J.; Li, T. A novel image encryption approach based on a hyperchaotic system, pixel-level filtering with variable kernels, and DNA-level diffusion. *Entropy* **2020**, *22*, 5. [CrossRef] [PubMed]
43. Yang, Q.; Zhang, K.; Chen, G. A modified generalized Lorenz-type system and its canonical form. *Int. J. Bifurc. Chaos* **2009**, *19*, 1931–1949. [CrossRef]
44. Yang, Q.; Bai, M. A new 5D hyperchaotic system based on modified generalized Lorenz system. *Nonlinear Dyn.* **2017**, *88*, 189–221. [CrossRef]
45. Wolf, A.; Swift, J.B.; Swinney, H.L.; Vastano, J.A. Determining Lyapunov exponents from a time series. *Phys. D Nonlinear Phenom.* **1985**, *16*, 285–317. [CrossRef]
46. Li, Y.; Wang, C.; Chen, H. A hyper-chaos-based image encryption algorithm using pixel-level permutation and bit-level permutation. *Opt. Lasers Eng.* **2017**, *90*, 238–246. [CrossRef]
47. Zhu, C.; Hu, Y.; Sun, K. New image encryption algorithm based on hyperchaotic system and ciphertext diffusion in crisscross pattern. *J. Electron. Inf. Technol.* **2012**, *34*, 1735–1743. [CrossRef]
48. Li, T.; Zhou, M. ECG classification using wavelet packet entropy and random forests. *Entropy* **2016**, *18*, 285. [CrossRef]
49. Li, T.; Hu, Z.; Jia, Y.; Wu, J.; Zhou, Y. Forecasting crude oil prices using ensemble empirical mode decomposition and sparse Bayesian learning. *Energies* **2018**, *11*, 1882. [CrossRef]
50. Song, Y.; Wu, D.; Deng, W.; Gao, X.Z.; Li, T.; Zhang, B.; Li, Y. MPPCEDE: Multi-population parallel co-evolutionary differential evolution for parameter optimization. *Energy Conv. Manag.* **2021**, *228*, 113661. [CrossRef]

51. Deng, W.; Shang, S.; Cai, X.; Zhao, H.; Song, Y.; Xu, J. An improved differential evolution algorithm and its application in optimization problem. *Soft Comput.* **2021**, *25*, 5277–5298. [CrossRef]
52. Li, T.; Qian, Z.; He, T. Short-term Load Forecasting with Improved CEEMDAN and GWO-based Multiple Kernel ELM. *Complexity* **2020**, *2020*, 1209547. [CrossRef]
53. Wang, Z.; Bovik, A.C.; Sheikh, H.R.; Simoncelli, E.P. Image quality assessment: from error visibility to structural similarity. *IEEE Trans. Image Process.* **2004**, *13*, 600–612. [CrossRef]
54. Wu, Y.; Noonan, J.P.; Agaian, S. NPCR and UACI randomness tests for image encryption. *Cyber J. Multidiscip. J. Sci. Technol. J. Sel. Areas Telecommun. (JSAT)* **2011**, *1*, 31–38.

entropy

Article

Automated Spleen Injury Detection Using 3D Active Contours and Machine Learning

Julie Wang [1], Alexander Wood [2], Chao Gao [2], Kayvan Najarian [1,2,3,4,5] and Jonathan Gryak [2,5,*]

1   Department of Electrical Engineering and Computer Science, University of Michigan,
    Ann Arbor, MI 48109, USA; wangjuli@umich.edu (J.W.); kayvan@med.umich.edu (K.N.)
2   Department of Computational Medicine and Bioinformatics, University of Michigan,
    Ann Arbor, MI 48109, USA; alexwood2@gmail.com (A.W.); gchao@umich.edu (C.G.)
3   Department of Emergency Medicine, University of Michigan, Ann Arbor, MI 48109, USA
4   Michigan Center for Integrative Research in Critical Care, University of Michigan, Ann Arbor, MI 48109, USA
5   Michigan Institute for Data Science, University of Michigan, Ann Arbor, MI 48109, USA
*   Correspondence: gryakj@med.umich.edu

Abstract: The spleen is one of the most frequently injured organs in blunt abdominal trauma. Computed tomography (CT) is the imaging modality of choice to assess patients with blunt spleen trauma, which may include lacerations, subcapsular or parenchymal hematomas, active hemorrhage, and vascular injuries. While computer-assisted diagnosis systems exist for other conditions assessed using CT scans, the current method to detect spleen injuries involves the manual review of scans by radiologists, which is a time-consuming and repetitive process. In this study, we propose an automated spleen injury detection method using machine learning. CT scans from patients experiencing traumatic injuries were collected from Michigan Medicine and the Crash Injury Research Engineering Network (CIREN) dataset. Ninety-nine scans of healthy and lacerated spleens were split into disjoint training and test sets, with random forest (RF), naive Bayes, SVM, k-nearest neighbors (k-NN) ensemble, and subspace discriminant ensemble models trained via 5-fold cross validation. Of these models, random forest performed the best, achieving an Area Under the receiver operating characteristic Curve (AUC) of 0.91 and an F1 score of 0.80 on the test set. These results suggest that an automated, quantitative assessment of traumatic spleen injury has the potential to enable faster triage and improve patient outcomes.

Keywords: image segmentation; computer-assisted diagnosis; machine learning; spleen injury detection

Citation: Wang, J.; Wood, A.; Gao, C.; Najarian, K.; Gryak, J. Automated Spleen Injury Detection Using 3D Active Contours and Machine Learning. *Entropy* **2021**, *23*, 382. https://doi.org/10.3390/e23040382

Academic Editor: Amelia Carolina Sparavigna

Received: 15 February 2021
Accepted: 22 March 2021
Published: 24 March 2021

Publisher's Note: MDPI stays neutral with regard to jurisdictional claims in published maps and institutional affiliations.

1. Introduction

Blunt spleen injuries account for up to half of all abdominal solid organ injuries. Common causes include road traffic accidents, falls, physical assaults, and sports-related injuries. Multiphasic contrast-enhanced computed tomography (CT) is the standard non-invasive diagnostic tool for injury evaluation of blunt spleen injuries [1], which include lacerations, subcapsular or parenchymal hematomas, active hemorrhage, and vascular injuries. The type and severity of spleen injuries are commonly described based on the Abbreviated Injury Scale (AIS) or the American Association for Trauma (AAST) Organ Injury Scale (OIS). Currently, detection and classification of spleen injuries rely on the manual review of radiologists. This manual process is not only inefficient but also subject to variability based on the reviewer [1,2].

Many computer-assisted diagnosis (CAD) systems have been developed to detect, locate, and assess potential anomalies or injuries to aid radiologists in the diagnostic process. Detection of pathology in the chest, breast, and colon has been the main focus of previous CAD studies [3]. Other extant CAD systems include those that target the brain, liver, skeletal, and vascular systems [3–5]. Although there have not been previous studies on CAD systems for the spleen, an automated method for localizing and segmenting the spleen [6] was previously developed by the co-authors of this study. This method can be

147

utilized to segment the region of interest in a CT volume, a prerequisite step to performing spleen injury detection.

Machine learning techniques, including Support Vector Machines (SVM), random forest (RF), logistic regression (LR), and deep learning methods, have been widely applied for analysis of medical images [5]. A critical step in application of machine learning to medical image analysis is the extraction and representation of features salient to the classification or detection task at hand. Different types of features and feature extraction methods have been employed based on the anomaly of interest. Common features used include histogram-based [7,8], shape-based [7,9–11], texture-based [12–14], region-based [10,15], and bag-of-words features [15], among others.

In this paper, we propose a supervised classification scheme to discriminate lacerated spleens from healthy controls, a schematic diagram that is presented in Figure 1. Lacerations were chosen for study as they are major types of blunt spleen injury that can be readily observed from contrast-enhanced CT, appearing as linear or branching regions extending from the capsular surface of the spleen and often disrupting the smooth splenic contour [1]. CT scans from patients experiencing traumatic injuries were collected from the Michigan Medicine and the Crash Injury Research Engineering Network (CIREN) dataset [16]. Healthy and lacerated spleens within CT scans from 99 patients were automatically segmented using a previously developed method [6]. From the segmented spleen region, various features were extracted: statistical histogram-based features including Rényi entropy; shape-based feature including fractal dimension [17], whose generalized version is directly related to Rényi entropy [18]; and texture-based features. The performance of five machine learning models: RF, naive Bayes, SVM, k-nearest neighbors (k-NN) ensemble, and subspace discriminant ensemble, were trained using 5-fold cross-validation. On a distinct test set, RF was the best performing classifier, achieving an Area Under the receiver operating characteristic Curve (AUC) of 0.91 and F1 score of 0.80. This study demonstrates the potential for such an automated injury assessment method to reduce physician workload and improve patient outcomes by enabling faster injury triage.

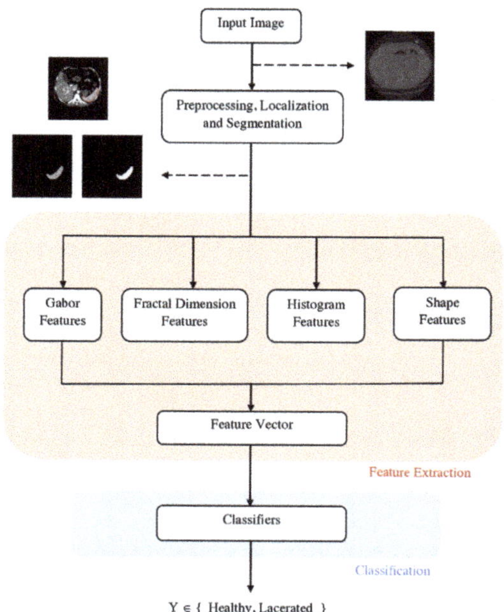

Figure 1. A schematic diagram of the proposed method.

2. Materials and Methods

2.1. Dataset

CT scans used in this study were obtained from Michigan Medicine patients who experienced traumatic abdominopelvic injuries under an IRB-approved retrospective study. Patient consent was waived by the IRB as the research involved no more than minimal risk to the subjects. Additional training data were obtained from the Crash Injury Research Engineering Network (CIREN) dataset [16] containing CT volumes for patients who experienced traumatic injuries in a motor vehicle accident. Each patient CT scan used in this study contained an axial abdominopelvic volume, comprised of between 42 and 122 slices of 5 mm thickness from the heart to the pelvic region. Samples with artifacts around the spleen region were removed.

A total of 99 CT scans, one per patient, were used in this study, consisting of 54 healthy spleen samples and 45 lacerated spleen samples. The lacerated samples are categorized by the Abbreviated Injury Scale (AIS) and the Organ Injury Scale (OIS). Of the 45 lacerated spleen samples, the distribution of injury is as follows: OIS grade I or II (AIS = 2): 15, OIS grade III (AIS = 3): 16, OIS grade IV (AIS = 4): 10, OIS grade V (AIS = 5): 4.

The previously developed spleen segmentation method utilized in this study [6] also made use of the Michigan Medicine and CIREN datasets. In that study, CT scans from 147 patients (one scan per patient) were used to train and test automated spleen segmentation on patients with healthy spleens. The training set was composed of 108 patients, 65 from Michigan Medicine, and 43 from CIREN, with a disjoint test set containing 39 CT scans, 21 from Michigan Medicine, and 18 from CIREN. The patients utilized for training in the prior segmentation study are distinct from those used in this study for training spleen injury detection.

2.2. Spleen Segmentation

Segmentations of the spleen were obtained from each abdominopelvic CT volume using a previously developed fully automated spleen localization and segmentation method [6]. Preprocessing was first applied to the images in order to remove noise through standard and local contrast adjustment, as well as the application of image denoising filters. Localization then utilized machine learning methods to identify a small region within the spleen as a seed mask. Segmentation was then performed via a series of reinitialized active contours using the established seed mask.

Segmentations that resulted in a total segmented spleen volume of less than 80 cm^2 were considered segmentation errors. This occurred in 6 out of 99 cases, and these samples were removed from the dataset. Manual annotations reviewed by an expert radiologist were obtained for 36 healthy samples as well as one lacerated sample. The segmentation method achieved an average Dice score of 0.87, excluding segmentation errors. Sample segmentations of healthy and lacerated spleens are illustrated in Figure 2.

2.3. Feature Extraction

In this study, four types of features—histogram features, fractal dimension features, Gabor features, and shape features—were extracted to train classifiers capable of discriminating injured spleens from healthy controls. Histogram and Gabor features were used to represent and discriminate textures within the spleen segmentation, while fractal and shape analyses were applied to characterize the spleen contour.

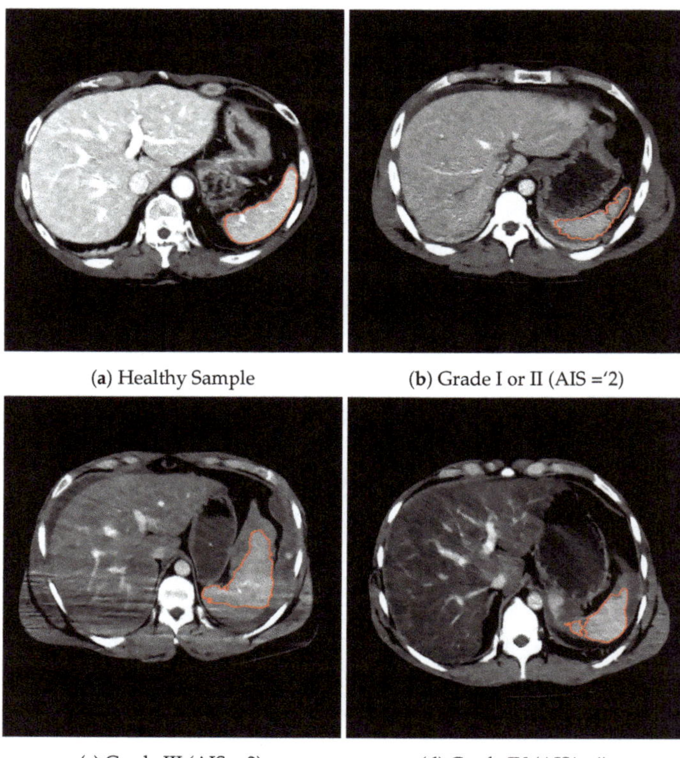

(**a**) Healthy Sample

(**b**) Grade I or II (AIS =′2)

(**c**) Grade III (AIS′= 3)

(**d**) Grade IV (AIS′= 4)

Figure 2. Segmentation of healthy and lacerated spleens.

2.3.1. Histogram Features

The histogram of an image is a plot of the intensity values of a color channel against the number of pixels at that value. The shape of the histogram provides information regarding the contrast and brightness of the image [19]. Five statistical and information-theoretic features of the histogram were extracted from the data for this analysis: mean, variance, skewness, kurtosis, and Rényi entropy. Mean denotes the average intensity level, while variance represents the variation of intensities around the mean. Skewness measures the asymmetry of the data about the mean and kurtosis specifies whether the distributions are flat or peaked relative to a normal distribution. Additionally, entropy measures the disorder in the image based on the distribution of intensity levels.

2.3.2. Fractal Dimension Analysis

Fractals are mathematical sets with high degrees of geometrical complexity capable of modeling irregular, complex shapes [20]. Fractal features have been widely applied in texture and shape analyses of images, including medical images [9,21] to characterize the irregularity of physical structures.

Fractal dimension (D_f) is one of the most important fractal features and provides a quantitative measure of the coarseness of an image. Since lacerated spleens generally display an irregular [2], jagged contour as compared to healthy spleens (see Figure 2), the fractal dimension of binary segmentation images was calculated as a shape-based feature. Both the fractal dimension of the segmentation perimeter as well as the segmentation area were extracted.

In this study, the widely used box counting method [17] was employed to estimate D_f for each binary image of segmentation, after which the fractal dimension D_f was calculated for each frame in the CT volume containing the segmented spleen. Let $N(r)$

denote the number of boxes with fixed side length r necessary to cover the positive pixels of the segmentation. The box-counting method iteratively calculates $N(r)$ for each r of $1, 2, 3, ..., 512$ pixels. D_f is then calculated by fitting $\log N(r)$ to a linear function of $\log r$ by the least squares error method.

2.3.3. Gabor Features

A Gabor filter is a linear filter often used for edge detection. Gabor filter-based features are commonly used to represent and discriminate textures in images and are captured from responses of images convolved with Gabor filters. A two-dimensional Gabor filter is a Gaussian kernel function modulated by a complex sinusoidal plane wave, and can be defined as follows:

$$g(x, y; \lambda, \theta, \psi, \sigma, \gamma) = \exp\left(-\frac{x'^2 + \gamma^2 y'^2}{2\sigma^2}\right) \exp\left(i(2\pi\frac{x'}{\lambda} + \psi)\right)$$
$$x' = x\cos\theta + y\sin\theta \tag{1}$$
$$y' = -x\sin\theta + y\cos\theta$$

In Equation (1), λ is the wavelength of the sinusoidal factor, θ is the orientation of the normal to the parallel stripes of a Gabor function, ψ is the phase offset, σ is the standard deviation of the Gaussian envelope, and γ is the spatial aspect ratio [22,23].

In this study, a filter bank of 40 Gabor filters in 5 scales and 8 orientations was employed. From the response matrices, two types of Gabor features were extracted: local energy and mean amplitude. Local energy is calculated by the sum of the squared values in each response matrix. Mean amplitude captures the response amplitude for each response matrix by taking the sum of absolute values in each matrix.

2.3.4. Shape Features

Values of circularity, eccentricity, orientation, and the difference between the segmented area and its convex area were extracted to characterize the shape of the segmented spleen. Circularity, calculated as

$$(4 * Area * \pi) / (Perimeter^2), \tag{2}$$

captures the roundness of objects; a perfect circle would have a circularity of 1. Eccentricity is the ratio of the distance between the foci of an ellipse and its major axis length. Orientation was calculated as the angle between the x-axis and the major axis of an ellipse. In addition, finally, the convex area is the area of the convex hull of the region, defined as the smallest convex set that contains the original region. The difference between this area and the original segmented area was also extracted.

2.4. Classification

2.4.1. Training

Ninety-three classification samples were randomly separated into training and test sets, respectively, comprising 80% and 20% of the samples, with the relative number of injury and healthy samples being balanced. As only one CT scan per patient was utilized in this study, patients and their respective scans were exclusively assigned to either the training or test set.

5-fold cross validation was employed during the training phase to select models with low variance and low bias. The training set was divided into 5 folds of roughly equal size. The classifier was then trained on 4 folds and tested on the remaining fold. Validation accuracy, AUC, and the associated standard deviations were used to select models.

2.4.2. Model Selection

Five models—RF, naive Bayes, SVM, k-NN ensemble, and subspace discriminant ensemble—were selected based on validation performance during the training phase as reported in Table 1. Of the five models, RF performed the best on the training set with an AUC of 0.91.

RF, naive Bayes, and SVM are all popular supervised learning models used for analysis on medical images [5]. Naive Bayes is a probabilistic classifier applying Bayes theorem with an assumption of feature pairwise independence given class values. Ensemble learning combines several classifiers to improve prediction performance. RF is an ensemble learner that leverages multiple decision trees to produce a more accurate and stable prediction. Subspace discriminant ensemble [24] employs the linear discriminant analysis (LDA) scheme for a specific discriminant subspace of low dimension. The k-NN ensemble employed in this study uses the Random Space method with k-NN learners.

Deep learning, and more specifically the application of convoluted neural networks (CNN) to image analysis, has achieved great success in recent years [25]. To assess the validity of the hand-crafted features proposed in this study, an end-to-end deep learning method was evaluated along with the traditional machine learning models. A pre-trained CNN, ResNet-50 [26], was used for feature extraction on the segmented CT volumes, with subsequent classification performed by a Long Short-term Memory (LSTM) artificial recurrent neural network (RNN). This combination of CNN for slice-wise feature extraction and LSTM for spatial information extraction across the CT volume has been successful in previous injury detection studies, including classification of intracranial hemorrhage [27,28], lung cancer [29], as well as liver and brain tumors [30]. The goal of this approach is to leverage 2D models pre-trained on the ImageNet dataset [31] while still accounting for spatial information between slices in the 3D volume. ResNet-50 was selected for feature extraction because of its relatively higher accuracy and lower number of parameters (23 M) compared to other architectures commonly used for medical image analysis, such as AlexNet (62 M parameters) and VGGNet (138 M parameters).

Feature extraction was performed by ResNet-50 on each slice of the segmented CT, which were cropped to reduce blank space surrounding the region of interest. An LSTM model was then employed to perform classification on the extracted features across each patient's CT sequence.

Table 1. Mean and standard deviation (SD) of performance metrics for spleen injury classification from 5-fold cross validation on the training set. The highest value for each performance metric is **bolded** while the lowest SD is *italicized*.

Metric	RF	Naive Bayes	SVM	k-NN	Subspace Discriminant
Accuracy	**0.83** *(0.10)*	0.71 (0.11)	0.73 (0.10)	0.73 (0.10)	0.67 (0.10)
Sensitivity	**0.77** *(0.16)*	0.66 (0.17)	0.61 (0.17)	0.56 (0.18)	0.44 (0.19)
Specificity	0.89 (0.12)	0.75 (0.15)	0.84 (0.13)	**0.89** *(0.11)*	0.87 (0.13)
F1	**0.81** *(0.12)*	0.68 (0.13)	0.67 (0.14)	0.65 (0.16)	0.54 (0.18)
AUC	**0.91** *(0.08)*	0.75 (0.12)	0.81 (0.10)	0.84 (0.10)	0.77 (0.13)

3. Results

3.1. Classifier Performance

The trained classifiers were evaluated on the test set, with the resulting accuracy, sensitivity, specificity, F1, and AUC reported in Table 2. The RF model achieved the best classification performance with an AUC of 0.91 and an F1 of 0.80. Overall, testing loss and accuracy are consistent with 5-fold cross validation results on the training set, demonstrating the generalizability of the proposed method on unseen data. No over-fitting seemed to occur on any of the models reported.

Table 2. Performance metrics for spleen injury classification on the test set. The highest value for each performance metric is **bolded**.

Metric	RF	Naive Bayes	SVM	*k*-NN	Subspace Discriminant
Accuracy	**0.83**	0.70	0.71	0.75	0.64
Sensitivity	**0.76**	0.63	0.56	0.59	0.40
Specificity	**0.89**	0.76	0.85	0.88	0.85
F1	**0.80**	0.66	0.64	0.68	0.50
AUC	**0.91**	0.74	0.80	0.84	0.76

3.2. Comparison against Deep Learning

A comparison between the RF classifier performance and the deep learning method is shown in Table 3. The RF classifier trained with hand-crafted features demonstrated better performance than the deep learning method, with RF achieving an AUC of 0.91 while the deep learning method achieved an AUC of 0.72. The lower deep learning performance is likely due to the small sample size available in this study, as deep learning methods require large datasets to minimize over-fitting and achieve good performance [25,32]. These results demonstrate that hand-crafted features using domain knowledge can overcome sample size limitations.

Table 3. Performance metrics for the RF classifier trained using hand-crafted features and for the deep learning method. The highest value for each performance metric is **bolded**.

Metric	RF (Hand-Crafted)	ResNet + LSTM (Deep Learning)
Accuracy	**0.83**	0.79
Sensitivity	**0.76**	0.67
Specificity	0.89	**0.90**
F1	**0.80**	0.75
AUC	**0.91**	0.72

3.3. Leave-One-Site-Out Analysis

This study utilizes two different datasets—the internal Michigan Medicine dataset and the public CIREN dataset. A leave-one-site-out analysis was performed to evaluate the cross-site generalizability of the proposed method.

To achieve an 80% to 20% training/test split, the Michigan Medicine dataset, containing a total of 54 samples, was used as the training set while 14 CIREN samples were used as the test set. The 14 CIREN test samples were randomly stratified based on injury grade. The best performing classifier from Section 2.4.2, RF, was trained on the Michigan Medicine samples and tested on the CIREN samples. Performance metrics of the classifier are reported in Table 4.

Table 4. Performance metrics for the RF classifier trained on Michigan Medicine samples and tested on CIREN samples.

Metric	RF
Accuracy	0.75
Sensitivity	0.59
Specificity	0.94
F1	0.71
AUC	0.91

The RF classifier achieved good performance on the cross-site generalizability assessment, with an AUC of 0.91 and a F1 of 0.71. Compared to the performance on the mixed-site test set, the classifier achieved the same AUC but lower F1, accuracy, and sensitivity. This performance difference is likely affected by the limited sample size used to train the classifier as only one dataset is utilized. Overall, the performance of the classifier demonstrates that the proposed method is relatively robust against variability stemming from differences in the data from two different sites.

4. Discussion

RF outperformed other classifiers on both the training and test set, which is consistent with its popularity among many previous medical image analysis studies [33–36]. Several features of RF may contribute to its higher performance on medical images—RFs are suited for high predictor dimension relative to sample size, they inherently perform feature selection, and they generalize well to regions of the feature space with sparse data [34,35].

Table 5 reports the classification accuracy of RF by injury grade on both the training and test sets. Although the RF classifier correctly classified the majority of samples across all injury grades, most incorrect classifications occurred within mildly or moderately injured samples (AIS = 2, 3). High classification accuracy is seen among healthy samples and more severe samples (AIS = 4, 5). Lower accuracy and higher variance were achieved for all injury grades as compared to healthy samples, likely due to the smaller number of samples within each individual injury grade compared to the healthy dataset. Despite the lower performance on less severe cases, the proposed method performs well on severe cases, demonstrating the potential to increase injury triage efficiency in real-world applications.

Table 5. RF classification accuracy by injury grades. The mean accuracy and standard deviation (SD) across 5-fold cross validation on the training set, as well as the mean accuracy on the test set are reported.

Injury Grade	Training Accuracy	Testing Accuracy
Healthy	0.89 (0.12)	0.89
AIS = 2	0.72 (0.30)	0.70
AIS = 3	0.74 (0.27)	0.78
AIS = 4, 5	0.88 (0.22)	0.79

Common misclassifications included classification of a mildly or moderately injured sample as healthy and classification of a healthy sample as lacerated, as illustrated in Figure 3. A lacerated sample with lower injury severity misclassified as healthy is likely caused by a relatively smooth segmentation contour (Figure 3c), which may be the result of imperfect segmentation of the lacerated region and/or a lower degree of laceration. Healthy samples misclassified as lacerated were often due to noise in the original image, which produces misleading segmentations or irregular contour shapes (Figure 3d,e).

Existence of a small portion of samples with localization errors likely led to lower model performance due to imperfect or erroneous segmentations. Image resolution and noise are likely contributing factors to imperfect localization and segmentation results. Previous studies have shown that image thickness is inversely related to image noise but directly related to image resolution [37]. In order, 5 mm CT slices were utilized in this study, which has worked relatively well due to its lower image noise compared to thinner slices. However, 5 mm slices have a lower resolution, decreasing diagnostic content and the proposed method's ability to detect small lesions. Although not available in the datasets utilized in this study, 3 mm slices may strike an ideal balance between minimizing image noise and maximizing image resolution and can be explored in the future [37].

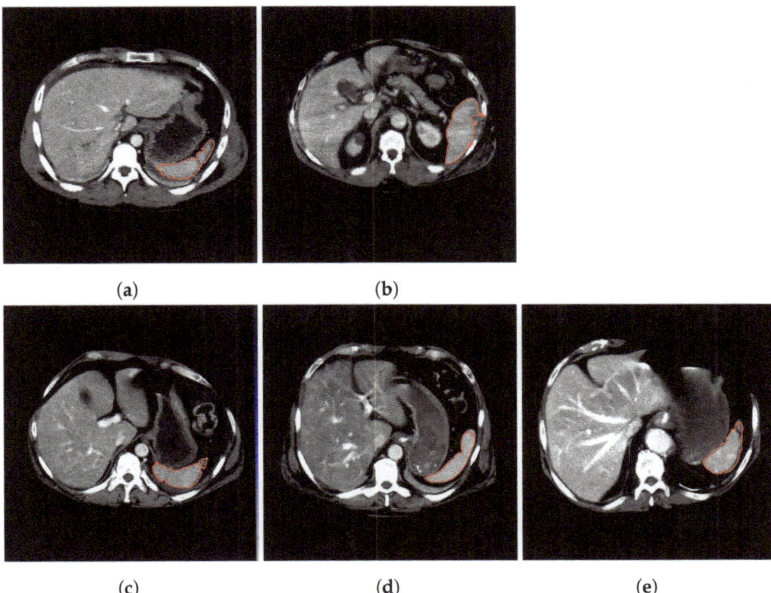

Figure 3. Classification results. (**a**,**b**) lacerated (AIS = 2) samples correctly classified as lacerated; (**c**) lacerated (AIS = 2) sample incorrectly classified as healthy; (**d**,**e**) healthy samples incorrectly classified as lacerated.

Future work will focus on refinement of the segmentation method to improve classification accuracy in lower severity cases. Additional pre- and post-processing of in the segmentation method can be introduced to reduce noise and increase discrimination between healthy and mildly lacerated spleen. Incorporation of more samples in each injury grade may increase classifier performance and support extension of the current binary classification to multi-class classification on different injury grades, providing additional clinical use cases. Finally, although this study focuses on spleen lacerations, future work should generalize to other blunt spleen injuries, including hematomas and hemorrhages.

5. Conclusions

In this study, an automated method for detecting spleen lacerations in CT scans was proposed. The classification scheme was built upon a previously developed localization and segmentation process [6], and used histogram, Gabor filters, fractal dimension, and shape features to distinguish lacerated spleens from healthy controls. Classifiers examined were RF, naive Bayes, SVM, k-NN ensemble, subspace discriminant ensemble, and a CNN-based architecture. The RF method outperformed other models in discriminating between lacerated and healthy spleens, achieving an AUC of 0.91 and an F1 of 0.80. Additionally, a leave-one-site-out analysis was performed that demonstrated the method's robustness against variability stemming from differences in the data from two different sites. Results from this study demonstrate the potential for automated, quantitative assessment of traumatic spleen injury to increase triage efficiency and improve patient outcomes. Future work will focus on improving classifier accuracy in less severe cases, extension of the method to support multi-class classification based on injury grade, and generalization to other types of blunt spleen injuries.

Author Contributions: Conceptualization, K.N. and J.G.; Methodology, J.W., A.W., C.G., K.N., and J.G.; Validation, J.W. and J.G.; Formal Analysis, J.W.; Data Curation, J.W., A.W., and J.G.; Writing—Original Draft Preparation, J.W.; Writing—Review and Editing, J.W. and J.G.; Supervision, J.G. All authors have read and agreed to the published version of the manuscript.

Funding: This research received no external funding.

Institutional Review Board Statement: The study was conducted according to the guidelines of the Declaration of Helsinki, and approved by the Institutional Review Board of University of Michigan (protocol code HUM00098656, approved 13 December 2020).

Informed Consent Statement: Patient consent was waived as the research involved no more than minimal risk to the subjects.

Data Availability Statement: Two datasets were employed in this study—the Crash Injury Research Engineering Network (CIREN) dataset and an internal dataset collected from Michigan Medicine. CIREN is a public dataset that is available for download at https://www.nhtsa.gov/research-data/crash-injury-research (accessed on 1 February 2021). Data collected from Michigan Medicine can be made available to external entities under a data use agreement with the University of Michigan.

Conflicts of Interest: The authors declare no conflict of interest.

References

1. Shi, H.; Teoh, W.; Chin, F.; Tirukonda, P.; Cheong, S.; Yiin, R. CT of blunt splenic injuries: What the trauma team wants to know from the radiologist. *Clin. Radiol.* **2019**, *74*, 903–911. [CrossRef]
2. Hassan, R.; Aziz, A.A.; Ralib, A.R.M.; Saat, A. Computed tomography of blunt spleen injury: A pictorial review. *Malays. J. Med. Sci. MJMS* **2011**, *18*, 60.
3. Zhang, Z.; Sejdić, E. Radiological images and machine learning: Trends, perspectives, and prospects. *Comput. Biol. Med.* **2019**, *108*, 354–370. [CrossRef]
4. Doi, K. Computer-aided diagnosis in medical imaging: Historical review, current status and future potential. *Comput. Med Imaging Graph.* **2007**, *31*, 198–211. [CrossRef]
5. Syeda-Mahmood, T. Role of big data and machine learning in diagnostic decision support in radiology. *J. Am. Coll. Radiol.* **2018**, *15*, 569–576. [CrossRef] [PubMed]
6. Wood, A.; Soroushmehr, S.R.; Farzaneh, N.; Fessell, D.; Ward, K.R.; Gryak, J.; Kahrobaei, D.; Najarian, K. Fully Automated Spleen Localization In addition, Segmentation Using Machine Learning In addition, 3D Active Contours. In Proceedings of the 2018 40th Annual International Conference of the IEEE Engineering in Medicine and Biology Society (EMBC), Honolulu, HI, USA, 18–21 July 2018; pp. 53–56.
7. Shi, X.; Cheng, H.D.; Hu, L.; Ju, W.; Tian, J. Detection and classification of masses in breast ultrasound images. *Digit. Signal Process.* **2010**, *20*, 824–836. [CrossRef]
8. Dhanalakshmi, K.; Rajamani, V. An intelligent mining system for diagnosing medical images using combined texture-histogram features. *Int. J. Imaging Syst. Technol.* **2013**, *23*, 194–203. [CrossRef]
9. Lee, W.L.; Chen, Y.C.; Hsieh, K.S. Ultrasonic liver tissues classification by fractal feature vector based on M-band wavelet transform. *IEEE Trans. Med. Imaging* **2003**, *22*, 382–392. [CrossRef]
10. Xu, Y.; Lin, L.; Hu, H.; Yu, H.; Jin, C.; Wang, J.; Han, X.; Chen, Y.W. Combined density, texture and shape features of multi-phase contrast-enhanced CT images for CBIR of focal liver lesions: A preliminary study. In *Innovation in Medicine and Healthcare 2015*; Springer: Cham, Switzerland, 2016; pp. 215–224.
11. Dhara, A.K.; Mukhopadhyay, S.; Dutta, A.; Garg, M.; Khandelwal, N. A combination of shape and texture features for classification of pulmonary nodules in lung CT images. *J. Digit. Imaging* **2016**, *29*, 466–475. [CrossRef]
12. Zhu, X.; He, X.; Wang, P.; He, Q.; Gao, D.; Cheng, J.; Wu, B. A method of localization and segmentation of intervertebral discs in spine MRI based on Gabor filter bank. *Biomed. Eng. Online* **2016**, *15*, 32. [CrossRef]
13. Wu, C.C.; Lee, W.L.; Chen, Y.C.; Lai, C.H.; Hsieh, K.S. Ultrasonic liver tissue characterization by feature fusion. *Expert Syst. Appl.* **2012**, *39*, 9389–9397. [CrossRef]
14. Lee, W.L. An ensemble-based data fusion approach for characterizing ultrasonic liver tissue. *Appl. Soft Comput.* **2013**, *13*, 3683–3692. [CrossRef]
15. Alkhawlani, M.; Elmogy, M.; Elbakry, H. Content-based image retrieval using local features descriptors and bag-of-visual words. *Int. J. Adv. Comput. Sci. Appl.* **2015**, *6*, 212–219. [CrossRef]
16. U.S. Department of Transportation, National Highway Traffic Safety Administration (NHTSA). Crash Injury Research Engineering Network. 2017. Available online: https://www.nhtsa.gov/research-data/crash-injury-research (accessed on 1 February 2021).
17. Keller, J.M.; Crownover, R.M.; Chen, R.Y. Characteristics of natural scenes related to the fractal dimension. *IEEE Trans. Pattern Anal. Mach. Intell.* **1987**, *9*, 621–627. [CrossRef] [PubMed]
18. Zmeskal, O.; Dzik, P.; Vesely, M. Entropy of fractal systems. *Comput. Math. Appl.* **2013**, *66*, 135–146. [CrossRef]
19. Sergyan, S. Color histogram features based image classification in content-based image retrieval systems. In Proceedings of the 2008 6th International Symposium on Applied Machine Intelligence and Informatics, Herlany, Slovakia, 21–22 January 2008; pp. 221–224.
20. Mandelbrot, B.B. *The Fractal Geometry of Nature*; WH freeman: New York, NY, USA, 1983; Volume 173.
21. Chen, C.C.; DaPonte, J.S.; Fox, M.D. Fractal feature analysis and classification in medical imaging. *IEEE Trans. Med. Imaging* **1989**, *8*, 133–142. [CrossRef] [PubMed]

22. Zheng, D.; Zhao, Y.; Wang, J. Features extraction using a Gabor filter family. In Proceedings of the Sixth IASTED International Conference, Signal and Image Processing, Honolulu, HI, USA, 23–25 August 2004.

23. Haghighat, M.; Zonouz, S.; Abdel-Mottaleb, M. CloudID: Trustworthy cloud-based and cross-enterprise biometric identification. *Expert Syst. Appl.* **2015**, *42*, 7905–7916. [CrossRef]

24. Ashour, A.S.; Guo, Y.; Hawas, A.R.; Xu, G. Ensemble of subspace discriminant classifiers for schistosomal liver fibrosis staging in mice microscopic images. *Health Inf. Sci. Syst.* **2018**, *6*, 21. [CrossRef]

25. Lee, J.G.; Jun, S.; Cho, Y.W.; Lee, H.; Kim, G.B.; Seo, J.B.; Kim, N. Deep learning in medical imaging: General overview. *Korean J. Radiol.* **2017**, *18*, 570. [CrossRef] [PubMed]

26. He, K.; Zhang, X.; Ren, S.; Sun, J. Deep Residual Learning for Image Recognition. In Proceedings of the 2016 IEEE Conference on Computer Vision and Pattern Recognition (CVPR), Las Vegas, NV, USA, 27–30 June 2016; pp. 770–778. [CrossRef]

27. Burduja, M.; Ionescu, R.T.; Verga, N. Accurate and Efficient Intracranial Hemorrhage Detection and Subtype Classification in 3D CT Scans with Convolutional and Long Short-Term Memory Neural Networks. *Sensors* **2020**, *20*, 5611. [CrossRef]

28. Nguyen, N.T.; Tran, D.Q.; Nguyen, N.T.; Nguyen, H.Q. A CNN-LSTM Architecture for Detection of Intracranial Hemorrhage on CT scans. *arXiv* **2020**, arXiv:2005.10992.

29. Marentakis, P.; Karaiskos, P.; Kouloulias, V.; Kelekis, N.; Argentos, S.; Oikonomopoulos, N.; Loukas, C. Lung cancer histology classification from CT images based on radiomics and deep learning models. *Med. Biol. Eng. Comput.* **2021**, *59*, 215–226. [CrossRef]

30. Kutlu, H.; Avcı, E. A novel method for classifying liver and brain tumors using convolutional neural networks, discrete wavelet transform and long short-term memory networks. *Sensors* **2019**, *19*, 1992. [CrossRef] [PubMed]

31. Russakovsky, O.; Deng, J.; Su, H.; Krause, J.; Satheesh, S.; Ma, S.; Huang, Z.; Karpathy, A.; Khosla, A.; Bernstein, M.; et al. Imagenet large scale visual recognition challenge. *Int. J. Comput. Vis.* **2015**, *115*, 211–252. [CrossRef]

32. Luo, C.; Li, X.; Wang, L.; He, J.; Li, D.; Zhou, J. How Does the Data set Affect CNN-based Image Classification Performance? In Proceedings of the 2018 5th International Conference on Systems and Informatics (ICSAI), Nanjing, China, 10–12 November 2018; pp. 361–366. [CrossRef]

33. Tang, T.T.; Zawaski, J.A.; Francis, K.N.; Qutub, A.A.; Gaber, M.W. Image-based classification of tumor type and growth rate using machine learning: A preclinical study. *Sci. Rep.* **2019**, *9*, 1–10. [CrossRef] [PubMed]

34. Nedjar, I.; EL HABIB DAHO, M.; Settouti, N.; Mahmoudi, S.; Chikh, M.A. Random forest based classification of medical x-ray images using a genetic algorithm for feature selection. *J. Mech. Med. Biol.* **2015**, *15*, 1540025. [CrossRef]

35. Geremia, E.; Clatz, O.; Menze, B.H.; Konukoglu, E.; Criminisi, A.; Ayache, N. Spatial decision forests for MS lesion segmentation in multi-channel magnetic resonance images. *NeuroImage* **2011**, *57*, 378–390. [CrossRef]

36. Lebedev, A.; Westman, E.; Van Westen, G.; Kramberger, M.; Lundervold, A.; Aarsland, D.; Soininen, H.; Kłoszewska, I.; Mecocci, P.; Tsolaki, M.; et al. Random Forest ensembles for detection and prediction of Alzheimer's disease with a good between-cohort robustness. *Neuroimage Clin.* **2014**, *6*, 115–125. [CrossRef]

37. Alshipli, M.; Kabir, N.A. Effect of slice thickness on image noise and diagnostic content of single-source-dual energy computed tomography. *J. Phys. Conf. Ser. IOP Publ.* **2017**, *851*, 012003. [CrossRef]

Article

Hyper-Chaotic Color Image Encryption Based on Transformed Zigzag Diffusion and RNA Operation

Duzhong Zhang, Lexing Chen and Taiyong Li *

School of Economic Information Engineering, Southwestern University of Finance and Economics, Chengdu 611130, China; zhangduzhong@swufe.edu.cn (D.Z.); clx220081203001@smail.swufe.edu.cn (L.C.)
* Correspondence: litaiyong@gmail.com

Abstract: With increasing utilization of digital multimedia and the Internet, protection on this digital information from cracks has become a hot topic in the communication field. As a path for protecting digital visual information, image encryption plays a crucial role in modern society. In this paper, a novel six-dimensional (6D) hyper-chaotic encryption scheme with three-dimensional (3D) transformed Zigzag diffusion and RNA operation (HCZRNA) is proposed for color images. For this HCZRNA scheme, four phases are included. First, three pseudo-random matrices are generated from the 6D hyper-chaotic system. Second, plaintext color image would be permuted by using the first pseudo-random matrix to convert to an initial cipher image. Third, the initial cipher image is placed on cube for 3D transformed Zigzag diffusion using the second pseudo-random matrix. Finally, the diffused image is converted to RNA codons array and updated through RNA codons tables, which are generated by codons and the third pseudo-random matrix. After four phases, a cipher image is obtained, and the experimental results show that HCZRNA has high resistance against well-known attacks and it is superior to other schemes.

Keywords: hyper-chaotic; ribonucleic acid; color image encryption; transformed Zigzag

Citation: Zhang, D.; Chen, L.; Li, T. Hyper-Chaotic Color Image Encryption Based on Transformed Zigzag Diffusion and RNA Operation. *Entropy* **2021**, *23*, 361. https://doi.org/10.3390/e23030361

Academic Editors: Amelia Carolina Sparavigna and Salim Lahmiri

Received: 27 January 2021
Accepted: 15 March 2021
Published: 17 March 2021

Publisher's Note: MDPI stays neutral with regard to jurisdictional claims in published maps and institutional affiliations.

1. Introduction

Nowadays, rapid developments of Internet and digital technologies have led to tremendous digital multimedia contents transmitting over Internet networks. Thus, protection on the contents of digital data has attracted serious concern from medical, military, and many other areas. Various image encryption methods have emerged by using cryptographic techniques [1–4]. Although there exists a view that AES is not suitable for image encryption, Zhang recently refuted it by using AES of cipher block chaining mode to encrypt images [5].

The chaos-based encryption method has become one of the most ideal methods, since it has a lot of appropriate characteristics, e.g. high sensitivity on initial conditions, mixing property, ergodicity, complex behavior, etc. [6–8]. As a result, a lot of researchers have presented plenty of image encryption schemes with a chaotic system [9–13]. In [14], Askar et al. proposed a chaotic economic map based image encryption method, whose simulation results indicated that the proposed algorithm could successfully encrypt and decrypt the images, and it had a good performance on security tests, except noise attacks analysis. By using a single round based hyper-chaotic system, Shaikh et al. presented a color image encryption method with bi-directional pixel diffusion [15]. Additionally, Li et al. presented a "transforming-scrambling-diffusion" model based color image encryption method with a four-dimensional (4D) hyper-chaotic system, which could convert pixel values to gray format before scrambling [16]. There is no doubt that some of the encryption methods in these chaos-based schemes still have weaknesses to some extent. However, different chaotic systems are neither superior nor inferior each other. A high-dimensional chaotic system has complex chaotic behaviors with high time cost, while a low-dimensional

159

chaotic system is opposite [17–19]. Hence, in this paper, a 6D hyper-chaotic system is employed as a pseudo-random numbers sequence generator for more complexity.

Zigzag is a common scrambling operation in image encryption [20,21]. In [22], Li et al. presented a 3D logistic map based color image encryption method with Zigzag scramble; the experiments showed that this method had brute-force attack and statistical attack resistance, but differential attacks analysis was missing. While, Wang et al. proposed a color encryption method with a Zigzag transformation, which could change the start pixel from upper left corner to the other three corners in an image [20]. Next year, Wang et al. [23] presented another image encryption method, which introduced an extended Zigzag confusion for a non-square image. Additionally, in [24], Zhao et al. proposed a novel color image encryption by combining Zigzag map and Hénon map together for permutation. However, these image encryption schemes implement Zigzag scramble on 2D images, which leads to some adjacent values in special positions of the image not being able to be scrambled, and different channels of a color image could not be scrambled, either. On the other hand, some image encryptions transformed 2D image to 3D cube [25], which gives out a new encryption inspiration on permutation, but most of them were focused on rotation, but not Zigzag. Therefore, Zigzag is utilized in diffusion on a 3D cube instead of scramble on 2D image to eliminate these drawbacks in this paper.

Deoxyribonucleic acid (DNA), a biological concept, has recently become a popular trend in the image encryption field [26,27]. By using DNA-based techniques, cipher images could obtain competitive entropy, correlation coefficients etc. [4,28–31]. In [29], Chai et al. presented a new diffusion mechanism that is based on the random numbers that are generated by plaintext image, and incorporated DNA encryption with four-wing hyper-chaotic system. Reference [32] proposed an image encryption method using a spatial map based DNA sequence matrix. In general, the DNA-based encryption mechanism includes two steps: use DNA operation rules to convert pixels of plaintext image to DNA codon matrix and change chaotic sequence to DNA keys to generate cipher image with DNA codon matrix.

While unlike the two strands structure of DNA sequences, Ribonucleic acid (RNA) is a single strand structure. RNA could form double helixes with complementary base pairing. By using this feature, some new image encryption methods have been proposed. In [33], Mahmud et al. presented an image encryption method by combining RNA with Genetic Algorithm (GA) through using a logistic map. In [34], Abbasi et al. employed Chen's chaotic system to encrypt an image with imperialist competition algorithm and RNA operations. Yadollahi et al. utilized the concepts of DNA and RNA to construct a two-phase image encryption method [35]. While an image encryption method is presented by Wang et al. through using an one-dimensional (1D) chaotic system combined from Logistic and Sine map, extended Zigzag confusion, and RNA operation [23]. However, all of these four schemes focus on gray image encryption. Although there is a color image experiment in [23], it is realized by running the scheme three times in three channels.

Being motivated by above discussions, a novel color image encryption method, called HCZRNA, is proposed in this paper. At the beginning, a 6D hyper-chaotic system is employed to generate three pseudo-random matrices. Subsequently, one of the pseudo-random matrices is used to permute plaintext color image. Additionally, 3D transformed Zigzag diffusion is implemented on initial cipher image with the second pseudo-random matrix. After diffusion, an RNA operation is used to convert the diffused image to RNA codons array, and update this array through RNA codons tables that are generated by the third pseudo-random matrix. Finally, a cipher image is obtained.

The main contributions of this work is listed as follows:

- A novel 6D hyper-chaotic system is employed in this paper to produce chaotic matrix for permutation, diffusion, and RNA operation.
- A new 3D transformed Zigzag diffusion scheme is proposed to encrypt color images.
- RNA operation is modified specifically for color images.

- Extensive experiments and analyses demonstrate that the proposed HCZRNA could resist various types of attacks.

The rest of this paper is structured as follows: Section 2 introduces the used 6D hyper-chaotic system, 3D Zigzag and RNA. Section 3 presents the HCZRNA scheme and explains how initial values and pseudo-random matrix are generated in detail. Section 4 reports and analyzes the experimental results. Finally, Section 5 concludes this paper.

2. Preliminaries

2.1. The 6D Hyper-Chaotic System

There are a lot of classical chaotic systems, e.g. Sine map, Logistic map, Tent map, etc., which have simple mathematical forms and can be implemented easily. However, they suffer from small key spaces, predictable orbits, limited ranges, etc. Existing research has shown that higher dimensional chaotic systems are much securer for image encryption [36]. Therefore, a novel 6D hyper-chaotic system is employed in this paper for chaotic sequences generation, which could be described as Equation (1) [37].

$$\begin{aligned}
\dot{x}_1 &= g(\omega + \beta x_6^2)x_2 - ax_1 \\
\dot{x}_2 &= cx_1 + dx_2 - x_1x_3 + x_5 \\
\dot{x}_3 &= -bx_3 + x_1^2 \\
\dot{x}_4 &= ex_2 + fx_4 \\
\dot{x}_5 &= -rx_1 \\
\dot{x}_6 &= x_2
\end{aligned} \tag{1}$$

where $a, b, c, d, e, f, g, r, \omega$, and β are controlling parameters, and $x_i(i = 1, 2, \cdots, 6)$ are state variables.

The fourth-order Runge–Kutta method is used to solve this hyper-chaotic system with step size $h = 0.001$. We set the controlling parameters as $(a, b, c, d, e, f, g, r, \omega, \beta) = (0.3, 1.5, 8.5, -2, 1, -0.1, 0.9, 1, 1, 0.2)$ and initial state variables as $(x_1, x_2, x_3, x_4, x_5, x_6) = (0.1, 0.6, 0.2, 0.02, 1, 0.5)$; Figure 1 shows this 6D hyper-chaotic system's attractors. Its Lyapunov exponents are $\lambda_1 = 7.340$, $\lambda_2 = 0.087$, $\lambda_3 = 0.006$, $\lambda_4 = -0.368$, $\lambda_5 = -1.349$, $\lambda_6 = -67.426$. Since this chaotic system has three positive Lyapunov exponents, its prediction time should be longer than other chaotic systems and it is hard to crack. Besides, this hyper-chaotic system exhibits limit cycles, quasiperiodic, and bursting behavior. Accordingly, it could generate effective a pseudo random sequence. More detailed demonstration could be found in reference [37].

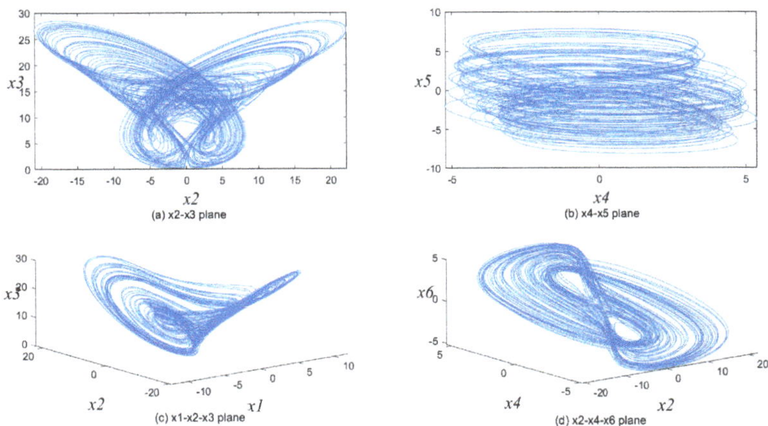

Figure 1. The attractors of six-dimensional (6D) hyper-chaotic system.

2.2. 3D Transformed Zigzag Diffusion

Zigzag is a scanning method that is used to scramble pixels in image encryption. By scanning and taking a pixel with left upper corner of image, then taking other pixels one-by-one through Zigzag path, the image could be converted to a matrix in a fixed way. Hence, the image's pixels could be scrambled.

Traditional Zigzag scrambling could only walk through numbers in $N \times N$ matrix with a fixed Zigzag path; an example of 4×4 matrix is shown in Figure 2. By this fixed path, different channels of color image could not be scrambled with each other. Due to its drawbacks, this paper proposes a novel 3D Zigzag transformation. Using this transformation, each channel of color image would be cut into two triangles through a diagonal line, and be placed on opposite surfaces of a cube, which is illustrated by an example of $4 \times 4 \times 3$ matrix, as shown in Figure 3. Subsequently, diffusion would start from origin vertex of the frontal side of the cube, and walk through every pixels on six surfaces at the front and back side synchronously with spiral Zigzag path, as in Figure 4. For all triangles that are placed on the cube, the order of diffusion is shown in Figure 5. In this way, different channels of color image could be diffused together.

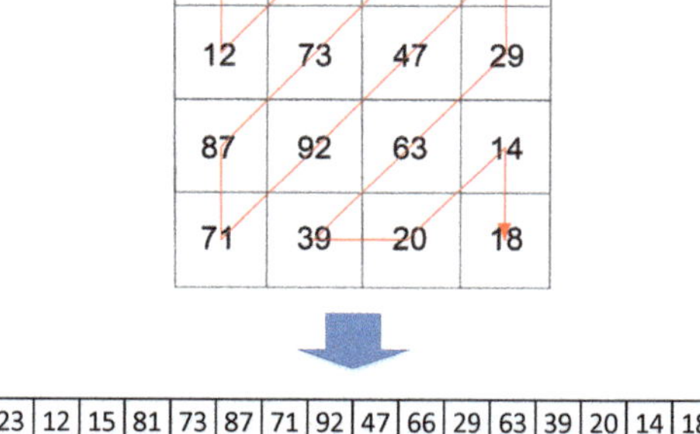

Figure 2. 4×4 traditional Zigzag scramble.

2.3. RNA Operation

RNA is one of the major macromolecules necessary for living organism. RNA has a single strand structure with four nitrogen bases: adnine (A), cytosine (C), guanine (G), and uracil (U). For these four units of RNA, a binary system could be employed for representation, which is shown in Table 1. According to the base pairing rules, four bases of RNA could be coded and constructed into three nucleotides that correspond to one amino acid called codon. Accordingly, there are 64 codons truth table of bases combinations, as shown in Table 2. Assuming that pixels in the image could transfer into six-bits format, a corresponding RNA codon could be found in Table 2.

Table 1. Binary representation of RNA.

RNA Bases	A	C	G	U
Binary	00	01	10	11

Table 2. RNA codon table.

#	Bin.	Codon	#	Bin.	Codon	#	Bin.	Codon	#	Bin.	Codon
0	000000	AAA	16	010000	CAA	32	100000	GAA	48	110000	UAA
1	000001	AAC	17	010001	CAC	33	100001	GAC	49	110001	UAC
2	000010	AAG	18	010010	CAG	34	100010	GAG	50	110010	UAG
3	000011	AAU	19	010011	CAU	35	100011	GAU	51	110011	UAU
4	000100	ACA	20	010100	CCA	36	100100	GCA	52	110100	UCA
5	000101	ACC	21	010101	CCC	37	100101	GCC	53	110101	UCC
6	000110	ACG	22	010110	CCG	38	100110	GCG	54	110110	UCG
7	000111	ACU	23	010111	CCU	39	100111	GCU	55	110111	UCU
8	001000	AGA	24	011000	CGA	40	101000	GGA	56	111000	UGA
9	001001	AGC	25	011001	CGC	41	101001	GGC	57	111001	UGC
10	001010	AGG	26	011010	CGG	42	101010	GGG	58	111010	UGG
11	001011	AGU	27	011011	CGU	43	101011	GGU	59	111011	UGU
12	001100	AUA	28	011100	CUA	44	101100	GUA	60	111100	UUA
13	001101	AUC	29	011101	CUC	45	101101	GUC	61	111101	UUC
14	001110	AUG	30	011110	CUG	46	101110	GUG	62	111110	UUG
15	001111	AUU	31	011111	CUU	47	101111	GUU	63	111111	UUU

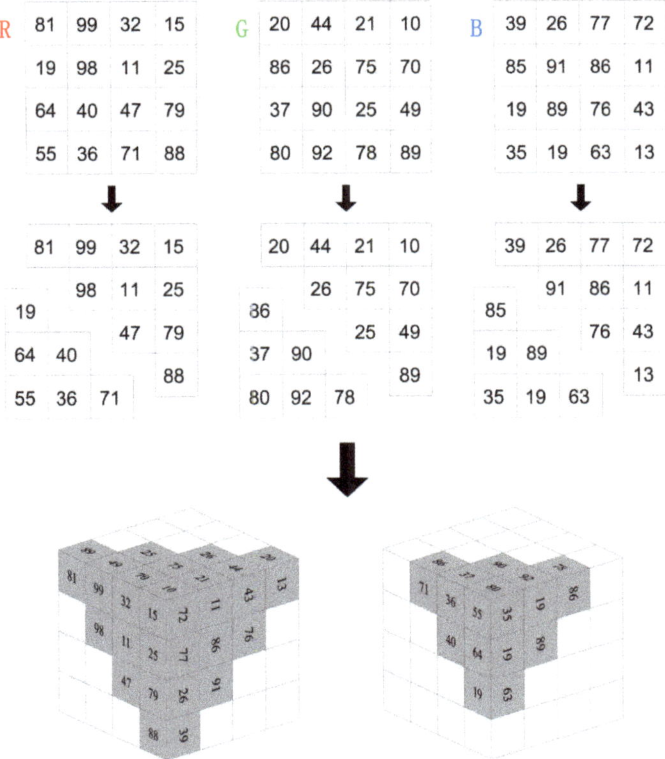

Figure 3. Color image to cube. The first row is three channels of a color image. The second row is the triangles generated from image. Additionally, the third row is the placement of triangles on a cube.

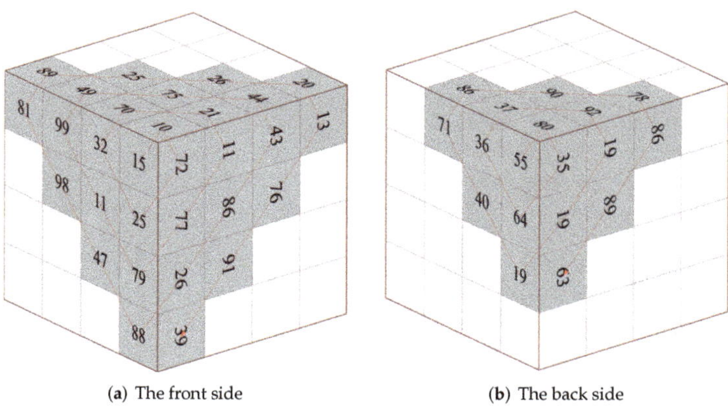

(**a**) The front side   (**b**) The back side

Figure 4. Three-dimensional (3D) transformed Zigzag diffusion. (**a**) is the Zigzag diffusion process on the front side of cube. (**b**) is the Zigzag diffusion process on the back side of cube.

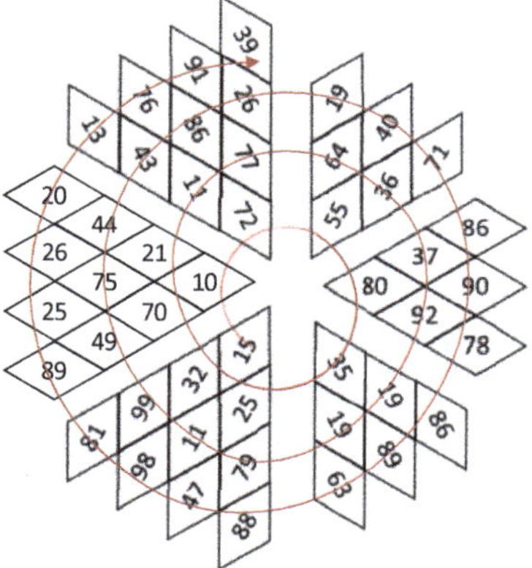

Figure 5. 3D transformed Zigzag path. For all triangles on the cube, 3D transformed Zigzag diffusion is implemented through this order.

3. Encryption and Decryption

In this paper, image encryption could be divided into three parts. Firstly, a 6D hyper-chaotic system is employed to generate chaotic matrices for encryption processes. Subsequently, three-dimensional (3D) transformed Zigzag diffusion is implemented on the permuted image. Finally, RNA concept is used for encoding and decoding.

3.1. Encryption Scheme

Suppose that plaintext image has N rows and N columns with RGB channels.

The flowchart of HCZRNA is described in Figure 6, and the specific operations are listed, as follows.

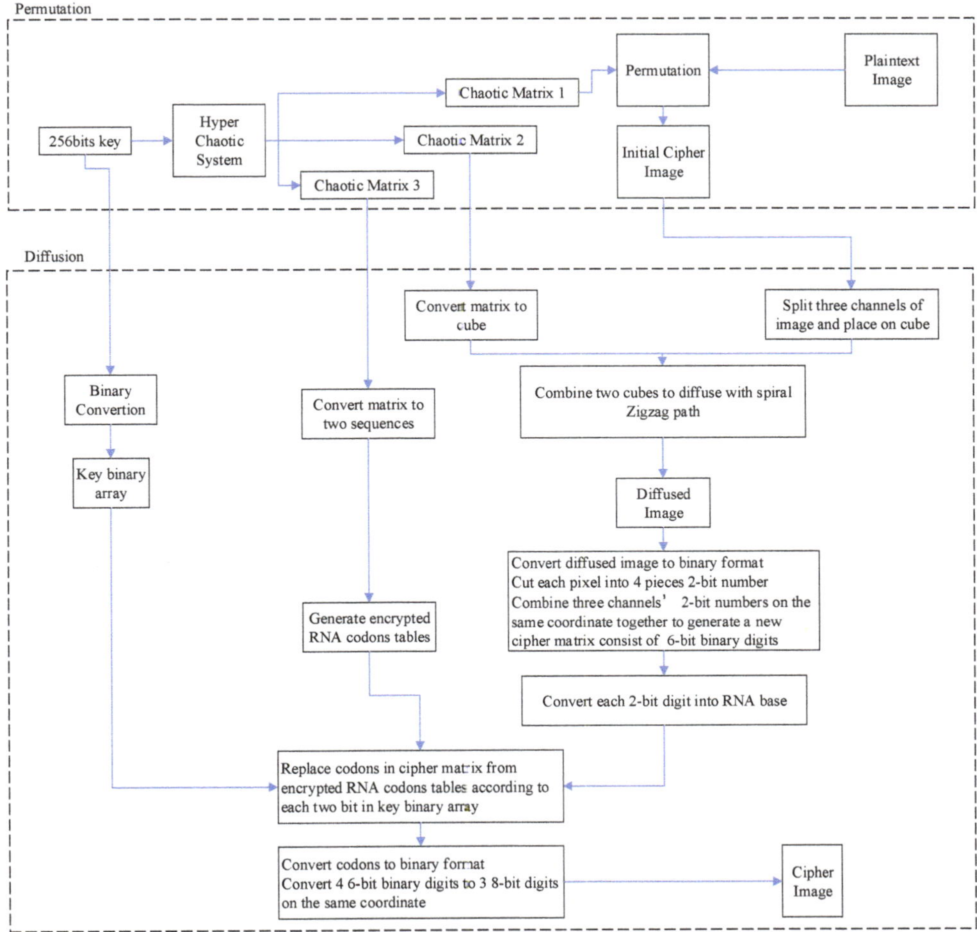

Figure 6. The process of encryption.

3.1.1. Initial Values Generation

The HCZRNA scheme uses a 256-bit key of different characters against attacks. The 256-bit long security key would be utilized in two parts, which are hyper-chaotic system initial values generation and RNA encryption.

At first, the initial values of hyper-chaotic system should be generated by a security key. Details of initial values generation is performed in three steps:

- Step 1: divide the secret key K into 32 blocks, which could be expressed as $K = \{k_1, k_2, \dots, k_{32}\}$, each k is a 8-bits number.

- Step 2: K array that is generated in step 1 is calculated into four intermediate parameters d_1, d_2, d_3, d_4 by Equation (2) with four user-defined constants c_1, c_2, c_3 and c_4.

$$\begin{cases} d_1 = c_1 + \dfrac{k_1 \oplus k_2 \oplus \cdots \oplus k_8}{256} \\[2mm] d_2 = c_2 + \dfrac{k_9 \oplus k_{10} \oplus \cdots \oplus k_{16}}{256} \\[2mm] d_3 = c_3 + \dfrac{k_{17} \oplus k_{18} \oplus \cdots \oplus k_{24}}{256} \\[2mm] d_4 = c_4 + \dfrac{k_{25} \oplus k_{26} \oplus \cdots \oplus k_{32}}{256} \end{cases} \tag{2}$$

where $\oplus$ represents bitwise XOR operation.

- Step 3: The initial values x_1 to x_6 of 6D hyper-chaotic system could be obtained from the 4 intermediate parameters by Equation (3).

$$\begin{cases} x_1 = \dfrac{((d_1 + d_2) \times 10^8) \bmod 256}{255} \\[2mm] x_2 = \dfrac{((d_2 + d_3) \times 10^8) \bmod 256}{255} \\[2mm] x_3 = \dfrac{((d_3 + d_4) \times 10^8) \bmod 256}{255} \\[2mm] x_4 = \dfrac{((d_1 + d_3) \times 10^8) \bmod 256}{255} \\[2mm] x_5 = \dfrac{((d_1 + d_4) \times 10^8) \bmod 256}{255} \\[2mm] x_6 = \dfrac{((d_2 + d_4) \times 10^8) \bmod 256}{255} \end{cases} \tag{3}$$

where *mod* means module operation.

3.1.2. Hyper-Chaotic Matrices Generation

With the initial values that are calculated in Section 3.1.1, chaotic matrices could be generated from 6D hyper-chaotic system. In HCZRNA, chaotic matrices would be utilized in three parts, which are permutation, 3D transformed Zigzag diffusion, and RNA operation. Suppose that the plaintext image has $N \times N \times 3$ pixels, an $N \times N \times 6$ chaotic matrix is needed for permutation, a $2 \times N \times N \times 6$ chaotic matrix for Zigzag, and 64×6 chaotic matrix for RNA.

Therefore, the 6D hyper-chaotic system utilizes initial values from Equation (3) to iterate for generating a $(3 \times N \times N + 64) \times 6$ matrix. Given that i^{th} iteration's state values could be described as $s^i = \{x_{1,i}, x_{2,i}, x_{3,i}, x_{4,i}, x_{5,i}, x_{6,i}\}$, a hyper-chaotic matrix S could be depicted as Equation (4) after all iterations.

$$S = \{s^1, s^2, \ldots, s^M\} = \begin{cases} x_{1,1}, x_{1,2}, \ldots, x_{1,M} \\ x_{2,1}, x_{2,2}, \ldots, x_{2,M} \\ x_{3,1}, x_{3,2}, \ldots, x_{3,M} \\ x_{4,1}, x_{4,2}, \ldots, x_{4,M} \\ x_{5,1}, x_{5,2}, \ldots, x_{5,M} \\ x_{6,1}, x_{6,2}, \ldots, x_{6,M} \end{cases}_{6 \times M} \tag{4}$$

where $M = 3 \times N \times N + 64$.

However, the numbers in matrix S are double-precision values, which are suitable for permutation but not for Zigzag and RNA, and color image only has three channels that are smaller than channels of S. Hence, matrix S should be separated into three pieces respectively.

For permutation, a matrix S_1 is calculated from the first $N \times N$ part of S by Equation (5).

$$S_1 = \left\{ \begin{array}{l} x_{1,1} + x_{2,1}, x_{1,2} + x_{2,2}, \ldots, x_{1,M'} + x_{2,M'} \\ x_{3,1} + x_{4,1}, x_{3,2} + x_{4,2}, \ldots, x_{3,M'} + x_{4,M'} \\ x_{5,1} + x_{6,1}, x_{5,2} + x_{6,2}, \ldots, x_{5,M'} + x_{6,M'} \end{array} \right\}_{3 \times M'} \tag{5}$$

where $M' = N \times N$.

While matrix S_2 is cut from $s^{M'+1}$ to $s^{3M'}$ in S for 3D transformed Zigzag diffusion. Additionally, because 8-bit integer digits are needed for diffusion, each item $x'_{i,j}$ in S_2 should be calculated by Equation (6).

$$\begin{aligned} Suppose \quad & \vec{s}_i = \{x_{i,(M'+1)}, x_{i,(M'+2)}, \ldots, x_{i,3M'}\} \\ & y_{i,j} = 2 \times x_{i,j} + \frac{max(\vec{s}_i) + min(\vec{s}_i)}{max(\vec{s}_i) - min(\vec{s}_i)} \\ & x'_{i,j} = ((\lfloor |y_{i,j}| \rfloor - \lfloor |y_{i,j}| \rfloor \rfloor) \times 10^{10}) \; mod \; 256 \end{aligned} \tag{6}$$

where *max* and *min* are maximum and minimum operations.

Matrix S_3 is the last part of matrix S and it is used to sort operation for encrypting RNA codons tables as indexes. Because there only needs two indexes sequences, matrix S_3 should be summarized as Equation (7).

$$\begin{aligned} index_1 &= \{x_{1,3M'+1} + x_{2,3M'+1} + x_{3,3M'+1}, x_{1,3M'+2} + x_{2,3M'+2} + x_{3,3M'+2}, \ldots, x_{1,3M'+64} + x_{2,3M'+64} + x_{3,3M'+64}\} \\ index_2 &= \{x_{4,3M'+1} + x_{5,3M'+1} + x_{6,3M'+1}, x_{4,3M'+2} + x_{5,3M'+2} + x_{6,3M'+2}, \ldots, x_{4,3M'+64} + x_{5,3M'+64} + x_{6,3M'+64}\} \end{aligned} \tag{7}$$

3.1.3. Permutation

In this part, matrix S_1 is used to permute plaintext image. At the beginning, each element in S_1 should be allocated to each pixel as index. Hence, an $N \times N \times 3$ matrix S'_1 is needed to be converted from S_1 by reshaping.

$$S'_1 = reshape(S_1, N, N, 3) \tag{8}$$

Afterwards, each pixel in plaintext image has a corresponding index in S'_1 at the same coordinate. Combine plaintext image with matrix S'_1, and take another reshaping operation to convert these two matrix into two sequences with a length of $N \times N \times 3$. After sorting S'_1 ascendingly with image sequence synchronously, pixels' orders in plaintext image sequence have been scrambled.

Finally, reshaping the sorted image sequence to an $N \times N \times 3$ matrix, the initial cipher image could be generated.

3.1.4. Diffusion

After permutation, a diffusion scheme by 3D Zigzag transformation is proposed, as follows. An initial cipher image would be split and placed on the surfaces of an $N \times N \times 6$ cube, termed as P, as described in Section 2.2. Additionally, chaotic matrix S_2 would also be placed on another two $N \times N \times 6$ cubes, since diffusion would implement two rounds. For the first $N \times N \times 6$ numbers in S_2, each number would be placed on a cube in order, which could be called cube SC_1. For the last $N \times N \times 6$ numbers in S_2, cube SC_2 could be generated by the same process.

Subsequently, diffusion would start from origin point of cube P on the front side, and its coordinate is $[1, 1, 1]$. At each iteration, the pixel's value $C_{i,j,m}$ is calculated by Equation (9).

$$C_{i,j,m} = (P_{i,j,m} \oplus (T + x_{i,j,m;1})) \bmod 256 \tag{9}$$

where i, j, m are the coordinates of pixel at the ith row, j^{th} column, and m^{th} side on the cube. T is the previous one diffused pixel's C value, if $i, j, m = 1, 1, 1$, T is a user-defined constant. $x_{i,j,m;1}$ is the corresponding coordinate's value in SC_1.

For the second round of diffusion, Equation (9) would change to Equation (10).

$$D_{i,j,m} = (C_{i,j,m} \oplus (T' + x_{i,j,m;2})) \bmod 256 \tag{10}$$

where D is result of diffusion, and T' is the previous one diffused pixel's D value, and, if $i, j, m = 1, 1, 1$, T' is the last pixel's C value after the first round diffusion. While $x_{i,j,m;2}$ is corresponding coordinate's value in SC_2.

Through these two round diffusions, D cube is generated. Additionally, recover the D's $N \times N \times 6$ matrix by reversing processes of image splitting and cube placement in Section 2.2. A diffused $N \times N \times 3$ matrix D_{mat} is obtained.

3.1.5. RNA Operation

The encryption from diffused matrix D_{mat} through RNA operation could be described, as follows:

- Step 1: RNA operation is initiated from creating two encrypted codons tables, called T_{00} and T_{01}. In which, T_{00} and T_{01} are shuffled tables from codons truth, as in Table 2. The shuffle orders are generated according to indexes sequences calculated from Equation (7). After sorting with these two indexes sequences, the original codons truth table could be shuffled to two different encrypted codons tables T_{00} and T_{01}. Subsequently, by the complementary rules of RNA, additional tables T_{10} and T_{11} could be generated from T_{00} and T_{01}. Hence, four encrypted codons tables are generated.
- Step 2: for each element in D_{mat}, binary number conversion is processed, which is recorded as B.

$$B = \{b_{i,j,m}\}. \quad i, j = 1, 2, \ldots, N; m = 1, 2, 3 \tag{11}$$

Each $b_{i,j,m}$ could be expressed as eight binary numbers, which could be depicted as $b_0^{i,j,m} b_1^{i,j,m} b_2^{i,j,m} b_3^{i,j,m} b_4^{i,j,m} b_5^{i,j,m} b_6^{i,j,m} b_7^{i,j,m}$.

- Step 3: divide $b_{i,j,m}$ into four pieces, each two bits are one piece, which are recorded as:

$$\begin{aligned} bt_1^{i,j,m} &= b_0^{i,j,m} b_1^{i,j,m} \\ bt_2^{i,j,m} &= b_2^{i,j,m} b_3^{i,j,m} \\ bt_3^{i,j,m} &= b_4^{i,j,m} b_5^{i,j,m} \\ bt_4^{i,j,m} &= b_6^{i,j,m} b_7^{i,j,m} \end{aligned} \tag{12}$$

Additionally, combine three channels' bts at the same coordinate together:

$$\begin{aligned} bt_1^{i,j} &= bt_1^{i,j,1} bt_1^{i,j,2} bt_1^{i,j,3} \\ bt_2^{i,j} &= bt_2^{i,j,1} bt_2^{i,j,2} bt_2^{i,j,3} \\ bt_3^{i,j} &= bt_3^{i,j,1} bt_3^{i,j,2} bt_3^{i,j,3} \\ bt_4^{i,j} &= bt_4^{i,j,1} bt_4^{i,j,2} bt_4^{i,j,3} \end{aligned} \tag{13}$$

Therefore, each $bt^{i,j}$ has six bits that could transfer to RNA codons according to Table 1. Exchange each two bits in bts to RNA base one-by-one according to the principle of row priority, *bts* could be coded to codons. And put them into a one-dimension sequence *BS* as Equation (14).

$$BS = \{bt_1^{1,1}, bt_2^{1,1}, bt_3^{1,1}, bt_4^{1,1}, bt_1^{1,2}, bt_2^{1,2}, \ldots, bt_4^{2,1}, bt_1^{2,2}, \ldots, bt_3^{N,N}, bt_4^{N,N}\}. \tag{14}$$

- Step 4: convert key to binary format. 256-bit key could be changed into a binary sequence *BK*.

$$
\begin{aligned}
key &= [key_0, key_1, \ldots, key_{31}] \\
key_i &= key_{i,0}, key_{i,1}, \ldots, key_{i,7} \\
BK &= [key_{1,0}, key_{1,1}, \ldots, key_{1,7}, key_{2,0}, key_{2,1}, \ldots, key_{31,7}]
\end{aligned}
\tag{15}
$$

Walk through sequence *BS*, and find corresponding index *id* of each codon in *BS* from Table 2. For each codon in *BS*, check 2-bits table number *z* in sequence *BK*.

$$z = BK_{n \bmod 2048} BK_{(n+1) \bmod 2048} \tag{16}$$

where *n* is the walking times.

Take the codon $T_z(id)$ to replace the origin codon $BS(n)$.

When iterations termination, an encrypted sequence is generated.

- Step 5: decode each base in encrypted sequence *BS* to binary format by Table 1, put all of the binary digits back to original coordinates by reversing operations in Step 3. Additionally, change binary matrix into 2-bit matrix. The cipher image is generated.

The HCZRNA encryption has four stages: hyper-chaotic matrices generation (Sections 3.1.1 and 3.1.2), hyper-chaotic permutation (Section 3.1.3), 3D transformed Zigzag diffusion on surfaces of cubes, which is generated from initial cipher image (Section 3.1.4), and a bit-level RNA operation (Section 3.1.5). The major steps of the HCZRNA are Sections 3.1.4 and 3.1.5, i.e., the transformed Zigzag diffusion on 3D cubes and bit-level RNA substitutions with hyper-chaotic matrix, respectively. The HCZRNA uses the strategy of "divide and conquer" that is widely used in various applications to decompose the original encryption task into a couple of simpler sub-tasks [38,39].

3.2. Decryption

In this paper, the encryption scheme has been depicted, and decryption is the inverse process of encryption. Details are proposed, as follows.

- Step 1: redo the processes that are listed in Sections 3.1.1 and 3.1.2 to generate hyper-chaotic matrices S_1, S_2, and S_3.
- Step 2: convert the cipher image to a binary format, and reconstruct three channels' pixels at each coordinate into four 6-bit binary arrays by using Euqation (12) and (13). Change 6-bit arrays into codons from codons truth Table 2, and put them in a one-dimension sequence BS' as Equation (14).
- Step 3: generate key binary sequence *BK* through Equation (15) and encrypted codons tables $\{T_{00}, T_{01}, T_{10}, T_{11}\}$ by redoing Step 1 in Section 3.1.5.
- Step 4: Check each 2-bits *z* in *BK* and find corresponding table T_z from $\{T_{00}, T_{01}, T_{10}, T_{11}\}$. Walk through BS' and find each codon's corresponding index id' in Table T_z. Replace codon in BS' to codon id' in codons truth Table 2. After all codons are replaced, convert them into binary formats and 8-bit numbers, matrix D'_{mat} is obtained.
- Step 5: split matrix D'_{mat} and place triangles on cube surfaces as the process shown in Section 2.2. Redo Section 3.1.4 with modified Equation (17) two rounds, and then walk through pixels with reversed Zigzag path. Take Figure 4 in Section 2.2 as an

example, the traversal road of decryption is shown in Figure 7. If we put all the pixels together, the order of traversal is depicted in Figure 8.

$$C'_{i,j,m} = (D'_{i,j,m} \oplus (T' + x_{i,j,m;2}))mod256$$
$$P'_{i,j,m} = (C'_{i,j,m} \oplus (T + x_{i,j,m;1}))mod256$$

(17)

where T' is the previous one pixel's D' value, and, if $i,j,m = 1,1,1$, T' is the last pixel's C' value after first round iteration. T is the previous one pixel's C' value and, if $i,j,m = 1,1,1$, T is the user-defined constant that is used in Section 3.1.4.

- Step 6: after the process in Step 5, return the triangles in the cube to theirs original coordinates on a image. Additionally, the reverse processes in Section 3.1.3, reshape S_1 to construct sorted sequence. Find image pixels' corresponding coordinates through sorted sequence and recover. The decrypted image is generated.

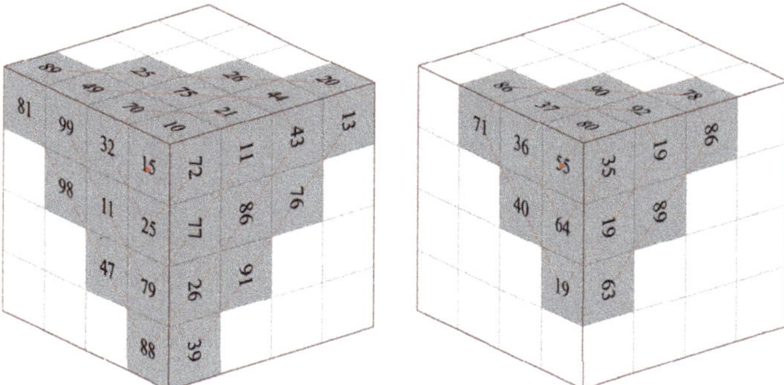

Figure 7. Reverse traversal.

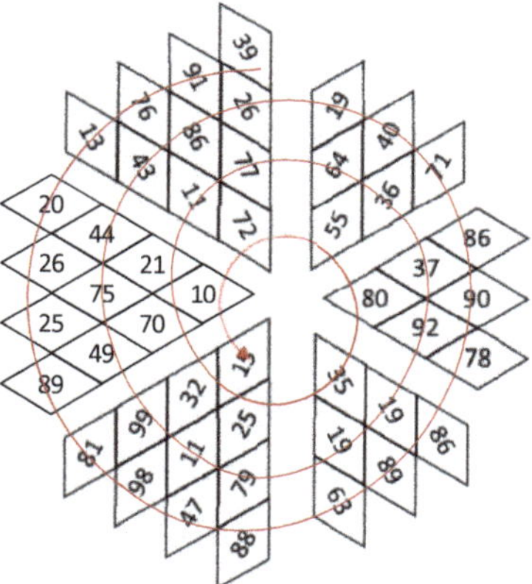

Figure 8. The order of Zigzag in decryption.

4. Experimental Results

The encryption and decryption schemes have been tested on four popular RGB color images in Table 3. All of the experiments are conducted by MATLAB R2019b on 64-bit Windows 10 system, and the main hardware includes an Xeon(R) W-2223 @ 3.60 GHz CPU as well as 32 GB RAM.

Table 3. Testing images.

Image	Size ($h \times w \times c$)	Image	Size ($h \times w \times c$)
Lena	$256 \times 256 \times 3$	Baboon	$512 \times 512 \times 3$
Peppers	$256 \times 256 \times 3$	Splash	$512 \times 512 \times 3$

For the controlling parameters setting in Equation (1), $(a, b, c, d, e, f, g, r, \omega, \beta) = (0.3, 1.5, 8.5, -2, 1, -0.1, 0.9, 1, 1, 0.2)$. Constants c_1, c_2, c_3, c_4 in Equation (2) are set as $(1, 1, 2, 2)$ and the initial constant of T in Equation (9) is 11. The security key can be set by users, so we set a 256-bit hexadecimal sequence that is shown below as the security key in all of the experiments. The key can also be optimized by some evolutionary optimizations, such as differential evolution and particle swarm optimization [40–43].

$$key = '743B5A203B1E8EDF6C0FB0D7497CB2E228689AD00F57F8953B5C6127E1C26053'$$

In order to demonstrate performance of proposed HCZRNA scheme, five state-of-the-art encryption schemes are employed for comparison: a Four-wing hyper-chaotic system based dynamic DNA encryption scheme [29], an extended Zigzag confusion and RNA encryption based scheme [23], a Hopfield chaotic neural network-based scheme [44], a scheme with utilization of differences between two 1D chaotic maps [45] and a scheme with 4D hyper-chaotic system and DNA encryption [46].

4.1. Key Space

For an image encryption system, large enough key space is necessary to withstand a brute-force attack. In HCZRNA, a 256-bit security key is used to calculate the initial values of the hyper-chaotic system to generate the pseudo random matrices that could affect the outputs of permutation, diffusion, and RNA operations. As we know, different initial values in a hyper-chaotic system would get different pseudo random sequences, and each bit has two states, the security key has 2^{256} different states, so it could generate 2^{256} results of a hyper-chaotic system. Therefore, the key space of HCZRNA could be calculated as 2^{256}. Theoretically, if the key space of an encryption scheme is larger than 2^{100}, this scheme could resist violent crack by modern computers [47]. Therefore, the proposed HCZRNA in this paper has a large enough key space to resist brute-force attack.

4.2. Sensitivity of Keys

The sensitivity test on keys refers to utilize slightly different keys to encrypt the same images. If an encryption is sensitive, the encryption with slight difference on keys would get completely different cipher images. To test the key sensitivity, we would use two different keys to encrypt four test images, one of these two keys is initial security key key_1, another key is key_2, which is one bit changed for key_1. These two keys are stated as follows, where the changed bits are shown in red:

$$key_1 = '743B5A203B1E8EDF6C0FB0D7497CB2E228689AD00F57F8953B5C6127E1C26053'$$

$$key_2 = '743B5A203B1F8EDF6C0FB0D7497CB2E228689AD00F57F8953B5C6127E1C26053'$$

By comparing two cipher images from the same plaintext image, the differences of cipher images that are encrypted from these two security keys are stated in Table 4.

Table 4. Differences between the cipher images.

Image	Lena	Pepper	Baboon	Splash
Difference	99.59%	99.62%	99.61%	99.60%

In the table, it is obvious that all of the differences between two cipher images are over 99%, which reveals that, even with tiny changes in security keys, encryption by HCZRNA would also lead to extremely different outputs. Hence, HCZRNA satisfies sensitivity requirements.

4.3. Histogram

Because a histogram reflects each pixel's times in an image, histograms of meaningful images are fluctuated, while cipher images' histogram should be flat and uniform. That is to say, if an encryption scheme is well-designed, the histograms of cipher images should be as flat as possible. For the proposed HCZRNA, a histogram of Baboon and its cipher image are placed in Figure 9.

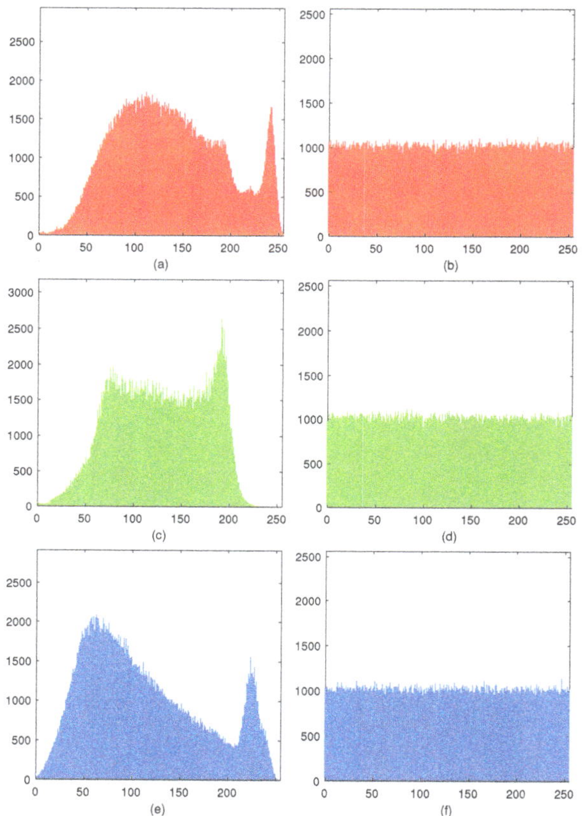

Figure 9. Histogram. image (**a,c,e**) are the histograms of three channels of Baboon, and image (**b,d,f**) are the histograms of corresponding channels of encrypted Baboon.

From this figure, it could find that histograms of all channels in plaintext image are fluctuated, while histograms of cipher image's different channels are almost distributed in a narrow range, and their values are around 1000. For more accurate results, histogram statistics are introduced to evaluate the variance and standard deviation of plaintext and cipher images [48,49]. Variance is used to calculate the average difference in each gray level frequency with respect to mean value $\bar{x}$, which could be formulated as Equation (18).

$$\alpha = \frac{1}{256} \sum_{i=1}^{256} (x_i - \bar{x})^2,$$
$$\bar{x} = \frac{h \times w}{256}$$

(18)

where h, w represent the image's height and width respectively, x is the frequency of different gray levels of pixels in a image, and the $\bar{x}$ is the mean value of xs. And α is the variance, the higher is α, the more fluctuate is the graphic histogram. Accordingly, if a encryption is well-designed, the α of encrypted image should be low.

As α is always very high in plaintext image, a standard deviation is used to evaluate histogram's fluctuations, which is stated as Equation (19).

$$\beta = \sqrt{\alpha} \tag{19}$$

where β is the standard deviation. For all test images, Table 5 describes the results of histogram statistics.

Table 5. Histogram statistics.

Image	Channels	Plaintext		Ciphertext	
		α	β	α	β
Lena	R	65,306	255	248	15
	G	30,665	175	258	16
	B	91,939	303	232	15
Pepper	R	57,413	239	249	15
	G	119,411	345	238	15
	B	151,644	389	237	15
Baboon	R	165,679	407	520	22
	G	285,616	534	532	23
	B	159,885	399	541	23
Splash	R	1,211,325	1100	566	23
	G	1,541,948	1241	495	22
	B	2,958,482	1720	504	22

In the table, the variances and standard deviations of plaintext images are very high, while they are extremely different in cipher images. All of these performances indicate that the proposed HCZRNA could effectively resist histogram attack.

4.4. Correlation

The correlation test refers to adjacent pixels' relationship. A meaningful image has high correlation because values of adjacent pixels are close to each other. This attribute could be utilized to crack. Therefore, a well-designed encryption scheme should have low enough correlations in three directions: horizontal, vertical, and diagonal directions. Given a pixel sequence that is represented by $X = \{x_1, x_2, \ldots, x_N\}$ and its adjacent pixel sequence $Y = \{y_1, y_2, \ldots, y_N\}$ in an image, correlation between X and Y could be denoted as $\gamma_{X,Y}$ in Equation (20).

$$\gamma_{X,Y} = \frac{\frac{1}{N} \sum_{i=1}^{N} (x_i - D(X))(y_i - D(Y))}{\sqrt{D(X)D(Y)}}$$
$$D(X) = \frac{\sum_{i=1}^{N} (x_i - E(X))^2}{N} \tag{20}$$
$$E(X) = \frac{\sum_{i=1}^{N} x_i}{N}$$

where $E(X)$ is X's mathematical expectation and $D(X)$ is standard deviation.

If X and Y are identical, $\gamma_{X,Y}$ would be a maximum of 1. On the contrary, $\gamma_{X,Y}$ would be close to 0 when X and Y have few correlations.

Figure 10 depicts the correlation test results. It is obvious that the adjacent pixels' distributions in plaintext images are concentrated, while the distributions in the cipher images are opposite.

More accurately, Table 6 provides correlation coefficients between plaintext images and cipher images. Additionally Table 7 demonstrates comparisons with references [44,45]. Through this test, it could find that the correlation coefficients of the proposed HCZRNA are extremely close to 0, which means that HCZRNA could effectively break correlations existing in plaintext images. While the comparisons show that the proposed HCZRNA achieves the best results with [44,45] in all cases. This reveals that HCZRNA outperforms when compared schemes in terms of reducing correlations.

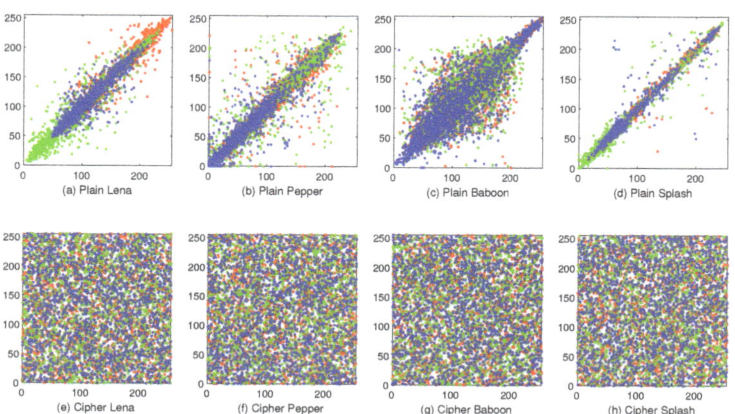

Figure 10. Correlations.The first row is correlations of plaintext images, and the second row is correlations of cipher images.

Table 6. The correlation coefficients of the testing images.

Image	Channels	Plaintext			Ciphertext		
		Horizontal	Vertical	Diagonal	Horizontal	Vertical	Diagonal
	R	0.9512	0.9755	0.9444	0.0046	0.0024	0.0051
Lena	G	0.9512	0.9679	0.9276	−0.0027	−0.0007	0.0002
	B	0.9512	0.9479	0.9021	−0.0023	0.0014	0.0004
	R	0.9218	0.8624	0.8531	0.0003	0.0001	0.0015
Baboon	G	0.9218	0.7591	0.7299	−0.0010	0.0004	0.0020
	B	0.9218	0.8782	0.8411	0.0005	−0.0022	0.0012

Table 7. Comparisons of correlation coefficients.

Image	Channels		Plaintext	Ciphertext		
				HCZRNA	Ref. [44]	Ref. [45]
		Horizontal	0.9218	**0.0003**	0.0054	−0.0073
	R	Vertical	0.8624	**0.0001**	−0.0042	−0.0059
		Diagonal	0.8531	**0.0015**	−0.0177	−0.0136
		Horizontal	0.9218	**−0.0010**	−0.0055	0.0046
Baboon	G	Vertical	0.7591	**0.0004**	0.0119	−0.0077
		Diagonal	0.7299	**0.0020**	0.0046	−0.0044
		Horizontal	0.9218	**0.0005**	−0.0021	−0.0067
	B	Vertical	0.8782	**−0.0022**	0.0104	−0.0111
		Diagonal	0.8411	**0.0012**	−0.0021	0.0122

4.5. Information Entropy

Information Entropy shows the randomness and uncertainty of image's pixels. If pixels in an image have uniform distribution, this image could resistant statistical attacks. Because there are 256 gray levels in each channel of color image, the Entropy calculation could be formulated as Equation (21):

$$H(C) = -\sum_{i=0}^{255} p(i)log_2 p(i) \tag{21}$$

where C denotes channels of color image and $p(i)$ is probability of gray level in whole channel.

The bigger $H(C)$, the bigger uncertainty of image. While the theoretical value of $H(C)$ is 8.

Table 8 shows the entropies of all channels of plaintext color images and corresponding cipher images through encryptions by proposed HCZRNA. It is obvious that cipher images have increased entropies a lot from plaintext images and their entropies are very close to the theoretical value. Moreover, a comparison is held between HCZRNA and Ref. [23,29,44–46], and the results are stated in Table 9. Among all of the encryption schemes, the proposed HCZRNA achieves the highest entropies in four out of six cases. It could conclude that HCZRNA has the ability to resist statistical attack.

Table 8. Information Entropies of testing images.

Image	Channels	Lena	Peppers	Baboon	Splash
Plaintext	R	7.2353	7.3369	7.7067	6.9481
	G	7.5683	7.4394	7.4744	6.8845
	B	6.9176	7.0219	7.7522	6.1265
Ciphertext	R	7.9973	7.9972	7.9993	7.9993
	G	7.9970	7.9970	7.9993	7.9994
	B	7.9972	7.9972	7.9993	7.9993

Table 9. Comparison of entropies.

Image	Channel	Plaintext	HCZRNA	Ref. [29]	Ref. [23]	Ref. [44]	Ref. [45]	Ref. [46]
Lena	R	7.2353	**7.9973**	7.9971	**7.9973**	-	-	7.9973
	G	7.5683	7.9970	7.9971	7.9972	-	-	7.9975
	B	6.9176	7.9972	7.9971	7.9971	-	-	7.9975
Baboon	R	7.7067	**7.9993**	7.9926	-	**7.9993**	**7.9993**	7.9970
	G	7.4744	**7.9993**	7.9926	-	**7.9993**	**7.9993**	7.9978
	B	7.7522	**7.9993**	7.9926	-	**7.9993**	7.9992	7.9987

4.6. Differential Attack

The differential attack test is an important security test for image encryption, which reveals the influence on the cipher image caused by a minor change in pixels of plaintext image. If a tiny change on pixels in plaintext image leads to significant different cipher image, that is to say the encryption scheme could resist differential attack.

Two important indices are introduced to measure the ability of differential attack resistance, which is called the number of pixel change rate (NPCR) and the unified average changing intensity (UACI). Additionally, they are defined as Equations (22) and (23):

$$NPCR = \frac{\sum_{i=0}^{h} \sum_{j=0}^{w} F(i,j) \times 100\%}{w \times h} \tag{22}$$

$$UACI = \frac{\sum_{i=0}^{h} \sum_{j=0}^{w} |e_1(i,j) - e_2(i,j)|}{255 \times w \times h} \tag{23}$$

where e_1 and e_2 are two cipher images, and $e(i,j)$ means the pixel's value at coordinate i, j in image e. $F(i,j)$ denotes whether the same coordinate's pixel values in e_1 and e_2 are independent or not, which could be formulated as Equation (24):

$$F(i,j) = \begin{cases} 0, & if \quad e_1(i,j) = e_2(i,j) \\ 1, & if \quad e_1(i,j) \neq e_2(i,j) \end{cases} \tag{24}$$

For two random images, NPCR and UACI's expected values are stated as: $NPCR = 99.6094\%$ and $UACI = 33.4635\%$ for an 8-bit gray image [30].

Hence, to realize the test, one bit would be changed on a random pixel in plaintext image. And both the plaintext image and changed image are encrypted to two different cipher images. Table 10 lists the average results of ten times tests. It could find that all NPCR values and UACI values of cipher images' different channels exceed the theoretical values. Additionally, comparisons with Ref. [23,29,44–46] are shown in Tables 11 and 12. Through the comparisons, the proposed HCZRNA encryption scheme has better performances on NPCR and UACI, which indicates that HCZRNA could resist differential attack well.

Table 10. The mean number of pixel change rate (NPCR) and unified average changing intensity (UACI) of cipher images.

Image	NPCR(%)			UACI(%)		
	R	G	B	R	G	B
Lena	99.6619	99.6272	99.6460	33.6177	33.6048	33.6422
Peppers	99.6481	99.6404	99.6239	33.7208	33.5701	33.6435
Baboon	99.6159	99.6769	99.6115	33.5196	33.5203	33.5049
Splash	99.6219	99.6934	99.6253	33.4983	33.5114	33.4816

Table 11. Average NPCR (%) of running the schemes 10 times.

Image	Channel	HCZRNA	Ref. [29]	Ref. [23]	Ref. [44]	Ref. [45]	Ref. [46]
Lena	R	**99.6619**	99.60	99.6323	-	-	99.615
	G	**99.6272**	99.61	99.6109	-	-	99.62
	B	**99.6460**	99.61	99.6338	-	-	99.617
Baboon	R	**99.6159**	99.6083	-	99.6037	99.6037	99.6140
	G	**99.6769**	99.6065	-	99.6048	99.6017	99.6073
	B	99.6115	99.6094	-	99.6059	99.6043	**99.6292**

Table 12. Average UACI (%) of running the schemes 10 times.

Image	Channel	HCZRNA	Ref. [29]	Ref. [23]	Ref. [44]	Ref. [45]	Ref. [46]
Lena	R	**33.6177**	33.56	33.4683	-	-	33.4732
	G	**33.6048**	33.45	33.4341	-	-	33.3428
	B	**33.6422**	33.49	33.4991	-	-	33.4647
Baboon	R	**33.5196**	33.4939	-	33.4427	29.9630	33.4843
	G	**33.5203**	33.4295	-	33.4605	28.5708	33.4690
	B	**33.5049**	33.4856	-	31.9747	31.2574	33.4965

4.7. Robustness

It is unavoidable that there data loss or noise attack occur when cipher images are transmitting. Hence, a well-designed encryption and decryption scheme should resist contamination on cipher images to recover plaintext images without great changes.

To demonstrate robustness of proposed HCZRNA scheme, 12.5%, 25%, and 50% data lose tests and 1%, 5%, and 10% salt and pepper noise tests would presented in Figures 11 and 12.

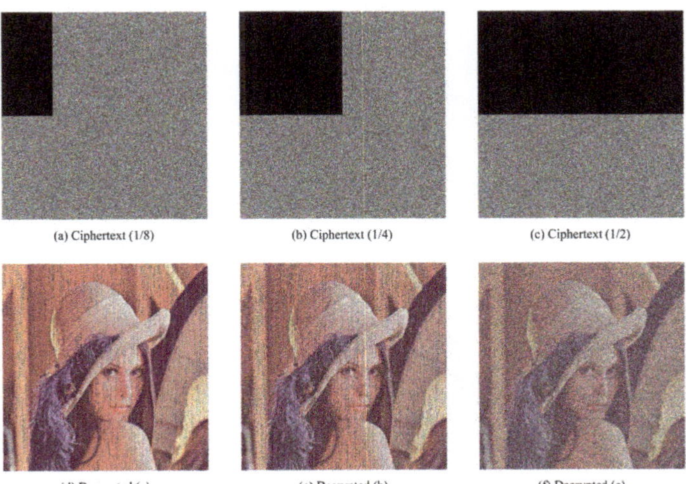

(a) Ciphertext (1/8) (b) Ciphertext (1/4) (c) Ciphertext (1/2)

(d) Decrypted (a) (e) Decrypted (b) (f) Decrypted (c)

Figure 11. Cropping attack tests. The first row is cipher images with 12.5%, 25% and 50% data loss, and the second row is decrypted images from the first row.

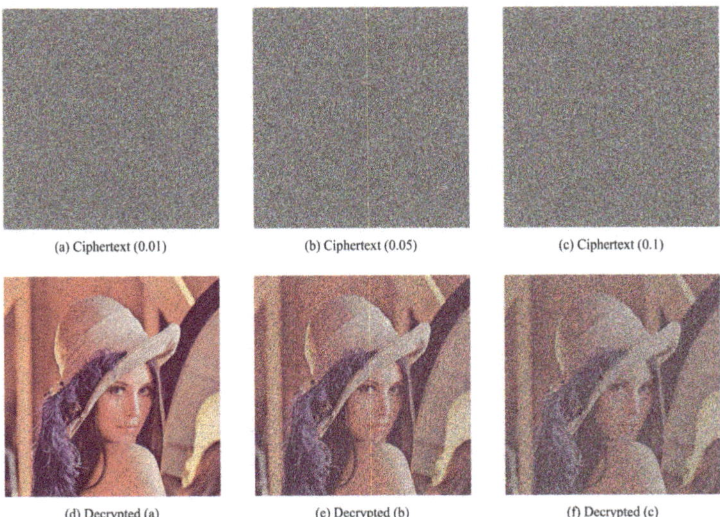

(a) Ciphertext (0.01) (b) Ciphertext (0.05) (c) Ciphertext (0.1)

(d) Decrypted (a) (e) Decrypted (b) (f) Decrypted (c)

Figure 12. Noise attack tests. The first row is cipher images with 1%, 5%, 10% salt and pepper noise, and the second row is decrypted images from the first row.

From the figures, the main information of plaintext images could be identified from decrypted images, which could conclude that HCZRNA has enough robustness for data loss and noise attacks. Here, the Mean Squared Error (MSE) and Peak Signal to Noise Ratio (PSNR) are also utilized to test robustness [48,49], which is formulated as Equation (25).

$$MSE = \frac{1}{h \times w} \sum_{i=1}^{h} \sum_{j=1}^{w} [P(i,j) - E(i,j)]^2,$$

$$PSNR = 20 log_{10}(\frac{255}{\sqrt{MSE}})$$

(25)

where P and E represent two different images. MSE is used to evaluate the difference between two images, and PSNR depicts the ratio between the maximum possible power of a signal and the power of distorting noise that affects the quality of its representation. The lower the MSE, the higher PSNR, which indicates that two images have high similarity. Hence, under noise attacks, if the PSNR between the plaintext image and decrypted image is high, the encryption and decryption schemes are good enough. Tables 13 and 14 present the results of plaintext image and decrypted image of Lena under data loss and salt and pepper noise attacks.

Table 13. Mean Squared Error (MSE) and Peak Signal to Noise Ratio (PSNR) under data loss.

Data Loss	MSE	PSNR
12.5%	1195	17.35
25%	2315	14.48
50%	4502	11.60

Table 14. MSE and PSNR under salt and pepper noise.

Salt and Pepper Noise	MSE	PSNR
1%	610	20.28
5%	2684	13.84
10%	4490	11.61

Through the results, we could find there are high MSEs and low PSNRs in these tables, which figures out that HCZRNA could resist attacks of data loss and noise.

4.8. Running Time

In HCZRNA, the pixels of image would be walked through multiple times in diffusion and RNA operation. Suppose that the size of RGB image is $N \times N \times 3$. For the five parts of encryption processes that are listed in Section 3.1, initial values of hyper-chaotic system are calculated from the security key, which costs $O(1)$ time complexity; the hyper-chaotic matrices are computed $3 \times N \times N + 64$ times iterations; for permutation, reshape and sort operations are implemented three times; while the diffusion process walks through each pixel two times, which costs $O(2 \times N \times N \times 3)$; at last, as RNA operation walks through all 6-bit codons that are transformed from 8-bit pixels, the times of iteration are increased to $\frac{4}{3} \times N \times N \times 3$. Hence, the time complexity of HCZRNA could be calculated as $O(1 + 3 \times N \times N + 64 + 3 + 2 \times N \times N \times 3 + \frac{4}{3} \times N \times N \times 3) = O(13N^2 + 68) = O(N^2)$. Using the experiment environment that is listed in this section, the running times of encryption and decryption could be stated in Table 15. Although the time costs of encryption and decryption are not very good, the time complexity is also a polynomial time, which could be tolerable. Additionally, the processes of RNA operation on different codons have no correlation with each other, which could improve computational time by computing RNA operation in parallel.

Table 15. Running time (unit: second).

Image Size	Encryption	Decryption
$64 \times 64 \times 3$	0.59	0.44
$128 \times 128 \times 3$	2.39	1.77
$256 \times 256 \times 3$	9.36	6.96
$512 \times 512 \times 3$	37.54	28.49

5. Conclusions

A novel hyper-chaotic system based image encryption scheme is proposed with 3D transformed Zigzag and RNA operation in this paper. By using the 6D hyper-chaotic system, three auxiliary matrices are generated, including one permutation index matrix, one mask matrix for Zigzag, and one codon table index matrix. Subsequently, two rounds 3D transformed Zigzag diffusion mechanism is proposed for pixels diffusion with each other. Nevertheless, additional encryption with RNA codons makes more reliable and secure results through employing codons tables and security keys. Through simulations, the proposed HCZRNA has better performances on the resistance of different types attacks than the compared encryption schemes, while the speed is not ideal, since it is a complex process. On the premise of ensuring performance, we would simplify diffusion and RNA operation processes and optimize the encryption steps for improving speed in the future.

Author Contributions: Conceptualization, D.Z. and T.L.; Data curation, L.C.; Formal analysis, D.Z.; Funding acquisition, T.L.; Investigation, T.L.; Methodology, D.Z. and T.L.; Project administration, T.L.; Resources, D.Z.; Software, D.Z.; Supervision, D.Z. and T.L.; Validation, L.C. and T.L.; Visualization, L.C.; Writing—original draft, D.Z.; Writing—review & editing, T.L. All authors have read and agreed to the published version of the manuscript.

Funding: This work was supported by the Ministry of Education of Humanities and Social Science Project (Grant No. 19YJAZH047) and the Scientific Research Fund of Sichuan Provincial Education Department (Grant No. 17ZB0433).

Institutional Review Board Statement: Not applicable.

Informed Consent Statement: Not applicable.

Data Availability Statement: Data available on request.

Acknowledgments: We gratefully acknowledge the reviewers for their comments and suggestions.

Conflicts of Interest: The authors declare no conflict of interest.

Abbreviations

The following abbreviations are used in this manuscript:

6D	6 Dimensional
RNA	Ribonucleic acid
HCZRNA	Hyper-chaotic color image encryption mechanism based on transformed Zigzag diffusion and RNA operations
RGB	Red Greeen Blue

References

1. Sneha, P.S.; Sankar, S.; Kumar, A.S. A chaotic colour image encryption scheme combining Walsh–Hadamard transform and Arnold–Tent maps. *J. Ambient. Intell. Humaniz. Comput.* **2020**, *11*, 1289–1308. [CrossRef]
2. Wang, H.; Xiao, D.; Chen, X.; Huang, H. Cryptanalysis and enhancements of image encryption using combination of the 1D chaotic map. *Signal Process.* **2018**, *144*, 444–452. [CrossRef]
3. Jeng, F.G.; Huang, W.L.; Chen, T.H. Cryptanalysis and improvement of two hyper-chaos-based image encryption schemes. *Signal Process. Image Commun.* **2015**, *34*, 45–51. [CrossRef]
4. Jithin, K.C.; Sankar, S. Colour image encryption algorithm combining Arnold map, DNA sequence operation, and a Mandelbrot set. *J. Inf. Secur. Appl.* **2020**, *50*, 102428. [CrossRef]
5. Zhang, Y. Test and verification of AES used for image encryption. *3D Res.* **2018**, *9*, 1–27. [CrossRef]

6. Xian, Z.H.; Sun, S.L. Image Encryption Algorithm Based on Chaos and S-Boxes Scrambling. *Adv. Mater. Res.* **2010**, *171–172*, 299–304. [CrossRef]
7. Zhou, Y.; Hua, Z.; Pun, C.M.; Chen, C.L.P. Cascade Chaotic System With Applications. *IEEE Trans. Cybern.* **2015**, *45*, 2001–2012. [CrossRef] [PubMed]
8. Liu, L.; Wang, Y.N.; Hou, L.; Feng, X.R. Easy encoding and low bit–error–rate chaos communication system based on reverse–time chaotic oscillator. *IET Signal Process.* **2017**, *11*, 869–876. [CrossRef]
9. Zhang, L.; Liao, X.; Wang, X. An image encryption approach based on chaotic maps. *Chaos Solitons Fractals* **2005**, *24*, 759–765. [CrossRef]
10. Bouslehi, H.; Seddik, H. Innovative image encryption scheme based on a new rapid hyperchaotic system and random iterative permutation. *Multimed. Tools Appl.* **2018**, *77*, 30841–30863. [CrossRef]
11. Zhang, Y.; Wen, W.; Su, M.; Li, M. Cryptanalyzing a novel image fusion encryption algorithm based on DNA sequence operation and hyper-chaotic system. *Optik* **2014**, *125*, 1562–1564. [CrossRef]
12. Mohammad Seyedzadeh, S.; Mirzakuchaki, S. A fast color image encryption algorithm based on coupled two-dimensional piecewise chaotic map. *Signal Process.* **2012**, *92*, 1202–1215. [CrossRef]
13. Li, T.; Shi, J.; Zhang, D. Color image encryption based on joint permutation and diffusion. *J. Electron. Imaging* **2021**, *30*, 013008. [CrossRef]
14. Askar, S.S.; Karawia, A.A.; Alshamrani, A. Image Encryption Algorithm Based on Chaotic Economic Model. *Math. Probl. Eng.* **2015**, *2015*, 1–10. [CrossRef]
15. Shaikh, N.; Chapaneri, S.; Jayaswal, D. Hyper chaotic color image cryptosystem. In Proceedings of the 2016 IEEE International Conference on Advances in Computer Applications (ICACA), Coimbatore, India, 24 October 2016; pp. 239–243. [CrossRef]
16. Li, C.; Zhao, F.; Liu, C.; Lei, L.; Zhang, J. A Hyperchaotic Color Image Encryption Algorithm and Security Analysis. *Secur. Commun. Netw.* **2019**, *2019*, 1–8. [CrossRef]
17. Zhou, Y.; Bao, L.; Chen, C.P. A new 1D chaotic system for image encryption. *Signal Process.* **2014**, *97*, 172–182. [CrossRef]
18. Kadir, A.; Aili, M.; Sattar, M. Color image encryption scheme using coupled hyper chaotic system with multiple impulse injections. *Opt. Int. J. Light Electron Opt.* **2017**, *129*, 231–238. [CrossRef]
19. Li, C.; Zhang, L.Y.; Ou, R.; Wong, K.W.; Shu, S. Breaking a novel colour image encryption algorithm based on chaos. *Nonlinear Dyn.* **2012**, *70*, 2383–2388. [CrossRef]
20. Xingyuan, W.; Junjian, Z.; Guanghui, C. An image encryption algorithm based on ZigZag transform and LL compound chaotic system. *Opt. Laser Technol.* **2019**, *119*, 105581. [CrossRef]
21. Xu, X.; Feng, J. Research and Implementation of Image Encryption Algorithm Based on Zigzag Transformation and Inner Product Polarization Vector. In Proceedings of the 2010 IEEE International Conference on Granular Computing, San Jose, CA, USA, 14–16 August 2010; pp. 556–561. [CrossRef]
22. Li, Y.; Li, X.; Jin, X.; Zhao, G.; Ge, S.; Tian, Y.; Zhang, X.; Zhang, K.; Wang, Z. An Image Encryption Algorithm Based on Zigzag Transformation and 3-Dimension Chaotic Logistic Map. In *Applications and Techniques in Information Security*; Niu, W., Li, G., Liu, J., Tan, J., Guo, L., Han, Z., Batten, L., Eds.; Springer: Berlin/Heidelberg, Germany, 2015; Volume 557, pp. 3–13. doi:10.1007/978-3-662-48683-2{\textunderscore }1. [CrossRef]
23. Wang, X.; Guan, N. A novel chaotic image encryption algorithm based on extended Zigzag confusion and RNA operation. *Opt. Laser Technol.* **2020**, *131*, 106366. [CrossRef]
24. Feixiang, Z.; Mingzhe, L.; Kun, W.; Hong, Z. Color image encryption via Hénon-zigzag map and chaotic restricted Boltzmann machine over Blockchain. *Opt. Laser Technol.* **2021**, *135*, 106610. [CrossRef]
25. Sahasrabuddhe, A.; Laiphrakpam, D.S. Multiple images encryption based on 3D scrambling and hyper-chaotic system. *Inf. Sci.* **2021**, *550*, 252–267. [CrossRef]
26. Hu, T.; Liu, Y.; Gong, L.H.; Ouyang, C.J. An image encryption scheme combining chaos with cycle operation for DNA sequences. *Nonlinear Dyn.* **2017**, *87*, 51–66. [CrossRef]
27. Li, T.; Yang, M.; Wu, J.; Jing, X. A novel image encryption algorithm based on a fractional-order hyperchaotic system and DNA computing. *Complexity* **2017**, *2017*, 9010251. [CrossRef]
28. Liu, Y.; Wang, J.; Fan, J.; Gong, L. Image encryption algorithm based on chaotic system and dynamic S-boxes composed of DNA sequences. *Multimed. Tools Appl.* **2016**, *75*, 4363–4382. [CrossRef]
29. Chai, X.; Fu, X.; Gan, Z.; Lu, Y.; Chen, Y. A color image cryptosystem based on dynamic DNA encryption and chaos. *Signal Process.* **2019**, *155*, 44–62. [CrossRef]
30. Hu, T.; Liu, Y.; Gong, L.H.; Guo, S.F.; Yuan, H.M. Chaotic image cryptosystem using DNA deletion and DNA insertion. *Signal Process.* **2017**, *134*, 234–243. [CrossRef]
31. Wu, J.; Shi, J.; Li, T. A novel image encryption approach based on a hyperchaotic system, pixel-level filtering with variable kernels, and DNA-level diffusion. *Entropy* **2020**, *22*, 5. [CrossRef] [PubMed]
32. Liu, P.; Zhang, T.; Li, X. A new color image encryption algorithm based on DNA and spatial chaotic map. *Multimed. Tools Appl.* **2019**, *78*, 14823–14835. [CrossRef]
33. Mahmud, M.; ur Rahman, A.; Lee, M.; Choi, J.Y. Evolutionary-based image encryption using RNA codons truth table. *Opt. Laser Technol.* **2020**, *121*, 105818. [CrossRef]

34. Abbasi, A.A.; Mazinani, M.; Hosseini, R. Chaotic evolutionary-based image encryption using RNA codons and amino acid truth table. *Opt. Laser Technol.* **2020**, *132*, 106465. [CrossRef]
35. Yadollahi, M.; Enayatifar, R.; Nematzadeh, H.; Lee, M.; Choi, J.Y. A novel image security technique based on nucleic acid concepts. *J. Inf. Secur. Appl.* **2020**, *53*, 102505. [CrossRef]
36. Li, T.; Shi, J.; Li, X.; Wu, J.; Pan, F. Image encryption based on pixel-level diffusion with dynamic filtering and DNA-level permutation with 3D Latin cubes. *Entropy* **2019**, *21*, 319. [CrossRef] [PubMed]
37. Mezatio, B.A.; Motchongom Tingue, M.; Kengne, R.; Tchagna Kouanou, A.; Fozin Fonzin, T.; Tchitnga, R. Complex dynamics from a novel memristive 6D hyperchaotic autonomous system. *Int. J. Dyn. Control.* **2020**, *8*, 70–90. [CrossRef]
38. Li, T.; Zhou, M. ECG Classification Using Wavelet Packet Entropy and Random Forests. *Entropy* **2016**, *18*, 285. [CrossRef]
39. Li, T.; Qian, Z.; He, T. Short-term load forecasting with improved CEEMDAN and GWO-based multiple kernel ELM. *Complexity* **2020**, *2020*, 1209547. [CrossRef]
40. Deng, W.; Shang, S.; Cai, X.; Zhao, H.; Song, Y.; Xu, J. An improved differential evolution algorithm and its application in optimization problem. *Soft Comput.* **2021**, 1–22. [CrossRef]
41. Song, Y.; Wu, D.; Deng, W.; Gao, X.; Li, T.; Zhang, B.; Li, Y. MPPCEDE: multi-population parallel co-evolutionary differential evolution for parameter optimization. *Energy Conv. Manag.* **2021**, *228*, 113661. [CrossRef]
42. Li, T.; Zhou, M.; Guo, C.; Luo, M.; Wu, J.; Pan, F.; Tao, Q.; He, T. Forecasting crude oil price using EEMD and RVM with adaptive PSO-based kernels. *Energies* **2016**, *9*, 1014. [CrossRef]
43. Deng, W.; Xu, J.; Cai, X.; Song, Y.; Zhao, H. Differential evolution algorithm with wavelet basis function and optimal mutation strategy for complex optimization problem. *Appl. Soft. Comput.* **2021**, *100*, 106724. [CrossRef]
44. Liu, L.; Zhang, L.; Jiang, D.; Guan, Y.; Zhang, Z. A Simultaneous Scrambling and Diffusion Color Image Encryption Algorithm Based on Hopfield Chaotic Neural Network. *IEEE Access* **2019**, *7*, 185796–185810. [CrossRef]
45. Pak, C.; Huang, L. A new color image encryption using combination of the 1D chaotic map. *Signal Process.* **2017**, *138*, 129–137. [CrossRef]
46. Mohamed, H.G.; ElKamchouchi, D.H.; Moussa, K.H. A Novel Color Image Encryption Algorithm Based on Hyperchaotic Maps and Mitochondrial DNA Sequences. *Entropy* **2020**, *22*, 158. [CrossRef]
47. Alvarez, G.; Li, S. Some basic cryptographic requirements for chaos-based cryptosystems. *Int. J. Bifurc. Chaos* **2006**, *16*, 2129–2151. [CrossRef]
48. Özkaynak, F. Brief review on application of nonlinear dynamics in image encryption. *Nonlinear Dyn.* **2018**, *92*, 305–313. [CrossRef]
49. Murillo-Escobar, M.A.; Meranza-Castillón, M.O.; López-Gutiérrez, R.M.; Cruz-Hernández, C. Suggested Integral Analysis for Chaos-Based Image Cryptosystems. *Entropy* **2019**, *21*, 815. [CrossRef] [PubMed]

*entropy*

MDPI

Article

Security Analysis of a Color Image Encryption Algorithm Using a Fractional-Order Chaos

Heping Wen [1,2,*], Chongfu Zhang [1,2], Lan Huang [1], Juxin Ke [3] and Dongqing Xiong [4]

1   School of Electronic and Information, Zhongshan Institute, University of Electronic Science and Technology of China, Zhongshan 528402, China; cfzhang@uestc.edu.cn (C.Z.); greentree_2001@163.com (L.H.)
2   School of Information and Communication Engineering, University of Electronic Science and Technology of China, Chengdu 611731, China
3   Center of Information and Technology, Dongguan Polytechnic, Dongguan 523808, China; kejx@dgpt.edu.cn
4   Guangdong Mechanical and Electronical College of Technology, Guangzhou 510515, China; xiongdongqing@gdmec.edu.cn
*   Correspondence: hepingwen@yeah.net

Abstract: Fractional-order chaos has complex dynamic behavior characteristics, so its application in secure communication has attracted much attention. Compared with the design of fractional-order chaos-based cipher, there are fewer researches on security analysis. This paper conducts a comprehensive security analysis of a color image encryption algorithm using a fractional-order hyperchaotic system (CIEA-FOHS). Experimental simulation based on excellent numerical statistical results supported that CIEA-FOHS is cryptographically secure. Yet, from the perspective of cryptanalysis, this paper found that CIEA-FOHS can be broken by a chosen-plaintext attack method owing to its some inherent security defects. Firstly, the diffusion part can be eliminated by choosing some special images with all the same pixel values. Secondly, the permutation-only part can be deciphered by some chosen plain images and the corresponding cipher images. Finally, using the equivalent diffusion and permutation keys obtained in the previous two steps, the original plain image can be recovered from a target cipher image. Theoretical analysis and experimental simulations show that the attack method is both effective and efficient. To enhance the security, some suggestions for improvement are given. The reported results would help the designers of chaotic cryptography pay more attention to the gap of complex chaotic system and secure cryptosystem.

Keywords: chaos; image encryption; cryptanalysis

Citation: Wen, H.; Zhang, C.; Huang, L.; Ke, J.; Xiong, D. Security Analysis of a Color Image Encryption Algorithm Using a Fractional-Order Chaos. *Entropy* **2021**, *23*, 258. https://doi.org/10.3390/e23020258

Academic Editor: Amelia Carolina Sparavigna

Received: 1 February 2021
Accepted: 13 February 2021
Published: 23 February 2021

1. Introduction

Nowadays, with the rapid development of optical fiber broadband access network, 5G and other communication technologies, the security of multimedia data, especially digital images, is of particular interest in communication networks [1]. As everyone knows, encryption is an effective means of achieving security enhancements [2]. However, traditional text encryption algorithms such as AES, DES, and IDEA are not suitable for digital images because they featured with strong correlation between adjacent pixels. To deal with the problem, various methodologies are introduced to design different image ciphers. Among them, chaos-based image encryption is the most popular one, because chaos has characteristics of sensitivity to initial values, dense periodic points, and long-term unpredictability of orbits [3–5]. In the past two decades, chaotic image encryption technology has been widely discussed and has become a research hotspot [6]. To improve the security performance of chaotic image encryption technology, various chaotic systems with resistance to dynamic degradation are studied, including quantum chaotic map [7], fractional-order chaos [8], non-degenerate hyperchaos [9], economic chaotic map [10], and cascaded chaotic systems [11], etc. However, chaotic cryptography still lacks authoritative metrics, especially in terms of security. Accordingly, many reported chaotic encryption algorithms have been broken [12–15]. As shown in Table 1, some previous chaos-based ciphers are vulnerable upon various attack methods, including chosen-ciphertext attack [16], chosen-/known-plaintext attack [12], differential cryptanalysis [17], even cipher-only attack [18]. Therefore, research on security is extremely important and has received much attention [19–33].

Table 1. Some chaos-based ciphers broken by various attack methods.

Ciphers	Broken by	Attack Methods
Fridrich et al. [34] in 1998	Xie et al. [16] in 2017	Chosen-ciphertext attack
Zhao et al. [35] in 2015	Norouzi et al. [36] in 2017	Chosen-plaintext attack
Ye [37] in 2010	Li et al. [18] in 2017	Cipher-only attack
Zhou [38] in 2015	Chen et al. [17] in 2016	Differential cryptanalysis
Song et al. [15] in 2015	Wen et al. [13] in 2019	Chosen-plaintext/cipertext attacks
Shafique et al. [14] in 2018	Wen et al. [12] in 2019	Chosen-plaintext attack

As described in Ref. [39], fractional-order chaotic systems have higher complexity and more optional key parameters and can be used as a competitive encryption scheme. Correspondingly, image encryption algorithms based on fractional-order chaotic systems have attracted the attention of researchers in recent years [35,40–42]. In 2013, Wang et al. [40] introduced a fractional-order chaos into image encryption for the first time, and gave some experiments to verify its performance. Since then, many image encryption schemes based on fractional-order chaotic systems have been proposed [35,41,42]. For example, in 2017, Zhang et al. [41] proposed a color image encryption scheme combing with fractional-order hyperchaotic system and DNA encoding. Yet, cryptanalysts have reported that some fractional-order chaotic image encryption algorithms have some fatal security issues. Exactly, Norouzi et al. [36] pointed out that the image cipher that using an improper fractional-order chaotic system was insecure, which was proposed in [35]. As far as we know, there are still few research studies concerning cryptanalysis on the ciphers based on fractional-order chaotic systems. Moreover, considering that each cryptosystem has its intrinsic characteristics, it is necessary and urgent to perform cryptanalysis on these existing ciphers.

In 2015, a color image encryption algorithm based on a fractional-order hyperchaotic system was proposed [42]. In color image encryption algorithm using a fractional-order hyperchaotic system (CIEA-FOHS), using the pseudo-random sequences generated by the fractional-order hyperchaotic system, RGB-inter permutation, RGB-intra permutation and pixel diffusion are successively performed to get cipher images from plain images. Meanwhile, the relevant pixel correlation, histogram and other experimental analysis are given to verify its security performance. However, from the perspective of cryptanalysis, we found some security defects as follows:

- The existence of an equivalent key. CIEA-FOHS encrypts the image using a pseudo-random sequence generated by fractional-order chaos. However, these sequences are not related to plaintext. Thus, these sequences can be considered as equivalent keys.
- Two-stage permutations can be equivalently simplified to only once. The reason is that the two permutations only change the position of the pixel without changing the value of the pixel.
- The paradigm of the diffusion part is insecure. According to the conclusion of Ref. [43], a class of diffusion encryption using module addition and XOR operations can be cracked with only two special plain images and their corresponding cipher images. Unfortunately, CIEA-FOHS is also the case.

Based on the three points, CIEA-FOHS cannot resist against a chosen-plaintext attack method with the divide-and-conquer strategy. More specifically, under the scenario of chosen-plaintext attack, firstly an equivalent diffusion key is obtained, and then an equivalent permutation key is achieved, and finally the original images can be restored from the encrypted images with the equivalent keys.

2. The Encryption Algorithm under Study

In this section, the fractional-order hyperchaotic system used in Reference [42] is presented, and then the specific steps of CIEA-FOHS are introduced.

2.1. Fractional-Order Hyperchaotic System

The fractional-order hyperchaotic system used in CIEA-FOHS is derived from Ref. [39], given as

$$\begin{cases} D_t^\alpha x(t) = -z - w \\ D_t^\alpha y(t) = 2y + z \\ D_t^\alpha z(t) = 14x - 14y \\ D_t^\alpha w(t) = 100(x - g(w)) \end{cases} \tag{1}$$

where x, y, z, w are the four state variables, $g(w) = w - (|w - 0.4| - |w - 0.8| - |w + 0.4| - |w + 0.8|)$, D_t^α is the fractional derivative under the definition of Caputo and α is the derivative order. The attractor of the fractional-order hyperchaotic system is shown in Figure 1.

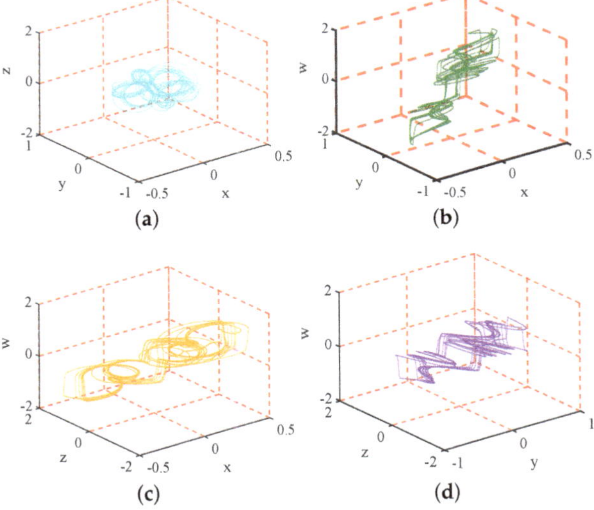

Figure 1. Attractor phase diagrams of the fractional-order hyperchaotic system with different variables: **(a)** (x, y, z); **(b)** (z, y, w); **(c)** (x, z, w); **(d)** (y, z, w).

2.2. Description of CIEA-FOHS

As shown in Figure 2, CIEA-FOHS consists of three main parts: inter-permutation, intra-permutation and pixel diffusion. It is noted that, a two-dimensional image is transformed into an one-dimensional sequence in raster scan order. Specifically, a color plain image I of size $H \times W \times 3$ is converted into three sequences of length $H \times W$ expressed as: IR, IG, and IB, which correspond to the three RGB channels of the image. The main contents are briefly introduced as follows:

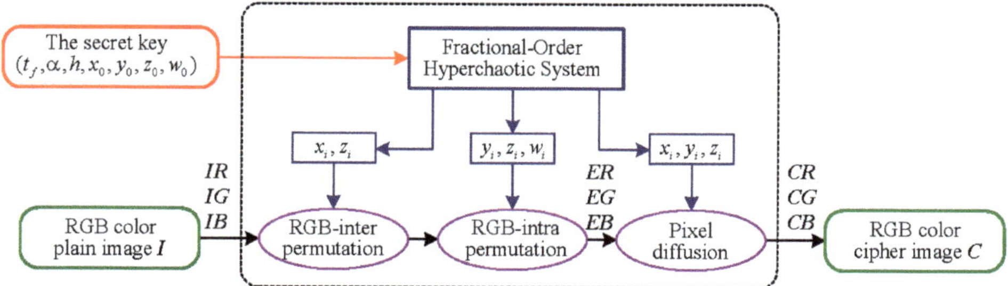

Figure 2. The block diagram of CIEA-FOHS.

- The Secret Key:
  The secret keys of CIEA-FOHS include $(t_f, \alpha, h, x_0, y_0, z_0, w_0)$, where t_f is the fractional derivative defined by Caputo definition, α is the dimension, h is the step size for discretization, and (x_0, y_0, z_0, w_0) are the four initial values of the fractional-order hyperchaotic system defined in Equation (1), respectively. In CIEA-FOHS, these keys are used to generate some chaos-based pseudo-random sequences for encryption [42].

- Initialization:
  In Equation (1), by selecting the secret key as the initial values and parameters and iterating L times, one gets four chaos-based pseudo-random sequences $\{x_i\}_{i=1}^L$, $\{y_i\}_{i=1}^L$, $\{z_i\}_{i=1}^L$ and $\{w_i\}_{i=1}^L$, where $L = H \times W$ represents the number of pixels in a single image channel.

- Stage 1. RGB-inter permutation:
  The RGB-inter permutation refers to the process of pixel replacement between channels. This stage is implemented by two control vectors $\{selE_i\}_{i=1}^L$ and $\{selLen_i\}_{i=1}^L$, which are given as

$$
\begin{cases}
selE_i = (|x_i| \times 10^{14}) \bmod 6 \\
selLen_i = (|z_i| \times 10^{14}) \bmod 3
\end{cases}
\tag{2}
$$

where $i = 1 \sim L$. More specifically, $\{selE_i\}_{i=1}^L$ is used to switch channels, as shown in Table 2, and $\{selLen_i\}_{i=1}^L$ is to control the position and length of the permutation pixel, given as

Table 2. The stutas of RGB-inter permutation under six rules.

Rule $selE(i)$	0	1	2	3	4	5
	$R \to R$	$R \to R$	$R \to G$	$R \to B$	$R \to G$	$R \to B$
Permutation status	$G \to G$	$G \to B$	$G \to R$	$G \to R$	$G \to B$	$G \to G$
	$B \to B$	$B \to G$	$B \to B$	$B \to G$	$B \to R$	$B \to R$

$$
\begin{cases}
length = (sum(ER(pos : pos + length - 1)) \bmod 64), \text{ if } selLen_i = 0 \\
length = (sum(EG(pos : pos + length - 1)) \bmod 64), \text{ if } selLen_i = 1 \\
length = (sum(EB(pos : pos + length - 1)) \bmod 64), \text{ if } selLen_i = 2
\end{cases}
\tag{3}
$$

where pos is the starting position, $length$ is the length of the permautation pixels, and sum is the cumulative function.

- Stage 2. RGB-intra permutation:
  Sort $\{y_i\}_{i=1}^L$, $\{z_i\}_{i=1}^L$ and $\{w_i\}_{i=1}^L$ to get three index sequences $\{IY_i\}_{i=1}^L$, $\{IZ_i\}_{i=1}^L$, and $\{IW_i\}_{i=1}^L$ respectively, and their values range $[1, L]$. Use $\{IY_i\}_{i=1}^L$, $\{IZ_i\}_{i=1}^L$, and

$\{IW_i\}_{i=1}^{L}$ to permute **ER**, **EG** and **EB** respectively, given as $ER_i = ER(IY_i)$, $EG_i = EG(IZ_i)$ and $EB_i = EB(IW_i)$.

- Stage 3. Pixel diffusion:
  Perform pixel diffusion on **ER**, **EG** and **EB**, and then get three channels of the cipher image **C**. Exactly, the three channels **CR**, **CG** and **CB** are defined as

$$
\begin{cases}
CR_i = SX_i \oplus ((ER_i + SX_i) \bmod 256) \oplus CR_{i-1} \\
CG_i = SY_i \oplus ((EG_i + SY_i) \bmod 256) \oplus CG_{i-1} \\
CB_i = SZ_i \oplus ((EB_i + SZ_i) \bmod 256) \oplus CB_{i-1}
\end{cases}
\tag{4}
$$

where $i = 1 \sim L$, $\oplus$ is bitwise XOR operation, mod represents modulo operation, and $CR_0 = SX_L$, $CG_0 = SY_L$, and $CB_0 = SZ_L$. Here, three diffusion sequences **SX**, **SY** and **SZ** are generated by $SX_i = round(x_i) \times 10^{14}$, $SY_i = round(y_i) \times 10^{14}$ and $SZ_i = round(z_i) \times 10^{14}$ respectively, where *round* is a rounding operation on real numbers.

Decryption is the inverse of encryption and is not described in detail here.

3. Security Analysis of CIEA-FOHS

3.1. Preliminary Analysis of CIEA-FOHS

Referring to the basic assumptions of cryptanalysis, everything about the cryptosystem is public and only the secret key is unknown for attackers [13]. Chosen-plaintext attack is a common and powerful method of cryptanalysis. It assumes that attackers can arbitrarily choose the plaintext that is conducive to deciphering and obtain the corresponding ciphertext [12]. Under the scenario of chosen-plaintext attack, attackers can construct special plain images, such as all black and all white, and obtain the corresponding cipher images to analyze the target cipher.

From the perspective of cryptanalysis, two-stage permutations of CIEA-FOHS can be treated as a global pixel permutation because they only change the pixels' position without their values. The difference is that the number of pixels performing the permutation is $3HW$ instead of HW. Then, the algorithm structure of CIEA-FOHS is actually a classic single-round permutation-diffusion. Moreover, the generation process of all chaos-based pseudo-random sequences is independent of the plain image, which means that these sequences can be regarded as an equivalent key. The reason is that, in the case of a given secret key, these sequences are fixed for encrypting different plain images with the same size. Then, CIEA-FOHS can be equivalently simplified as Figure 3, where **PM** is an equivalent permutation key and three diffusion sequences **SX**, **SY** and **SZ** serve as an equivalent diffusion key.

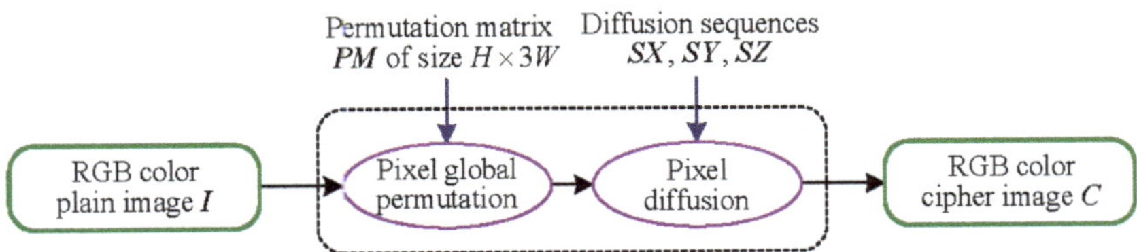

Figure 3. The block diagram of an equivalent simplified CIEA-FOHS.

Based on the above, under the scenario of chosen-plaintext attack and the strategy of divide and conquer, one can get the equivalent keys and then recover the original plain images. Specifically, firstly choose some plain images with same pixel values to cancel the permutation and get the corresponding plain images to obtain the diffusion key; then achieve the permutation key by the method of Reference [12]; finally, recover the images by the equivalent keys.

3.2. Analysis on the Diffusion Part

In this section, based on chosen-plaintext attack, it is assumed that the plaintext image with the same pixel value is selected, and the corresponding ciphertext image is obtained.

- *Step 1.* Choose the all-zero plain image $I^{(0)}$ and get the corresponding cipher image $C^{(0)}$ to determine SX_L, SY_L, SZ_L.

  The reason for choosing the all-zero image is that the permutation is invalid at this time, and the diffusion can be eliminated to the greatest extent. Then, Equation (4) becomes

$$\begin{cases} CR_i^{(0)} = CR_{i-1}^{(0)} \\ CG_i^{(0)} = CG_{i-1}^{(0)} \\ CB_i^{(0)} = CB_{i-1}^{(0)} \end{cases} \tag{5}$$

when $i = 1$, one has $CR_1^{(0)} = CR_0$. Since $CR_0 = SX_L$, thus $SX_L = CR_i^{(0)}$. Similarly, one further gets $SY_L = CG_i^{(0)}$ and $SZ_L = CB_i^{(0)}$.

- *Step 2.* Choose two special plain images and get the corresponding cipher images to determine SX_i, SY_i, SZ_i for $i = 1 \sim L-1$.

  Referring to [43,44], the two chosen plaintexts are pure-color images with pixel values of 85 and 170, represented as $I^{(85)}$ and $I^{(170)}$, respectively. Because for the combined operation of module addition and bitwise XOR, choosing these two plain images can minimize the number of solutions for SX, SY, SZ. Under the plain image $I^{(85)}$ and its corresponding cipher image $C^{(85)}$, one gets

$$\begin{cases} CR_i^{(85)} = SX_i \oplus ((85 + SX_i) \bmod 256) \oplus CR_{i-1}^{(85)} \\ CG_i^{(85)} = SY_i \oplus ((85 + SY_i) \bmod 256) \oplus CG_{i-1}^{(85)} \\ CB_i^{(85)} = SZ_i \oplus ((85 + SZ_i) \bmod 256) \oplus CB_{i-1}^{(85)} \end{cases} \tag{6}$$

Similarly, given the plain image $I^{(170)}$ and its corresponding cipher image $C^{(170)}$, one has

$$\begin{cases} CR_i^{(170)} = SX_i \oplus ((170 + SX_i) \bmod 256) \oplus CR_{i-1}^{(170)} \\ CG_i^{(170)} = SY_i \oplus ((170 + SY_i) \bmod 256) \oplus CG_{i-1}^{(170)} \\ CB_i^{(170)} = SZ_i \oplus ((170 + SZ_i) \bmod 256) \oplus CB_{i-1}^{(170)} \end{cases} \tag{7}$$

By performing bitwise on Equations (6) and (7), one further gets

$$\begin{cases} (85 \dotplus SX_i) \oplus (170 \dotplus SX_i) = CR_i^{(85)} \oplus CR_{i-1}^{(85)} \oplus CR_i^{(170)} \oplus CR_{i-1}^{(170)} \\ (85 \dotplus SY_i) \oplus (170 \dotplus SY_i) = CG_i^{(85)} \oplus CG_{i-1}^{(85)} \oplus CG_i^{(170)} \oplus CG_{i-1}^{(170)} \\ (85 \dotplus SZ_i) \oplus (170 \dotplus SZ_i) = CB_i^{(85)} \oplus CB_{i-1}^{(85)} \oplus CB_i^{(170)} \oplus CB_{i-1}^{(170)} \end{cases} \tag{8}$$

where $\dotplus$ is defined as $a \dotplus b \triangleq \bmod(a+b, 256)$. It is worth pointing out that the reason why 85 and 170 are chosen as the attack images is that their binary are 01010101 and 10101010 respectively. At this time, the number of possible solutions of SX_i, SY_i, SZ_i is the smallest, which is two. More precisely, the difference between the two solutions is 128. Then, based on Equation (8), we propose Alogrithm 1 to determine SX_i, SY_i, SZ_i, where $i = 1 \sim L-1$.

- *Step 3.* Eliminate the diffusion part by SX, SY, SZ.

Corresponding to Equation (4), the decryption process of diffusion is given as

$$\begin{cases} ER_i = (SX_i \oplus CR_i \oplus CR_{i-1} - SX_i) \bmod 256 \\ EG_i = (SY_i \oplus CG_i \oplus CG_{i-1} - SY_i) \bmod 256 \\ EB_i = (SZ_i \oplus CB_i \oplus CB_{i-1} - SZ_i) \bmod 256 \end{cases} \tag{9}$$

Thus, **ER, EG, EB** can be restored from **CR, CG, CB** with **SX, SY, SZ**, respectively.

Algorithm 1: Determining SX_i, SY_i, SZ_i for $i = 1 \sim L - 1$

Input: SX_L, SY_L, SZ_L: two chosen plain images $I^{(85)}$ and $I^{(170)}$, and their corresponding cipher images $C^{(85)}$ and $C^{(170)}$.

Output: SX_i, SY_i, SZ_i for $i = 1 \sim L - 1$

1   $i \leftarrow 1$;
2   **for** $x \leftarrow 0$ *to* 255 **do**
3   **if** $(85 \dotplus x) \oplus (170 \dotplus x) = CR_1^{(85)} \oplus CR_1^{(170)}$ **then**
4 | $SX_1 \leftarrow x$;
5   **end**
6   **if** $(85 \dotplus x) \oplus (170 \dotplus x) = CG_1^{(85)} \oplus CG_1^{(170)}$ **then**
7 | $SY_1 \leftarrow x$;
8   **end**
9   **if** $(85 \dotplus x) \oplus (170 \dotplus x) = CB_1^{(85)} \oplus CB_1^{(170)}$ **then**
10 | $SZ_1 \leftarrow x$;
11   **end**
12 **end**
13 **for** $i \leftarrow 2$ *to* $L - 1$ **do**
14   **for** $x \leftarrow 0$ *to* 255 **do**
15   **if** $(85 \dotplus x) \oplus (170 \dotplus x) = CR_i^{(85)} \oplus CR_{i-1}^{(85)} \oplus CR_i^{(170)} \oplus CR_{i-1}^{(170)}$ **then**
16 | $SX_i \leftarrow x$;
17   **end**
18   **if** $(85 \dotplus x) \oplus (170 \dotplus x) = CG_i^{(85)} \oplus CG_{i-1}^{(85)} \oplus CG_i^{(170)} \oplus CG_{i-1}^{(170)}$ **then**
19 | $SY_i \leftarrow x$;
20   **end**
21   **if** $(85 \dotplus x) \oplus (170 \dotplus x) = CB_i^{(85)} \oplus CB_{i-1}^{(85)} \oplus CB_i^{(170)} \oplus CB_{i-1}^{(170)}$ **then**
22 | $SZ_i \leftarrow x$;
23   **end**
24   **end**
25 **end**
26 **return** SX_i, SY_i, SZ_i for $i = 1 \sim L - 1$

3.3. Analysis on the Permutation Part

Once the diffusion part is broken, CIEA-FOHS degenerates into a permutation-only cipher. Based on existing research, it cannot resist a chosen-plaintext attack. The basic idea of attacking permutation-only is to construct a special plain image with unequal element values, and get the corresponding permuted image. Taking $2 \times 2 \times 3$ as an example, the process of solving **PM** is described below. First, a chosen plain image and the corresponding permuted image are given as

$$IR = \begin{bmatrix} 0 & 1 \\ 2 & 3 \end{bmatrix}; IG = \begin{bmatrix} 4 & 5 \\ 6 & 7 \end{bmatrix}; IB = \begin{bmatrix} 8 & 9 \\ 10 & 11 \end{bmatrix}$$

$$ER = \begin{bmatrix} 5 & 8 \\ 3 & 11 \end{bmatrix}; EG = \begin{bmatrix} 1 & 10 \\ 2 & 9 \end{bmatrix}; EB = \begin{bmatrix} 6 & 4 \\ 0 & 7 \end{bmatrix}$$

For ease of explanation, a matrix of size $H \times 3W$ is obtained by connecting three channels of size $H \times W$ in a row connection manner. Then, the permutation process can be described by

$$\begin{bmatrix} 0 & 1 & 4 & 5 & 8 & 9 \\ 2 & 3 & 6 & 7 & 10 & 11 \end{bmatrix} \xrightarrow{PM} \begin{bmatrix} 5 & 8 & 1 & 10 & 6 & 4 \\ 3 & 11 & 2 & 9 & 0 & 7 \end{bmatrix}$$

where PM is the permutation matrix of size $H \times 3W$. Finally, PM is determined as

$$PM = \begin{bmatrix} (2,5) & (1,3) & (1,6) & (1,1) & (1,2) & (2,4) \\ (2,3) & (2,1) & (1,5) & (2,6) & (1,4) & (2,2) \end{bmatrix} \tag{10}$$

Obviously, one can recover (IR, IG, IB) from (ER, EG, EB) with PM. However, the situation may be more complicated for large size images. For an 8-bit image, the pixel value range is $[0, 255]$. Thus, when $3HW > 256$, PM cannot be determined by only one chosen plain image and its corresponding cipher image. Fortunately, this problem has been solved in our latest research [12,13]. The basic idea is to combine multiple chosen plain images in a weighted manner to form a matrix with different elements, and the number of chosen plain images required for attacking permutation is $\lceil \log_{256}(3HW) \rceil$, where $\lceil \cdot \rceil$ is the rounding up operation.

Based on the above, the steps for attacking permutation are briefly summarized as follows:

- *Step 1.* Choose some special plain images and get their corresponding cipher images to determine the permutation matrix PM;
- *Step 2.* Use the permutation matrix PM to recover the original images from the permuted images.

3.4. The Proposed Chosen-Plaintext Attack Method

Following the above-mentioned discussion, CIEA-FOHS cannot resist the attack method proposed in this paper. The flowchart of the attack method is shown in Figure 4, and the specific steps based on chosen-plaintext attack are given as: firstly, get an equivalent diffusion key (SX, SY, SZ) by the method in Section 3.2; secondly, achieve the permutation matrix PM by the method in Section 3.3; finally, recover the original images with the equivalent keys.

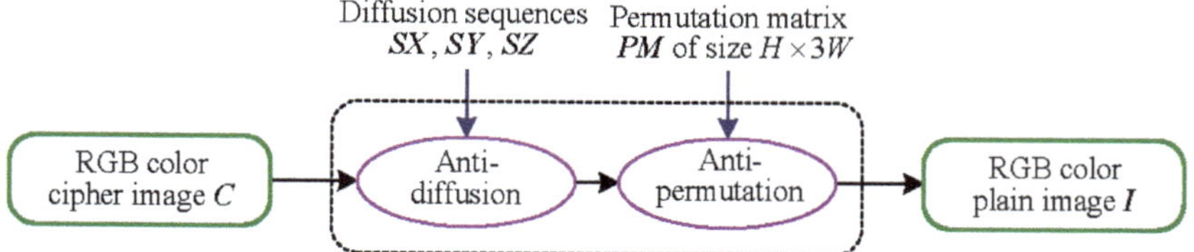

Figure 4. The overall flowchart of attacking CIEA-FOHS.

Moreover, the complexity required for the attack method is discussed here. In terms of data complexity, for color images of size $H \times W \times 3$, the number of chosen plain images required to decipher diffusion and permutation is 3 and $\lceil \log_{256}(3HW) \rceil$, respectively. Hence, the total data complexity required is $O(3 + \lceil \log_{256}(3HW) \rceil)$.

4. Experimental Verifications and Discussions

To verify our security analysis, the algorithm steps of CIEA-FOHS strictly follow Ref. [42]. Although Due to the complexity of fractional-order chaos, some parameters may not be completely consistent, but this does not affect the effectiveness of security analysis. We conduct simulation verification on the proposed image cryptosystem based on a PC (personal computer) with MATLAB r2018b. The running PC is installed with Windows 10 64-bit OS (operating system), Intel(R) Core(TM) i5-8265U CPU @ 1.60 GHz and 8 GB memory. We select some typical images listed in Table 3 for experiments. Among them, the image "Lenna" of size $256 \times 256 \times 3$ given in Ref. [42] is also included. In Equation (1), we set the experimental secret key parameters for $h = 0.001$, $\alpha = 104$, $t_f = 100$, $x_0 = 1.002$, $y_0 = 0.949$, $z_0 = 0.997$ and $w_0 = 1.103$.

- *Case 1.* Breaking CIEA-FOHS with an image of size $2 \times 2 \times 3$:

    In order to better illustrate the attack process, we first adopt an extremely simple image with a size of $2 \times 2 \times 3$. A pair of the given target plain and cipher images I and C is shown in Figure 5a,c respectively, and their histograms are shown in Figure 5b,d respectively. Accordingly, the numerical matrices of I and C are:

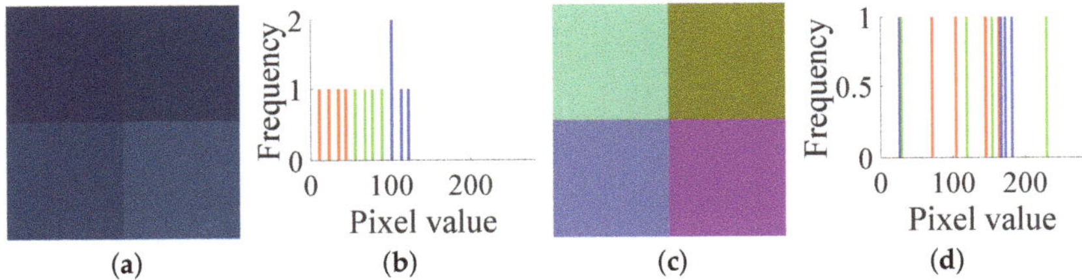

(a)   (b)   (c)   (d)

Figure 5. A pair of plain and cipher images of size $2 \times 2 \times 3$: (**a**) plain image I; (**b**) histogram of I; (**c**) cipher image C; (**d**) histogram of C.

$$IR = \begin{bmatrix} 11 & 22 \\ 33 & 44 \end{bmatrix}; IG = \begin{bmatrix} 55 & 66 \\ 77 & 88 \end{bmatrix}; IB = \begin{bmatrix} 99 & 100 \\ 111 & 122 \end{bmatrix}$$

$$CR = \begin{bmatrix} 70 & 165 \\ 103 & 145 \end{bmatrix}; CG = \begin{bmatrix} 231 & 154 \\ 118 & 28 \end{bmatrix}; CB = \begin{bmatrix} 181 & 24 \\ 171 & 165 \end{bmatrix}$$

Firstly, following Step 1 in Section 3.2, choose the all-zero plain image $I^{(0)}$ shown in Figure 6a and temporarily use the encryption machine of CIEA-FOHS, and then get the corresponding cipher image $C^{(0)}$, as shown in Figure 6c. The all-zero plain image $I^{(0)}$ and the corresponding cipher image $C^{(0)}$ and their histograms are shown in Figure 6b,d, respectively. Similarly, the numerical matrices of $I^{(0)}$ and $C^{(0)}$ are:

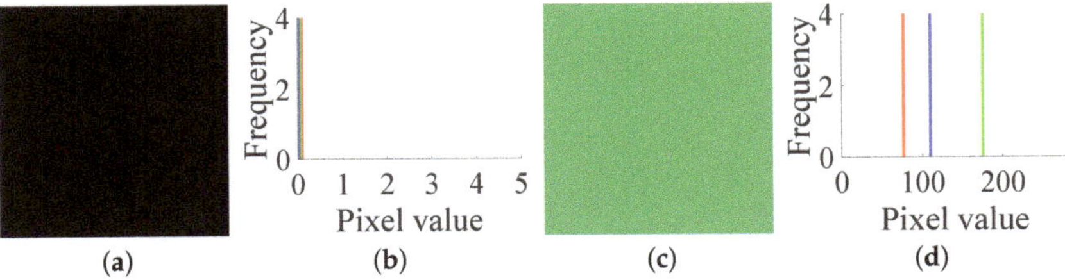

(a)   (b)   (c)   (d)

Figure 6. The all-zero chosen plain image $I^{(0)}$ and its corresponding cipher image $C^{(0)}$ of size $2 \times 2 \times 3$: (**a**) $I^{(0)}$; (**b**) histogram of $I^{(0)}$; (**c**) $C^{(0)}$; (**d**) histogram of $C^{(0)}$.

$$IR^{(0)} = \begin{bmatrix} 0 & 0 \\ 0 & 0 \end{bmatrix}; IG^{(0)} = \begin{bmatrix} 0 & 0 \\ 0 & 0 \end{bmatrix}; IB^{(0)} = \begin{bmatrix} 0 & 0 \\ 0 & 0 \end{bmatrix}$$

$$CR^{(0)} = \begin{bmatrix} 77 & 77 \\ 77 & 77 \end{bmatrix}; CG^{(0)} = \begin{bmatrix} 174 & 174 \\ 174 & 174 \end{bmatrix}; CB^{(0)} = \begin{bmatrix} 109 & 109 \\ 109 & 109 \end{bmatrix}$$

Then, one has $SX_L = 77$, $SY_L = 174$ and $SZ_L = 109$ because $SX_L = CR_0$, $SY_L = CG_0$ and $SZ_L = CB_0$, where $L = 2 \times 2 = 4$.

Secondly, based on Step 2 in Section 3.2, choose the two plain images $I^{(85)}$ and $I^{(170)}$, and get the corresponding cipher images, $C^{(85)}$ and $C^{(170)}$, which are shown in Figure 7a–d, respectively. The values of their RGB three channels are:

$$IR^{(85)} = \begin{bmatrix} 85 & 85 \\ 85 & 85 \end{bmatrix}; IG^{(85)} = \begin{bmatrix} 85 & 85 \\ 85 & 85 \end{bmatrix}; IB^{(85)} = \begin{bmatrix} 85 & 85 \\ 85 & 85 \end{bmatrix}$$

$$CR^{(85)} = \begin{bmatrix} 176 & 186 \\ 77 & 85 \end{bmatrix}; CG^{(85)} = \begin{bmatrix} 5 & 181 \\ 110 & 24 \end{bmatrix}; CB^{(85)} = \begin{bmatrix} 184 & 94 \\ 229 & 241 \end{bmatrix}$$

$$IR^{(170)} = \begin{bmatrix} 170 & 170 \\ 170 & 170 \end{bmatrix}; IG^{(170)} = \begin{bmatrix} 170 & 170 \\ 170 & 170 \end{bmatrix}; IB^{(170)} = \begin{bmatrix} 170 & 170 \\ 170 & 170 \end{bmatrix}$$

$$CR^{(170)} = \begin{bmatrix} 231 & 235 \\ 177 & 81 \end{bmatrix}; CG^{(170)} = \begin{bmatrix} 120 & 24 \\ 174 & 238 \end{bmatrix}; CB^{(170)} = \begin{bmatrix} 199 & 123 \\ 45 & 1 \end{bmatrix}$$

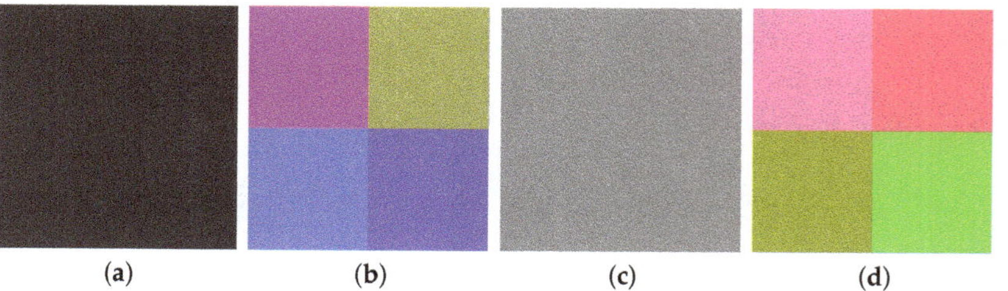

Figure 7. The two chosen plain images $I^{(85)}$, $I^{(170)}$ and their corresponding cipher images $C^{(85)}$, $C^{(170)}$ of size $2 \times 2 \times 3$: (**a**) $I^{(85)}$; (**b**) $C^{(85)}$; (**c**) $I^{(170)}$; (**d**) $C^{(170)}$.

Then, combining Algorithm 1, we determine $SX\ SY\ SZ$ as

$$SX = \begin{bmatrix} 84 & 86 & 89 & 77 \end{bmatrix}; SY = \begin{bmatrix} 63 & 31 & 71 & 46 \end{bmatrix}; SZ = \begin{bmatrix} 64 & 36 & 119 & 109 \end{bmatrix}$$

or

$$SX = \begin{bmatrix} 212 & 214 & 217 & 205 \end{bmatrix}; SY = \begin{bmatrix} 191 & 159 & 199 & 174 \end{bmatrix}; SZ = \begin{bmatrix} 192 & 164 & 247 & 237 \end{bmatrix}$$

Thirdly, by Step 3 in Section 3.2, the corresponding permuted image shown in Figure 8c can be restored from the targeted cipher image Figure 8a with $SX\ SY\ SZ$. Fourthly, following Step 1 in Section 3.3, construct some special attack images to obtain the permutation matrix PM. For images of size $2 \times 2 \times 3$, the process of solving PM is exactly the same as Section 3.3. Then, we determine the PM as Equation (10). Fifth, by Step 2 in Section 3.3, recover (IR, IG, IB) from (ER, EG, EB) with PM. Thus, the original plain image shown in Figure 8e can be recovered.

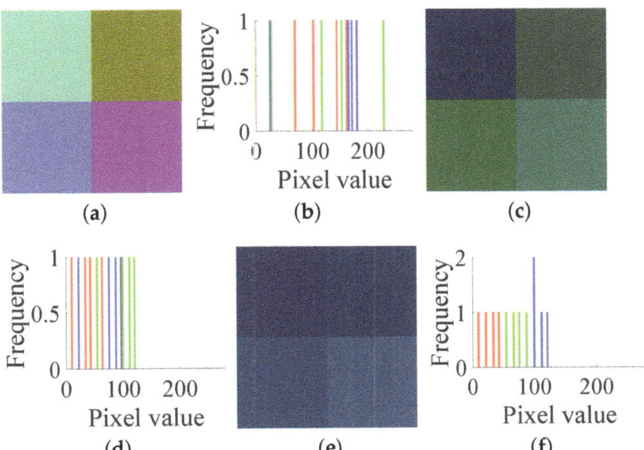

Figure 8. A target cipher image, the permuted image, the original plain image and their histograms of size $2 \times 2 \times 3$: (**a**) a target cipher image; (**b**) histogram of (**a**); (**c**) its permuted image; (**d**) histogram of (**c**); (**e**) its plain image; (**f**) histogram of (**e**).

- *Case 2.* Breaking CIEA-FOHS with "Lenna" of size $256 \times 256 \times 3$:

  Firstly, following Step 1 in Section 3.2, choose the all-zero plain image $I^{(0)}$ shown in Figure 9a and temporarily use the encryption machine of CIEA-FOHS, and then get the corresponding cipher image $C^{(0)}$, as shown in Figure 9b, and the corresponding three channel images and their histograms of $C^{(0)}$ are shown in Figure 9c,d, respectively. Exactly, one has $SX_L = 238$, $SY_L = 168$ and $SZ_L = 91$ owing to $SX_L = CR_0$, $SY_L = CG_0$ and $SZ_L = CB_0$.

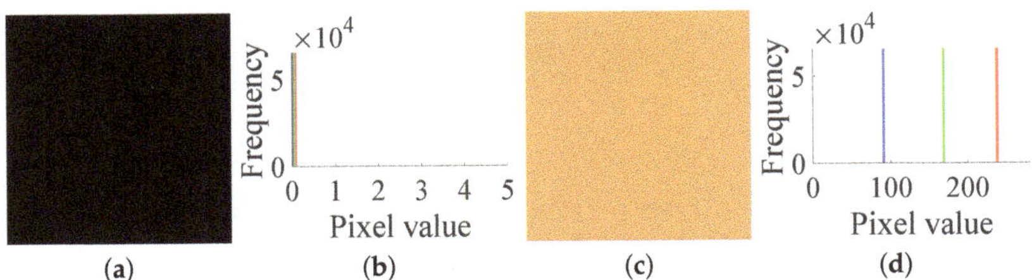

Figure 9. The all-zero chosen plain image $I^{(0)}$ and its corresponding cipher image $C^{(0)}$ of size $256 \times 256 \times 3$: (**a**) $I^{(0)}$; (**b**) histogram of $I^{(0)}$; (**c**) $C^{(0)}$; (**d**) histogram of $C^{(0)}$.

Secondly, based on Step 2 in Section 3.2, choose the two plain images, $I^{(85)}$ and $I^{(170)}$, and get the corresponding cipher images, $C^{(85)}$ and $C^{(170)}$, which are shown in Figure 10a–d, respectively.

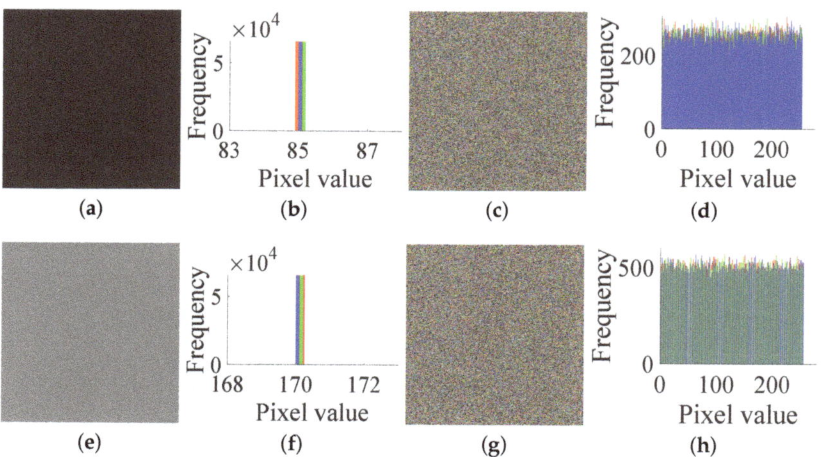

Figure 10. The two chosen plain images $I^{(85)}$, $I^{(170)}$ and their corresponding cipher images $C^{(85)}$, $C^{(170)}$ of size $256 \times 256 \times 3$: (**a**) $I^{(85)}$; (**b**) histogram of $I^{(85)}$; (**c**) $C^{(85)}$; (**d**) histogram of $C^{(85)}$; (**e**) $I^{(170)}$; (**f**) histogram of $I^{(170)}$; (**g**) $C^{(170)}$; (**h**) histogram of $C^{(170)}$.

Furthermore, one determines SX_i, SY_i, SZ_i for $i = 1 \sim L - 1$ by Algorithm 1. Thirdly, by the method in Section 3.3, choose the three plain images (shown in Figure 11a–f) and get the corresponding cipher images (shown in Figure 11g–l), and then use Algorithm 1 again to obtain their corresponding permuted images (shown in Figure 11m–r). Then, we can get **PM**.

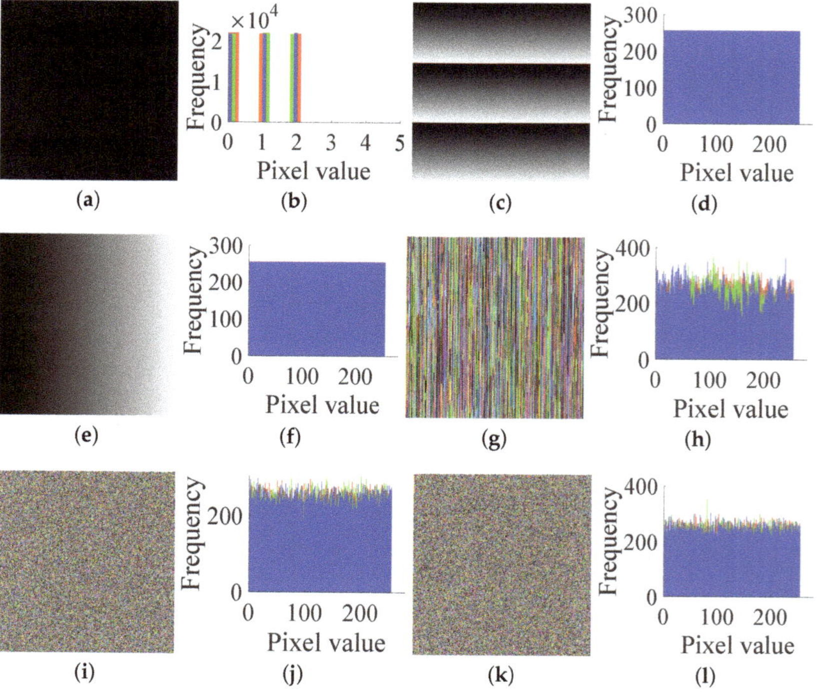

Figure 11. *Cont.*

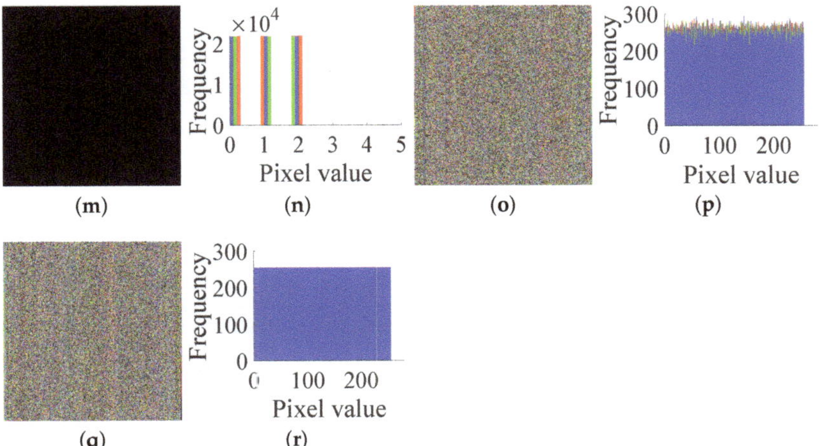

Figure 11. Three chosen plain images, the corresponding cipher and permuted images for attacking permutation: (**a**) $1^{\#}$ plain image; (**b**) The histogram of (**a**); (**c**) $2^{\#}$ plain image; (**d**) The histogram of (**c**); (**e**) $3^{\#}$ plain image; (**f**) The histogram of (**e**); (**g**) $1^{\#}$ cipher image; (**h**) The histogram of (**g**); (**i**) $2^{\#}$ cipher image; (**j**) The histogram of (**i**); (**k**) $3^{\#}$ cipher image; (**l**) The histogram of (**k**); (**m**) $1^{\#}$ permuted image; (**n**) The histogram of (**m**); (**o**) $2^{\#}$ permuted image; (**p**) The histogram of (**o**); (**q**) $3^{\#}$ permuted image; (**r**) The histogram of (**q**).

Finally, we recover the original image from the cipher image of "Lenna" shown in Figure 12a. First, the permuted image shown in Figure 12c is obtained from the cipher image with (SX, SY, SZ). Then, the plain image is restored by PM, which is shown in Figure 12e.

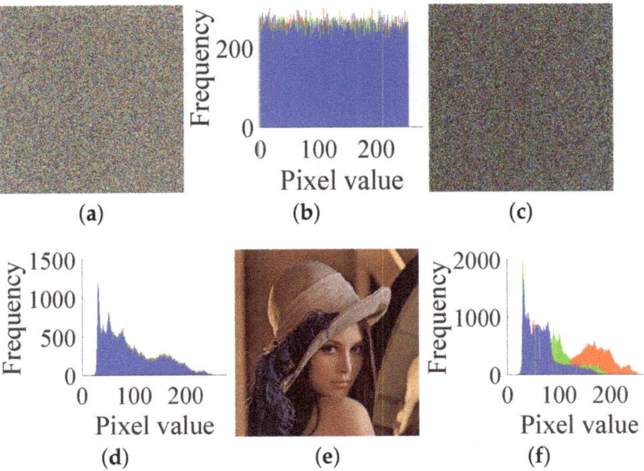

Figure 12. The cipher image, the permuted image, the original plain image of "Lenna" and their histograms of size $256 \times 256 \times 3$: (**a**) the cipher image; (**b**) histogram of (**a**); (**c**) its permuted image; (**d**) histogram of (**c**); (**e**) its plain image; (**f**) histogram of (**e**).

Without loss of generality, we do the experiments based on other images with different sizes. The experimental results are shown in Table 3 and Figure 13. They both verify the effectiveness of our attack method. Besides, it can be seen from Table 3 that the proposed attack is efficient. Taking the image "Lenna" of size $256 \times 256 \times 3$ as an example, when the encryption time is 0.6391 s, the time needed for the corresponding attack is just 129.4039 s. Even if the image size increases, the time required for the attack is still within an acceptable range. Thus, it verifies that our method is computationally feasible.

Moreover, we verified the data complexity required for the attack. As discussed in Section 3.4, the total data complexity required for breaking CIEA-FOHS is $O(3 + \lceil \log_{256}(3HW) \rceil)$. In our experiment with chosen-plaintext attack, the number of attack images required for sizes $2 \times 2 \times 3$ and $100 \times 100 \times 3$ are 4 and 5, respectively. And for sizes $300 \times 200 \times 3$, $256 \times 256 \times 3$ and $512 \times 512 \times 3$, the number of attack images required are all 6. Therefore, the experimental verification is consistent with the theoretical calculation.

Table 3. The time required for breaking CIEA-FOHS by our proposed attack method (unit: second).

Images	Sizes	Encryption Time	Attacking Diffusion			Attacking Permutation		Totol Attacking Time
			Step 1	*Step* 2	*Step* 3	*Step* 1	*Step* 2	
Figure 5a	$2 \times 2 \times 3$	0.0280	0.1559	0.1811	1.0297	0.0244	2.7151	4.1502
Figure 13b	$100 \times 100 \times 3$	0.1539	0.0920	19.6092	1.1407	0.2764	2.7102	24.0427
Figure 13d	$300 \times 200 \times 3$	0.3280	0.5092	101.7737	0.7872	0.9055	2.4353	106.8545
Figure 12e	$256 \times 256 \times 3$	0.6391	0.6913	120.4768	1.6147	1.9642	3.7725	129.4039
Figure 13f	$512 \times 512 \times 3$	3.5386	2.8134	988.3704	1.9930	4.2884	5.0459	1004.4617

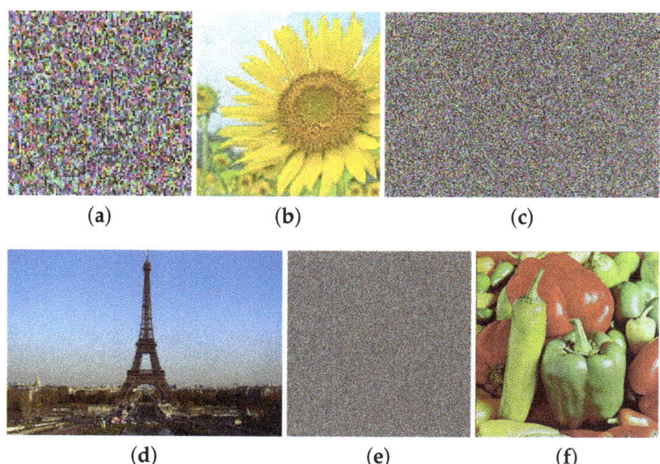

Figure 13. Attacking results with three images of size $100 \times 100 \times 3$, $300 \times 200 \times 3$ and $512 \times 512 \times 3$ respectively: (**a**) cipher image of size $100 \times 100 \times 3$; (**b**) plain image of (**a**); (**c**) cipher image of size $300 \times 200 \times 3$; (**d**) plain image of (**c**); (**e**) cipher image of size $512 \times 512 \times 3$; (**f**) plain image of (**e**).

5. Suggestions for Improvement

On the basis of the above, CIEA-FOHS is insecure against a chosen-plaintext attack method because of its inherent security defects. To enhance the security, some suggestions for improvement are listed below:

- Suggestion 1. Ensuring the substantial security contribution of the fractional-order chaos to the corresponding cipher. The attractor phase diagram of the fractional-order hyperchaotic system is shown in Figure 1, which shows the extremely complex dynamics. Undoubtedly, fractional-order chaos is one of the preferred sources of entropy for encryption. However, due to the negligence of algorithm design, CIEA-FOHS has serious security defects and is attacked.

- Suggestion 2. Security analysis should be implemented from the perspective of cryptography, not limited to numerical statistical verification. As Ref. [45] points out, many encryption algorithms have excellent statistical analysis results, but they are still insecure. In fact, good statistical analysis results are only a necessary and not a sufficient condition for security. Some security flaws are difficult to reflect with numerical statistical results, but they can be clearly revealed by theoretical security analysis. For example, the existence of an equivalent key makes CIEA-FOHS vulnerable to cryptographic attacks. Given the implementation of detailed cryptographic security analysis, these flaws can be avoided, thereby improving security.

6. Conclusions

In this paper, a detailed security analysis of a color image encryption algorithm named CIEA-FOHS using a fractional-order chaos was performed. From the perspective of cryptanalysis, this paper found that CIEA-FOHS can be broken by a chosen-plaintext attack method, owing to its some inherent security defects. Theoretical analysis and experimental simulations show that the attack method is both effective and efficient for attacking CIEA-FOHS. Although the fractional-order chaotic system has complex dynamics, the algorithm defects may cause insecurity. The reported results would help the designers of chaotic cryptography pay more attention to the gap between complex chaotic system and secure cryptosystem.

Author Contributions: Methodology, H.W.; Software, H.W. and J.K.; Validation, H.W., L.H. and C.Z.; Supervision, C.Z.; Project Administration, C.Z. and D.X.; Funding Acquisition, C.Z., H.W. and D.X. All authors have read and agreed to the published version of the manuscript.

Funding: This work was supported partly by the National Science Foundation of China (62071088, 61571092), Project for National Key RD Program of China (2018YFB1801302), Project for Innovation Team of Guangdong University (2018KCXTD033), Project for Zhongshan Social Public Welfare Science and Technology (2019B2007), Zhongshan Innovative Research Team Program (180809162197886), Research Project for Talent of UESTC Zhongshan Institute (418YKQN07, 419YKQN23), Natural Science Project for Young Innovative Talents by the Department of Education of Guangdong Province (2019KQNCX191), Characteristic Innovation Project of Department of Guangdong Province (2017GWTSCX010).

Data Availability Statement: No applicable.

Conflicts of Interest: The authors declare no conflict of interest.

References

1. Wu, T.; Zhang, C.; Chen, C.; Hou, H.; Wei, H.; Hu, S.; Qiu, K. Security enhancement for OFDM-PON using Brownian motion and chaos in cell. *Opt. Express* **2018**, *26*, 22857–22865. [CrossRef]
2. Wu, T.; Zhang, C.; Chen, Y.; Cui, M.; Huang, H.; Zhang, Z.; Wen, H.; Zhao, X.; Qiu, K. Compressive sensing chaotic encryption algorithms for OFDM-PON data transmission. *Opt. Express* **2021**, *29*, 3669–3684. [CrossRef]
3. Wen, H.; Zhang, C.; Chen, P.; Chen, R.; Xu, J.; Liao, Y.; Liang, Z.; Shen, D.; Zhou, L.; Ke, J. A quantum chaotic image cryptosystem and its application in IoT secure communication. *IEEE Access* **2021**, *1*. [CrossRef]
4. Li, C.; Feng, B.; Li, S.; Kurths, J.; Chen, G. Dynamic analysis of digital chaotic maps via state-mapping networks. *IEEE Trans. Circuits Syst. I Regul. Pap.* **2019**, *66*. [CrossRef]
5. Li, C.; Tan, K.; Feng, B.; Lu, J. The Graph Structure of the Generalized Discrete Arnold's Cat Map. *IEEE Trans. Comput.* **2021**, *1*. [CrossRef]
6. Li, C.; Zhang, Y.; Xie, E.Y. When an attacker meets a cipher-image in 2018: A Year in Review. *J. Inf. Secur. Appl.* **2019**, *48*. [CrossRef]
7. Akhshani, A.; Akhavan, A.; Mobaraki, A.; Lim, S.C.; Hassan, Z. Pseudo random number generator based on quantum chaotic map. *Commun. Nonlinear Sci. Numer. Simul.* **2014**, *19*, 101–111. [CrossRef]
8. He, S.; Sun, K.; Wang, H. Complexity analysis and DSP implementation of the fractional-order lorenz hyperchaotic system. *Entropy* **2015**, *17*, 8299–8311. [CrossRef]
9. Shen, C.; Yu, S.; Lü, J.; Chen, G. Designing Hyperchaotic Systems With Any Desired Number of Positive Lyapunov Exponents via A Simple Model. *IEEE Trans. Circuits Syst. I Regul. Pap.* **2014**, *61*, 2380–2389. [CrossRef]
10. Askar, S.S.; Karawia, A.; Al-Khedhairi, A.; Al-Ammar, F.S. An algorithm of image encryption using logistic and two-dimensional chaotic economic maps. *Entropy* **2019**, *21*, 44. [CrossRef]

11. Zhou, Y.; Hua, Z.; Pun, C.; Philip Chen, C.L. Cascade Chaotic System with Applications. *IEEE Trans. Cybern.* **2015**, *45*, 2001–2012. [CrossRef] [PubMed]
12. Wen, H.; Yu, S. Cryptanalysis of an image encryption cryptosystem based on binary bit planes extraction and multiple chaotic maps. *Eur. Phys. J. Plus* **2019**, *134*, 1–16. [CrossRef]
13. Wen, H.; Yu, S.; Lü, J. Breaking an Image Encryption Algorithm Based on DNA Encoding and Spatiotemporal Chaos. *Entropy* **2019**, *21*, 246. [CrossRef]
14. Shafique, A.; Shahid, J. Novel image encryption cryptosystem based on binary bit planes extraction and multiple chaotic maps. *Eur. Phys. J. Plus* **2018**, *133*, 331. [CrossRef]
15. Song, C.; Qiao, Y. A novel image encryption algorithm based on DNA encoding and spatiotemporal chaos. *Entropy* **2015**, *17*, 6954–6968. [CrossRef]
16. Xie, Y.; Li, C.; Yu, S.; Lü, J. On the cryptanalysis of Fridrich's chaotic image encryption scheme. *Signal Process.* **2017**, *132*, 150–154. [CrossRef]
17. Chen, L.; Ma, B.; Zhao, X.; Wang, S. Differential cryptanalysis of a novel image encryption algorithm based on chaos and Line map. *Nonlinear Dyn.* **2016**, *87*, 1797–1807. [CrossRef]
18. Li, C.; Lin, D.; Lü, J. Cryptanalyzing an Image-Scrambling Encryption Algorithm of Pixel Bits. *IEEE Multimed.* **2017**, *3*, 64–71. [CrossRef]
19. Wang, L.; Sun, K.; Peng, Y.; He, S. Chaos and complexity in a fractional-order higher-dimensional multicavity chaotic map. *Chaos Solitons Fractals* **2020**, *131*, 109488. [CrossRef]
20. Peng, D.; Sun, K.; He, S.; Zhang, L.; Alamodi, A.O.A. Numerical analysis of a simplest fractional-order hyperchaotic system. *Theor. Appl. Mech. Lett.* **2019**, *9*, 220–228. [CrossRef]
21. He, S.; Sun, K.; Wang, H. Dynamics and synchronization of conformable fractional-order hyperchaotic systems using the Homotopy analysis method. *Commun. Nonlinear Sci. Numer. Simul.* **2019**, *73*, 146–164. [CrossRef]
22. Chai, X.; Bi, J.; Gan, Z.; Liu, X.; Zhang, Y.; Chen, Y. Color image compression and encryption scheme based on compressive sensing and double random encryption strategy. *Signal Process.* **2020**, *176*, 107684. [CrossRef]
23. Chai, X.; Wu, H.; Gan, Z.; Han, D.; Zhang, Y.; Chen, Y. An efficient approach for encrypting double color images into a visually meaningful cipher image using 2D compressive sensing. *Inf. Sci.* **2020**, doi:10.1016/j.ins.2020.10.007. [CrossRef]
24. Wang, X.; Chen, S.; Zhang, Y. A chaotic image encryption algorithm based on random dynamic mixing. *Opt. Laser Technol.* **2021**, *138*, 106837. [CrossRef]
25. Hua, Z.; Zhu, Z.; Yi, S.; Zhang, Z.; Huang, H. Cross-plane colour image encryption using a two-dimensional logistic tent modular map. *Inf. Sci.* **2021**, *546*, 1063–1083. [CrossRef]
26. Kamal, F.M.; Elsonbaty, A.; Elsaid, A. A novel fractional nonautonomous chaotic circuit model and its application to image encryption. *Chaos Solitons Fractals* **2021**, *144*, 110686. [CrossRef]
27. Mani, P.; Rajan, R.; Shanmugam, L.; Joo, Y.H. Adaptive control for fractional order induced chaotic fuzzy cellular neural networks and its application to image encryption. *Inf. Sci.* **2019**, *491*, 74–89. [CrossRef]
28. Yang, F.; Mou, J.; Liu, J.; Ma, C.; Yan, H. Characteristic analysis of the fractional-order hyperchaotic complex system and its image encryption application. *Signal Process.* **2020**, *169*, 107373. [CrossRef]
29. Lahdir, M.; Hamiche, H.; Kassim, S.; Tahanout, M.; Kemih, K.; Addouche, S. A novel robust compression-encryption of images based on SPIHT coding and fractional-order discrete-time chaotic system. *Opt. Laser Technol.* **2019**, *109*, 534–546. [CrossRef]
30. Yang, F.; Mou, J.; Ma, C.; Cao, Y. Dynamic analysis of an improper fractional-order laser chaotic system and its image encryption application. *Opt. Lasers Eng.* **2020**, *129*, 106031. [CrossRef]
31. Yu, S.; Zhou, N.; Gong, L.; Nie, Z. Optical image encryption algorithm based on phase-truncated short-time fractional Fourier transform and hyper-chaotic system. *Opt. Lasers Eng.* **2020**, *124*, 105816. [CrossRef]
32. Sayed, W.S.; Radwan, A.G. Generalized switched synchronization and dependent image encryption using dynamically rotating fractional-order chaotic systems. *AEU–Int. J. Electron. Commun.* **2020**, *123*, 153268. [CrossRef]
33. Yang, Y.; Guan, B.; Li, J.; Li, D.; Zhou, Y.; Shi, W. Image compression-encryption scheme based on fractional order hyper-chaotic systems combined with 2D compressed sensing and DNA encoding. *Opt. Laser Technol.* **2019**, *119*, 105661. [CrossRef]
34. Fridrich, J. Symmetric Ciphers Based On Two-Dimensional Chaotic Maps. *Int. J. Bifurc. Chaos* **1998**, *8*, 1259–1284. [CrossRef]
35. Zhao, J.; Wang, S.; Chang, Y.; Li, X. A novel image encryption scheme based on an improper fractional-order chaotic system. *Nonlinear Dyn.* **2015**, *80*, 1721–1729. [CrossRef]
36. Norouzi, B.; Mirzakuchaki, S. Breaking a novel image encryption scheme based on an improper fractional order chaotic system. *Multimed. Tools Appl.* **2017**, *76*, 1817–1826. [CrossRef]
37. Ye, G. Image scrambling encryption algorithm of pixel bit based on chaos map. *Pattern Recognit. Lett.* **2010**, *31*, 347–354. [CrossRef]
38. Zhou, G.; Zhang, D.; Liu, Y.; Yuan, Y.; Liu, Q. A novel image encryption algorithm based on chaos and Line map. *Neurocomputing* **2015**, *169*, 150–157. [CrossRef]
39. Huang, X.; Zhao, Z.; Wang, Z.; Li, Y. Chaos and hyperchaos in fractional-order cellular neural networks. *Neurocomputing* **2012**, *94*, 13–21. [CrossRef]
40. Wang, Z.; Huang, X.; Li, Y.; Song, X. Image encryption based on a delayed fractional-order chaotic logistic system. *Chin. Phys. B* **2013**, *22*, 010504. [CrossRef]

41.	Zhang, L.; Sun, K.; Liu, W.; He, S. A novel color image encryption scheme using fractional-order hyperchaotic system and DNA sequence operations. *Chin. Phys. B* **2017**, *26*, 100504. [CrossRef]
42.	Huang, X.; Sun, T.; Li, Y.; Liang, J. A Color Image Encryption Algorithm Based on a Fractional-Order Hyperchaotic System. *Entropy* **2015**, *17*, 28–38. [CrossRef]
43.	Li, C.; Liu, Y.; Zhang, L.Y.; Chen, M.Z.Q. Breaking a chaotic image encryption algorithm based on modulo addition and XOR operation. *Int. J. Bifurc. Chaos* **2013**, *23*, 1350075. [CrossRef]
44.	Zhang, L.Y.; Liu, Y.; Pareschi, F.; Zhang, Y.; Wong, K.; Rovatti, R.; Setti, G. On the Security of a Class of Diffusion Mechanisms for Image Encryption. *IEEE Trans. Cybern.* **2018**, *48*, 1163–1175. [CrossRef] [PubMed]
45.	Preishuber, M.; Hütter, S.K.T.; Uhl, A. Depreciating Motivation and Empirical Security Analysis of Chaos-Based Image and Video Encryption. *IEEE Trans. Inf. Forensics Secur.* **2018**, *13*, 2137–2150. [CrossRef]

Article

Cryptographic Algorithm Using Newton-Raphson Method and General Bischi-Naimzadah Duopoly System

Abdelrahman Karawia

Mathematics Department, Faculty of Science Mansoura University, Mansoura 35516, Egypt; abibka@mans.edu.eg

Abstract: Image encryption is an excellent method for the protection of image content. Most authors used the permutation-substitution model to encrypt/decrypt the image. Chaos-based image encryption methods are used in this model to shuffle the rows/columns and change the pixel values. In parallel, authors proposed permutation using non-chaotic methods and have displayed good results in comparison to chaos-based methods. In the current article, a new image encryption algorithm is designed using combination of Newton-Raphson's method (non-chaotic) and general Bischi-Naimzadah duopoly system as a hyperchaotic two-dimensional map. The plain image is first shuffled by using Newton-Raphson's method. Next, a secret matrix with the same size of the plain image is created using general Bischi-Naimzadah duopoly system. Finally, the XOR between the secret matrix and the shuffled image is calculated and then the cipher image is obtained. Several security experiments are executed to measure the efficiency of the proposed algorithm, such as key space analysis, correlation coefficients analysis, histogram analysis, entropy analysis, differential attacks analysis, key sensitivity analysis, robustness analysis, chosen plaintext attack analysis, computational analysis, and NIST statistical Tests. Compared to many recent algorithms, the proposed algorithm has good security efficiency.

Keywords: Newton-Raphson's method; chaos; image encryption/decryption; security analysis

---

Citation: Karawia, A. Cryptographic Algorithm Using Newton-Raphson Method and General Bischi-Naimzadah Duopoly System. *Entropy* **2021**, *23*, 57. https://doi.org/10.3390/e23010057

Received: 7 November 2020
Accepted: 26 December 2020
Published: 31 December 2020

Publisher's Note: MDPI stays neutral with regard to jurisdictional claims in published maps and institutional affiliations.

1. Introduction

Digital images play a critical role in the world today. Digital images make up 70% of the transmitted data via the Internet [1]. They often contain sensitive and valuable information which requires protection against unauthorised access in various applications such as military images, medical images and Satellite images. Therefore, researchers have been designing methods to protect digital images from piracy while they are transferred from one place to another such as encryption algorithms via chaos [2–6], DNA coding [7], and wavelets [8]. Also S-boxes play an excellent role in confirming the resistance of block ciphers against cryptanalysis [9]. In Reference [10], the authors presented an efficient algorithm based on a class of Mordell elliptic curves to generate S-boxes. One of the most stable and powerful public key cryptosystems has been proven to be the Elliptic Curve Cryptography, which is popular for its high performance. But improving protection by increasing the duration of the key is inefficient [11,12].

Among the many ways of image cryptography, the image cryptography based on chaotic map will selected over the past two decades. This is because the chaotic mappings have necessary proprieties such as high sensitivity to the initial conditions and the parameters, nonlinearity, non-periodicity, and pseudorandomness [13–17]. Numerous researchers have presented image cryptography algorithms via chaotic maps. Some of these algorithms have limited key space, weak keys, vulnerability to chosen plaintext/ciphertext attacks [18–20]. Almost all the authors used the permutation-substitution (confusion-diffusion) model to encrypt/decrypt the image. There are different permutation methods, from performing a shuffling to rows/columns to performing more complicated iterative processes. For example, in Reference [3], the authors proposed rows/columns shuffling

---

algorithm using the logistic map to get permutation. Karawia in Reference [6] suggested an image encryption algorithm using Fisher-Yates shuffling to obtain permutation while Shakiba in Reference [21] performed cyclic shifts to the rows/columns via Chebyshev mapping to achieve permutation. Xiao et al. in Reference [22] used switch control mechanism to perform permutation for rows and columns of plain image. For substitution, majority of the authors applied the XOR processes [3–6,21,23], or addition modular 256 during the substitution stage of encryption [24]. There are many maps (chaotic and hyperchaotic) utilized to design encryption algorithms, for example, 1D chaotic map in Reference [25], 2D generalized Arnold map in Reference [26], 3D Cat chaotic map in Reference [27,28], and 4D chaotic map in Reference [29].

Many of the known chaotic image encryption algorithms are resistanceless for chosen plaintext attacks(CPA). These image encryption algorithms are broken by Li et al., algorithm [30], such as References [25,31,32]. To avoid this, the image encryption algorithm must be dependent on the plain image and randomized [21,33]. Based on the dimension of the chaotic map, most of 1D-chaotic maps have simple forms and simple chaotic orbits and can be guessed. So image encryption based on 1D chaotic maps are low secure [19,34]. On the contrary, the hyperchaotic maps have more complicated form and complicated chaotic performance which make expectation of their chaotic orbits is difficult [35].

In the current article, we design an image encryption algorithm that uses Newton-Raphson's method, to shuffle the rows/columns of the plain image, and the general Bischi-Naimzadah duopoly system, to diffuse the pixels of the shuffled image. The general Bischi-Naimzada is selected to solve three essential problems: (i) the randomness of the chaotic sequences, (ii) the space of the secret key, and (iii) improving the security compared with the algorithms in literature. The chaotic sequence generated from it is extremely random. Also, it has eight parameters and two initial values and thus increasing the secret key space for the image encryption algorithm. In this algorithm, the key mixing proportion factor K is utilized to generate the secret key [36]. So, the proposed algorithm depends on the plain image and it can provide CPA-security. For more details about chaos based image encryption techniques, see Reference [37].

The main contributions of the current article are: (i) using a 2D chaotic map (the general Bischi-Naimzadah duopoly system) with a large positive Lyapunov exponent, wide and uniform distribution, (ii) Performing rows/columns shuffle for the plain image using pseudo-random sequence generation based on Newton-Raphson's method, (iii) performing pixel diffusion to the shuffled image, and (iv) offering CPA-security for our algorithm.

This article is prepared as follows. In Section 2, the general Bischi-Naimzadah duopoly system is presented. The proposed algorithm is introduced in Section 3. In Section 4, security experimental results and comparative analyses are given. Finally, conclusions are mentioned in Section 5.

2. General Bischi-Naimzadah Duopoly System (GBNDS)

The image encryption needs a sequences of random numbers to generate a good secret image. The current paper takes advantage of the effectiveness of the general Bischi-Naimzadah duopoly system to generate pseudorandom numbers. The general Bischi-Naimzada game is a market vying between two companies based on sales constraints with the aim of maximising profits. The general Bischi-Naimzadah duopoly system is mathematically defined as [38]:

$$q_1(t+1) = q_1(t) + v_1 q_1(t)[(1 - \mu_1)(a - 2bq_1(t) - bq_2(t)) - c_1]$$
$$q_2(t+1) = q_2(t) + v_2 q_2(t)[(1 - \mu_2)(a - 2bq_2(t) - bq_1(t)) - c_2], \tag{1}$$

where

q_i: the output of company $i = 1, 2$,
$a > 0$: constant price,
$b > 0$: the market price slope,
c_i: the marginal cost, $i = 1, 2$,

$\mu_i > 0$: associated with the sales constraint, $i = 1, 2$,
$\nu_i > 0$: the adjustment speed of company $i = 1, 2$.

The chaotic behavior of system (1) is observed by the values of the parameters: $a = 11.25$, $b = 0.5$, $c_1 = 0.20$, $c_2 = 0.30$, $\mu_1 = 0.002$, $\mu_2 = 0.60$, $\nu_1 = 0.20$, $\nu_2 = 0.70$ and initial values $q_{10} = 0.10$, $q_{20} = 0.20$. Figure 1a displays the bifurcation diagram of system (1) regarding the parameter μ_1. Lyapunov exponent of system (1) regarding the parameter μ_1 is shown in Figure 1b. Figure 2 displays phase diagram of system (1). It presents four unconnected chaotic areas. Whereas the phase diagram of system (1) at $\mu_1 = 0.97$ is given in Figure 3 and it presents a chaotic attractor. The main advantages of the proposed coding scheme compared to other systems in the literature is that the chaotic coding sequence extracted from the **GBNDS** is extremely random. This is because this it contains many chaotic regions for different values of the parameters. The proposed system also shows a great positive feature, which is the emergence of a very wide range of chaos and complex dynamics with the parameter μ_1 in which the system (1) shows very complex chaotic behavior [38]. Moreover, incorporating the effects of sales constraints into the form has the advantage of increasing the number of parameters in the form and thus expanding the secret key space for the cryptography process. Also, a stable coexistence of multiple chaotic attractions is observed in this case [38].

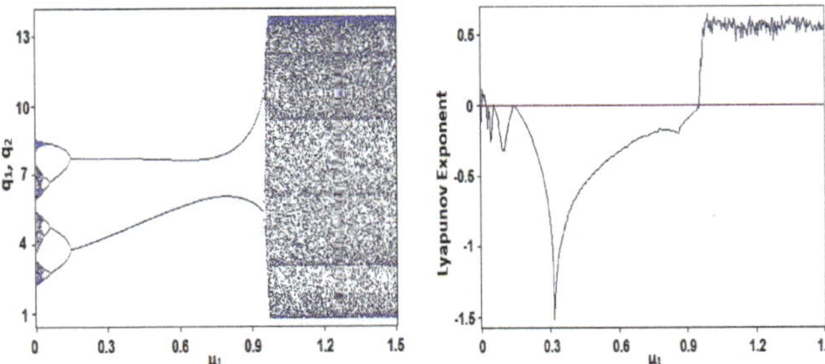

Figure 1. (**left**) Bifurcation diagram of system (1) regarding μ_1, (**right**) Lyapunov exponent of system (1) regarding μ_1.

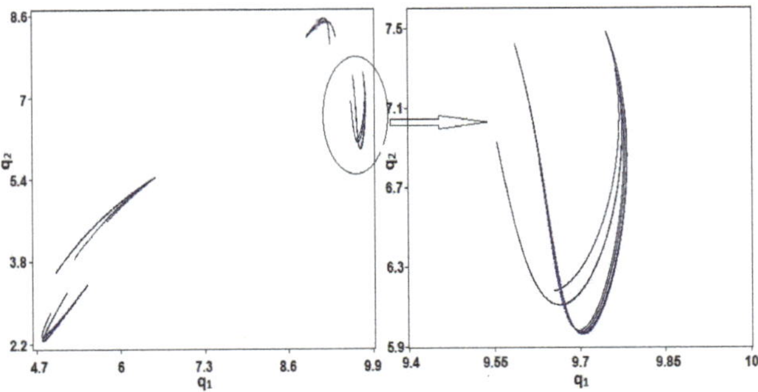

Figure 2. Phase diagram of system (1) for $a = 11.25$, $b = 0.5$, $c_1 = 0.20$, $c_2 = 0.30$, $\mu_1 = 0.002$, $\mu_2 = 0.60$, $\nu_1 = 0.20$, $\nu_2 = 0.70$, $q_{10} = 0.10$, and $q_{20} = 0.20$.

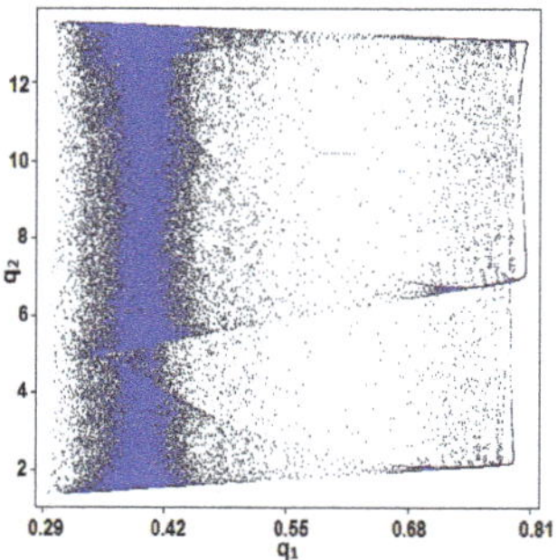

Figure 3. Phase diagram of system (1) for $a = 11.25, b = 0.5, c_1 = 0.20, c_2 = 0.30, \mu_1 = 0.97,$ $\mu_2 = 0.60, \nu_1 = 0.20, \nu_2 = 0.70, q_{10} = 0.10,$ and $q_{20} = 0.20$.

3. The Proposed Algorithm

The proposed method applies confusion-diffusion model to encrypt the plain image. Sequences are generated by the points, calculated by Newton-Raphson's method, on a polynomial function. Based on these sequences, the rows/columns of the plain image are shuffled (confusion phase). By applying *XOR* between the shuffled image and the generated values of the chaotic system (1), the diffusion stage modifies the pixel values. In the current section, the key generation, rows/columns Shuffling, and an image encryption/decryption algorithms are presented.

3.1. The Key Generation

Suppose that $\mathbf{O} = (o_{ij})$, $i = 1, 2, ..., M$, and $j = 1, 2, ..., N$, is the plain image. The secret key is generated by using the key mixing proportion factor K as follows [36]:

$$K_s = \frac{1}{256}mod\left(\sum_{i=\lceil\frac{(s-1)M}{2}\rceil+1}^{\lceil\frac{sM}{2}\rceil}\sum_{j=1}^{N}o_{ij}, 256\right), \quad s = 1, 2, \tag{2}$$

and, the key values ζ_s is changed via the following formula:

$$\zeta_s \leftarrow \frac{(\zeta_s + K_s)}{2}, \quad s = 1, 2, \tag{3}$$

where $[x]$ denoted to the nearest integer and ζ_s denoted to q_{s0}, $s = 1, 2$.

The key space of the proposed algorithm consists of the polynomial function with degree k and limits of $[\alpha, \beta]$ for the Newton-Raphson's method, two initial values and eight parameters for *GBNDS*. Then, for the confusion phase, select two values, α, β, and one polynomial function based on Newton-Raphson's method, and for diffusion phase, two initial values, q_{10}, q_{20}, and eight parameters $a, b, c_1, c_2, \mu_1, \mu_2, \nu_1, \nu_2$ for the System (1).

3.2. Rows/Columns Shuffling (Confusion Phase)

In this section, we design a technique for generating a random permutation of the integers $\{1, 2, \ldots, n\}$. Then, we shuffle the rows/columns of the plain image via the random permutation sequences.

Suppose a polynomial function of degree s, $p(x) = \sum_{i=1}^{s} a_i x^i$, where $a_s \neq 0$ and $s > 1$, is defined on the interval $[\alpha, \beta]$. Take $x_0 = (\alpha + \beta)/2$ and Newton-Raphson's method generates the sequence $\{x_i\}_{i=0}^{\infty}$ by the following formula:

$$x_i = x_{i-1} - \frac{p(x_{i-1})}{p'(x_{i-1})}, \quad p'(x_{i-1}) \neq 0 \quad \forall \quad i = 1, 2, 3, \ldots \tag{4}$$

Suppose, the Newton-Raphson's method generates the points $\{x_1, x_2, x_3, \ldots, x_n\}$. To get more randomness, instead of $p(x_i)$, the sequence is defined as the fraction part of $p(x_i)$. This sequence depends on the polynomial $p(x)$ and the interval $[\alpha, \beta]$.

The standard NIST SP300-22 test is used to assess the efficiency of the pseudorandom number generator(PRNG) of Newton-Raphson's method, and Table 1 gives the test results. In Table 1, the random number generator has passed all the tests. So, it has a good randomness.

Table 1. NIST statistical test for PRNG-Newton-Raphson's method.

Statistical Test	PRNG	Result
Frequency monobit test	100/100	PASS
Block frequency test	99/100	PASS
Rank test	99/100	PASS
Runs test	97/100	PASS
Longest runs test	99/100	PASS
Cumulative sums test	100/100	PASS
Discrete Fourier transform	100/100	PASS
Random excursion test	56/58	PASS
Random excursion variant test	57/58	PASS
Universal test	96/100	PASS
Approximate entropy	97/100	PASS
Linear complexity test	99/100	PASS
Serial	99/100	PASS
Non Overlapping templates test	97/100	PASS
Overlapping templates test	100/100	PASS

Algorithm 1 is proposed to generate a random permutation of the integers $\{1, 2, \ldots, n\}$ based on Newton-Raphson's method as follows:

Algorithm 1 Random-Permutation algorithm

Input: Size of random numbers, n, the polynomial $p(x)$, α, and β.
Output: S, the random permutation of the integers $\{1, 2, \ldots, n\}$.
Step 1: Set $S = 1$, $x_0 = (\alpha + \beta)/2$, $x = p(x_0) - fix(p(x_0))$
Step 2: For $i = 2$ to n, compute
$\qquad S = [S \quad i]$
$\qquad k = ceil(i * x)$
$\qquad S([k \quad i]) = S([i \quad k])$
$\qquad x_1 = x_0 - p(x_0)/p'(x_0)$
$\qquad x = p(x_1) - fix(p(x_1))$
$\qquad x_0 = x_1$
$\quad$ End For
Step 3: S

Suppose the size of the plain image is $M \times N$. Algorithm 2 is designed to shuffle the plain image based on the random permutation sequences of Algorithm 1. It may be processed as in Algorithm 2.

3.3. Diffusion Phase

The system (1) is utilized to generate a chaotic sequence of size $M \times N$. Then, reshape it to be of size $1 \times MN$, $Q = \{q_1, q_2, \ldots, q_{MN}\}$. The sequence Q is modified using the following formula:

$$q_i = mod(ceil(q_i \times 10^{14}), 256), i = 1, 2, \ldots, MN. \tag{5}$$

Algorithm 2 Row/Columns shuffling algorithm

Input: The plain image, O, the polynomial $p(x)$, $\alpha_1, \alpha_2, \beta_1$, and β_2.
Output: H, the shuffled image.
Step 1: Set $[M, N] = size(O)$
Step 2: Use Algorithm 1, with polynomial $p(x)$, and interval $[\alpha_1, \beta_1]$, to generate a random
permutation of size M for shuffling the rows, say S_{Rows}.
Step 3: Use Algorithm 1, with polynomial $p(x)$, and interval $[\alpha_2, \beta_2]$, to generate a random
permutation of size N for shuffling the columns, say $S_{Columns}$.
Step 4: For $i = 1$ to M, compute
For $j = 1$ to N, compute
$$H(i, j) = O(S_{Rows}(i), S_{Columns}(j))$$
End For j
End For i
Step 5: H, the shuffled image.

Moreover, the shuffled image H is reshaped to be of size $1 \times MN$, $H = \{h_1, h_2, \ldots, h_{MN}\}$. Finally, XOR is applied between each pixel in H and corresponding chaotic value of X, $D = XOR(H, X)$ (diffusion phase). The algorithm of diffusion phase may be processed as follows:

Algorithm 3 Diffusion algorithm

Input: The shuffled image, H, $q_{10}, q_{20}, a, b, c_1, c_2, \mu_1, \mu_2, \nu_1$, and ν_2.
Output: D, the diffusion vector.
Step 1: Reshape H, $H = \{h_1, h_2, \ldots, h_{MN}\}$.
Step 2: Covert H to binary, H_b.
Step 3: Set $q_1(0) = q_{10}, q_2(0) = q_{20}$.
Step 4: Perform initial iterations,
For $t = 0$ to 999
$$q_1(t+1) = q_1(t) + \nu_1 q_1(t)[(1 - \mu_1)(a - 2bq_1(t) - bq_2(t)) - c_1]$$
$$q_2(t+1) = q_2(t) + \nu_2 q_2(t)[(1 - \mu_2)(a - 2bq_2(t) - bq_1(t)) - c_2]$$
End For
Step 5: Set $q_1(0) = q_1(1000), q_2(0) = q_2(1000)$.
Step 6: For $t = 0$ to $MN - 1$
$$q_1(t+1) = q_1(t) + \nu_1 q_1(t)[(1 - \mu_1)(a - 1(t) - bq_2(t)) - c_1]$$
$$q_2(t+1) = q_2(t) + \nu_2 q_2(t)[(1 - \mu_2)(a - 2bq_2(t) - bq_1(t)) - c_2]$$
$$q(t+1) = (q_1(t+1) + q_2(t+1))/2$$
End For
Step 7: Preprocess the values of $Q = \{q(1), q(2), \ldots, q(MN)\}$ as follows:
$q(t) = mod(ceil(q(t) * 10^{14}), 256), t = 1, 2, \ldots, MN$.
Step 8: Covert Q to binary, Q_b.
Step 9: Perform XOR between H_b and Q_b, say $D = XOR(H_b, Q_b)$.

3.4. The Encryption/Decryption Algorithm

The encrypted image is produced from Algorithm 3 by reshape diffusion vector D to be of size $M \times N$, say E. The whole image encryption algorithm may be processed as in the Algorithm 4.

The Algorithm 4 (Image Encryption based on General Bischi-Naimzadah Duopoly System) will be referred to as **IEGBNDS** algorithm. Indeed, **IEGBNDS** algorithm can be applied to encrypt the color images. We can decompose color images into three grayscale images of red, green and blue colors (R, G, B components). After that we can encrypt them into their corresponding cipher images by applying the proposed algorithm. Then by re-joining the three cipher images of the R, G, B components, the color cipher image can be obtained.

Algorithm 4 Image encryption algorithm

Input: The plain image, O, the polynomial $p(x)$, α, β, q_{10}, q_{20}, a, b, c_1, c_2, μ_1, μ_2, ν_1, and ν_2.
Output: E, the encrypted image.
Step 1: Read the plain image, O.
Step 2: Generate the secret key by using the key mixing proportion factor.
Step 3: Call Algorithm 2 to get the shuffled image $\bar{H}$.
Step 4: Call Algorithm 3 to get the diffusion vector D.
Step 5: Covert D to decimal, say D_d.
Step 6: Change the dimension of D_d to $M \times N$, say E.
Step 7: E is the encrypted image.

The decryption algorithm is the inverse steps of **IEGBNDS** algorithm. Figure 4 displays the block diagram of **IEGBNDS** algorithm.

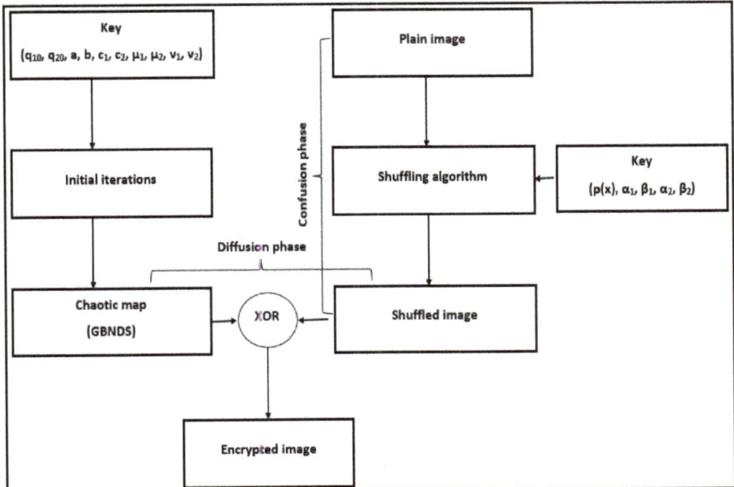

Figure 4. Block diagram of the proposed algorithm.

4. Experimental Results

The **IEGBNDS** algorithm has been applied to several 512×512 pixel gray-scale images and very promising results have been accomplished. All codes are accomplished on a Windows 10 Laptop with Intel(R) Core(TM) i7 2.40 GHz, CPU with 12 GB RAM using MATLAB R2016b.

4.1. Key Space Analysis

The key space must be large enough to hold out against brute-force attack. It must be above the value 2^{100} [39]. The key space of the **IEGBNDS** algorithm consists of the polynomial function with degree k and limits of $[\alpha, \beta]$ for the Newton-Raphson's method, two initial values and eight parameters for $GBNDS$. If the accuracy 10^{-14} has been used then it will be equal to $10^{14(k+1)} + 10^{168} (>> 2^{100})$. Table 2 gives the key space of the **IEGBNDS** algorithm compared to some recent algorithms in literature.

Table 2. Key space of the **IEGBNDS** algorithm compared to some recent algorithms in literature.

Algorithm	IEGBNDS Algorithm	[5]	[23]	[40]
Key space	$(10^{14(k+1)} + 10^{168}) > 2^{605}$	$10^{140} \approx 2^{466}$	$>10^4 \times 2^{208}$	2^{256}

4.2. Histogram Analysis

In a good encryption algorithms, the distribution of the pixel intensity values within a cipher image should be as similar to the uniform distribution as possible. Figures 5 and 6 show that the histograms of the cipher image is very similar to the uniform distribution. As the χ^2 statistical test is used to measure the nearness of produced histograms to the uniform histogram. The statistical χ^2-value is evaluated by [6]:

$$\chi^2 = \sum_{i=1}^{256} \frac{(E_i - e_i)^2}{e_i},$$ (6)

where the length of all possible values in an image is 256, E_i is the observed event frequencies of $i - 1$ and e_i is the expected event frequencies of $i - 1$, $i = 1, 2, ..., 256$. By evaluating the χ^2-value with the level of significance $\alpha = 0.05$, we got $\chi_{0.05}(255) = 293.25$. So, Both distributions are nearly equal if $\chi^2(255) < 293.25$. Table 3 shows that all tested images are smaller than 293.25. Therefore, the cipher images histograms are close to the uniform distributions. In other words, an attacker cannot retrieve any valuable information from them.

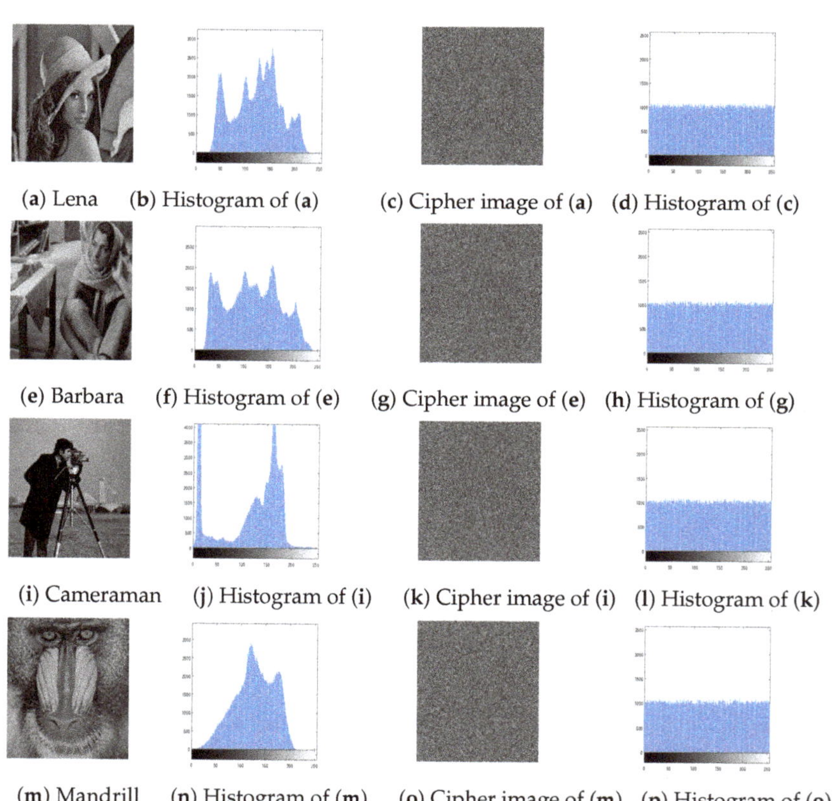

(a) Lena   **(b)** Histogram of **(a)**   **(c)** Cipher image of **(a)**   **(d)** Histogram of **(c)**

(e) Barbara   **(f)** Histogram of **(e)**   **(g)** Cipher image of **(e)**   **(h)** Histogram of **(g)**

(i) Cameraman   **(j)** Histogram of **(i)**   **(k)** Cipher image of **(i)**   **(l)** Histogram of **(k)**

(m) Mandrill   **(n)** Histogram of **(m)**   **(o)** Cipher image of **(m)**   **(p)** Histogram of **(o)**

Figure 5. Plain images, cipher images and their corresponding histograms.

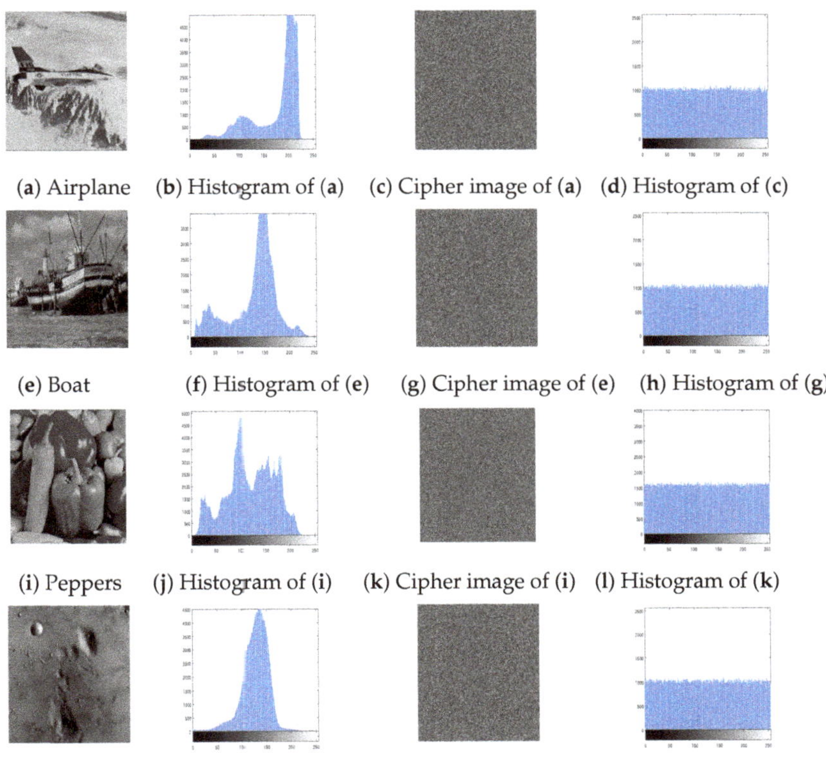

(**a**) Airplane (**b**) Histogram of (**a**) (**c**) Cipher image of (**a**) (**d**) Histogram of (**c**)

(**e**) Boat (**f**) Histogram of (**e**) (**g**) Cipher image of (**e**) (**h**) Histogram of (**g**)

(**i**) Peppers (**j**) Histogram of (**i**) (**k**) Cipher image of (**i**) (**l**) Histogram of (**k**)

(**m**) Moon_surface (**n**) Histogram of (**m**) (**o**) Cipher image of (**m**) (**p**) Histogram of (**o**)

Figure 6. Plain images, cipher images and their corresponding histograms.

Table 3. χ^2-values of the histograms of the cipher images at $a = 11.25, b = 0.5, c_1 = 0.20, c_2 = 0.30,$ $\mu_1 = 0.002, \mu_2 = 0.60, \nu_1 = 0.20, \nu_2 = 0.70, q_{10} = 0.10,$ and $q_{20} = 0.20$.

Image	χ^2-Value
Lena	286.52
Barbara	256.74
Cameraman	252.88
Mandrill	275.24
Airplane	279.64
Boat	260.23
Peppers	288.74
Moon_surface	249.12

Histogram Statistics

The variance and standard deviation are dispersion metrics applied in graphic histograms to help the effects of visual inspection. They calculate how often the elements of a dataset differ across the average with respect to each other. The same average value (mean) can be in two datasets, but the differences may be dramatically different. If the histogram has the lower variance then it has the more uniform of the graphic histogram, which is calculated by the following formula:

$$V = \frac{1}{256} \sum_{i=1}^{256} (\theta_i - \bar{\theta})^2, \tag{7}$$

where

$$\bar{\theta} = \frac{M \times N}{256},\tag{8}$$

θ_i is the frequency for each pixel's value from $0-255$ of the histogram, $i = 1, 2, \ldots, 256$, $\bar{\theta}$ is the histogram mean.

The standard deviation helps us to know the arithmetic average of the dataset's variations relative to the mean. It is calculated as follows:

$$S = \sqrt{V},\tag{9}$$

where V is the histogram variance.

Table 4 presents the histogram statistics for the plain and cipher images of the tested images for the **IEGBNDS** algorithm and the encryption algorithm in Reference [41].

Table 4. Histogram statistics for the **IEGBNDS** algorithm and the encryption algorithm in Reference [41].

| Image | Plain Image | | Cipher Image | | | |
| | | | IEGBNDS | | [41] | |
	V	S	V	S	V	S
lena (256×256)	38451	196.1	396	19.9	414	20.3
lena (512×512)	633397	795.9	3171	56.3	3340	57.8

4.3. Entropy Analysis

Information entropy [7] is utilized to detect the randomness of the cipher image. It is computed as follows:

$$H = \sum_{i=0}^{255} P_i log_2\left(\frac{1}{P_i}\right),\tag{10}$$

where P_i is the probability associated with gray level i. The largest value of the entropy reflects the randomness of the encrypted image. The maximum value of the entropy in our case is 8. Table 5 gives the information entropy for the plain and cipher images of the tested images. All values of entropy based on our algorithm are close to 8. In addition, the **IEGBNDS** algorithm gives average better than most averages of the listed recent algorithms. Based on the results of entropy, the **IEGBNDS** algorithm has reasonable protection.

Table 5. Information entropy analysis of the **IEGBNDS** algorithm compared to some recent algorithms in literature.

| Image | Information Entropy | |
	Plain Image	Encrypted Image
Lena	7.4475	7.9992
Barbara	7.6338	7.9993
Cameraman	7.0518	7.9993
Mandrill	7.2933	7.9992
Airplane	6.6823	7.9992
Boat	7.2151	7.9993
Peppers	7.4849	7.9992
Moon_surface	6.6974	7.9993
Average	7.1883	7.99925
[5] (Average)	–	7.99867
[23] (Average)	–	7.90252
[40] (Average)	7.266297	7.999224

4.4. Correlation Coefficients Analysis

In the plain image, adjacent pixels have strong relationships. So, reducing these relationships is required to hold out against statistical attacks. The correlation coefficient between two adjacent pixels, θ and ϕ, is defined as [6]:

$$r_{\theta\phi} = \frac{Cov(\theta, \phi)}{\sqrt{(D(\theta)D(\phi))}}, \tag{11}$$

where

$$Cov(\theta, \phi) = \frac{1}{N} \sum_{m=1}^{N} (\theta_m - E(\theta))(\phi_m - E(\phi)), \tag{12}$$

$$E(\theta) = \frac{1}{N} \sum_{m=1}^{N} \theta_m, \tag{13}$$

and

$$D(\theta) = \frac{1}{N} \sum_{m=1}^{N} (\theta_m - E(\theta))^2, \tag{14}$$

where θ and ϕ are selected randomly. 3000 pairs of adjacent pixels are chosen randomly from the plain and cipher images. Figure 7 displays the pixel intensity value's distribution of 3000 pairs for the Barbara image and its encrypted image in the three directions, diagonal, horizontal, and vertical. The correlation coefficients of the three directions for the **IEGBNDS** algorithm compared to some recent encryption algorithms based on the average of the correlation coefficients are given in Table 6. Table 6 shows that the **IEGBNDS** algorithm outperforms all of them at least in one direction. Also all values of $r_{\theta\phi}$ for the cipher images are close to zero. So, it can protect the image information.

Table 6. Correlation coefficient of the cipher images based on the **IEGBNDS** algorithm compared to some recent encryption algorithms in literature.

Image	Correlation Coefficient		
	Horizontal	Vertical	Diagonal
Lena	0.0011	−0.0026	−0.0015
Barbara	−0.0006	−0.0015	−0.0010
Cameraman	−0.0035	−0.0029	−0.0015
Mandrill	−0.0002	−0.0005	−0.0026
Airplane	0.0029	−0.0020	−0.0049
Boat	−0.0034	−0.0019	0.0010
Peppers	−0.0035	0.0032	0.0017
Moon_surface	−0.0013	0.0000	0.0011
Average	0.002063	0.001825	0.001913
[5] (Average)	0.007067	0.007867	0.014567
[23] (Average)	0.001544	0.001772	0.002678
[40] (Average)	0.003134	0.006602	0.004525

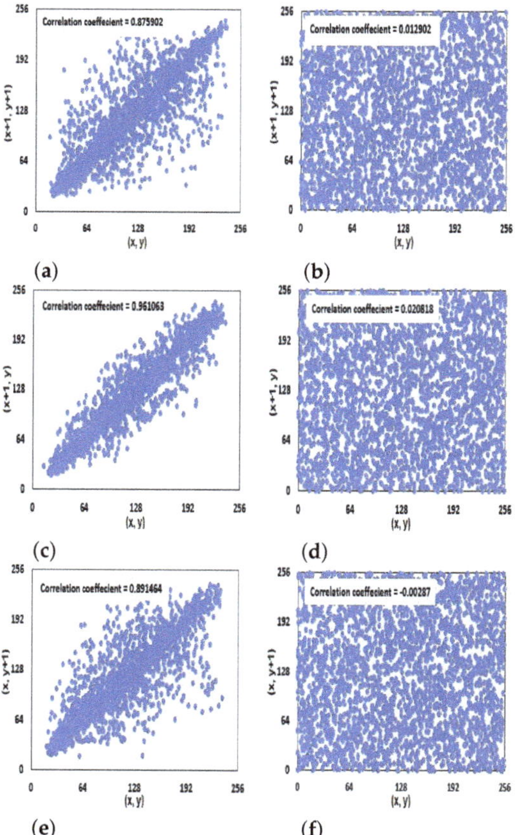

Figure 7. Distribution of adjacent pixels in the plain image (**a,c,e**) and the cipher image (**b,d,f**) for Barbara image in the three directions, diagonal, horizontal, and vertical.

4.5. Differential Attack Analysis

The protection against differential attacks is required for any image encryption algorithm. There are two main measurements, (1) $NPCR$ (Number of Pixel Change Rate), and (2) $UACI$ (Unified Average Changing Intensity). These measurements evaluated the amount of differences between two images, and can be defined as [5]:

$$NPCR = \frac{\sum_{m,n} D(m,n)}{M \times N} \times 100\%,$$ (15)

$$UACI = \frac{1}{M \times N} \left[\sum_{m,n} \frac{|O(m,n) - E(m,n)|}{255} \right] \times 100\%,$$ (16)

where

$$D(m,n) = \begin{cases} 0 & \text{if } O(m,n) = E(m,n), \\ 1 & \text{otherwise.} \end{cases}$$ (17)

A single pixel of the plain image is selected randomly and it modified to $255 - v$, where v is the original intensity value of pixel. The same key is utilized to encrypt the modified image and the plain image. Then, $NPCR$, and $UACI$ are calculated using the two cipher images. Table 7 shows $NPCR$ and $UACI$ for the tested images and compared them to some recent algorithms in literature. The **IEGBNDS** algorithm offers a good level of security. Based on the averages of $NPCR$ and $UACI$, the **IEGBNDS** algorithm outperforms all of them at least in one of the two measures. So, the **IEGBNDS** algorithm can be useful against differential attacks.

Table 7. NPCR and UACI of the tested images using the **IEGBNDS** algorithm and the recent algorithms.

Image	NPCR (%)	UACI%
Ideal value [23]	99.6094	33.4635
Lena	99.6326	33.4584
Barbara	99.6082	33.5339
Cameraman	99.6044	33.5797
Mandrill	99.6204	33.4392
Airplane	99.5907	33.4608
Boat	99.6086	33.4599
Peppers	99.6033	33.4868
Moon_surface	99.5998	33.4590
Average	99.6085	33.48471
[5] (Average)	99.6067	33.4267
[23] (Average)	99.6083	33.4521
[40] (Average)	99.6060	33.4646

4.6. Key Sensitivity Analysis

The sensitivity to the secret key is one of the important features of an excellent encryption algorithm. During the restoring plain image (decryption process), small changes in one of the initial values cr parameter are made and we will observe the restoring image via the modified secret key. Table 8 shows the restoring images using the true secret key and the modified secret keys. The plain image cannot be restored by any of modified secret keys. Therefore, the **IEGBNDS** algorithm is highly sensitive to any changes of the secret key.

Table 8. The result cf key sensitivity analysis.

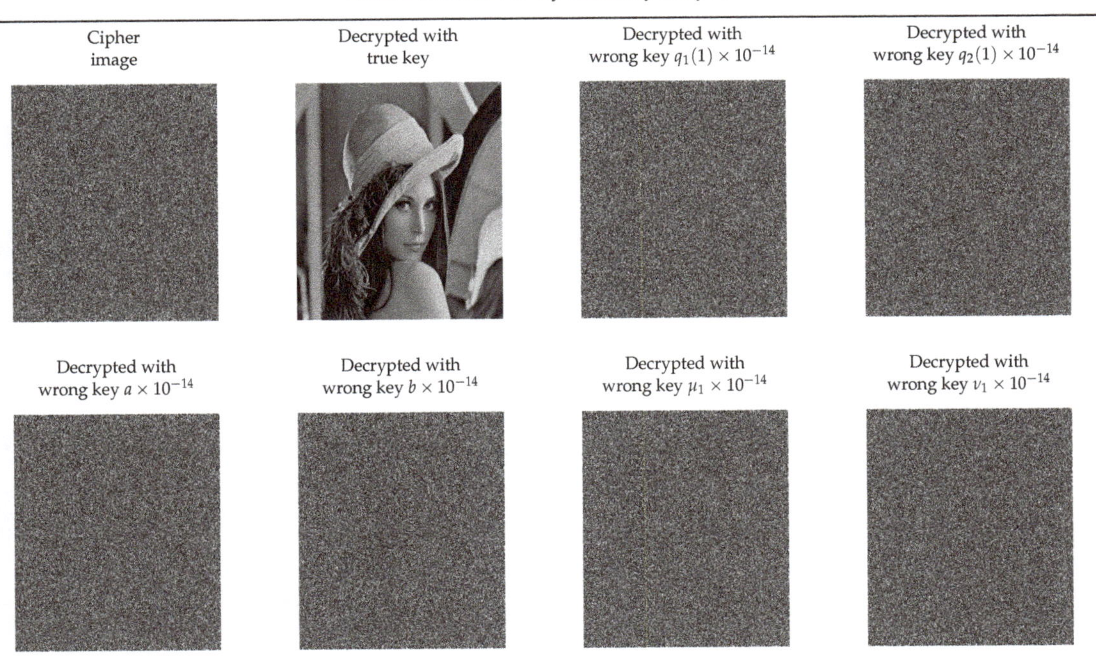

4.7. Robustness Analysis

In real life, noise or data loss is occurred and the **IEGBNDS** algorithm is tested against these problems. Salt&Pepper noise with different densities are added to the cipher image of lena with size 512×512. Table 9 shows the decrypted images of the noisy encrypted images. Moreover, the decryption image of the encryption image with some data loss is shown in Table 9. Based on the result of Table 9, The **IEGBNDS** algorithm can be robust against the noise and data loss attacks.

Table 9. Robustness analysis of the **IEGBNDS** algorithm for lena image with size 512×512.

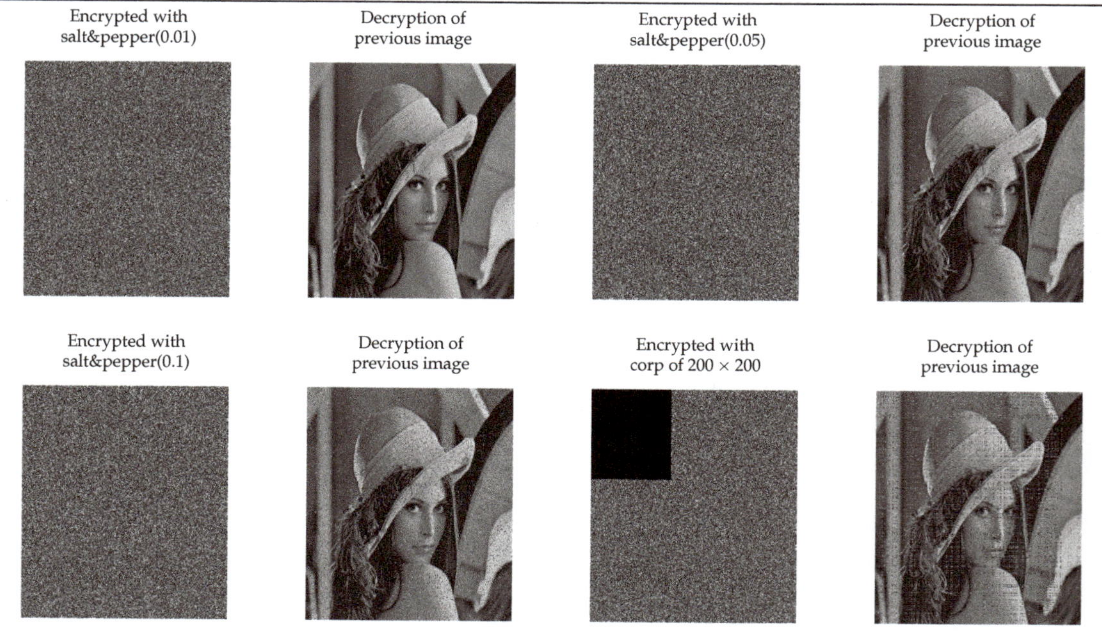

4.8. Chosen Plaintext Attack Analysis

The **IEGBNDS** algorithm is sensitive to the key generation, K_s, in Equation (2) and different sequences will be generated by small changes in the plain image. So, the **IEGBNDS** algorithm can hold out against the plaintext attacks. Now, we will examine the **IEGBNDS** algorithm against the chosen plaintext attack. Suppose the attacker has the encrypted image and the running of the **IEGBNDS** algorithm for a short time. The algorithm of Reference [42] will be used to examine our algorithm against chosen plaintext attack. In this algorithm, the following notations will be used:

P: plain image,

E: encrypted image of P,

D: designed image, where $d_{mn} = 0, \quad m = 1, 2, \ldots, M, \quad n = 1, 2, \ldots, N,$

E_D: encrypted image of D,

D_E: decrypted image of E.

The XOR operations between the pixels of E and E_D are performed to obtain the plain image P. Based on the result of Figure 8, the decrypted image is totally unlike the plain image. Therefore, the **IEGBNDS** algorithm can resist chosen plaintext attack.

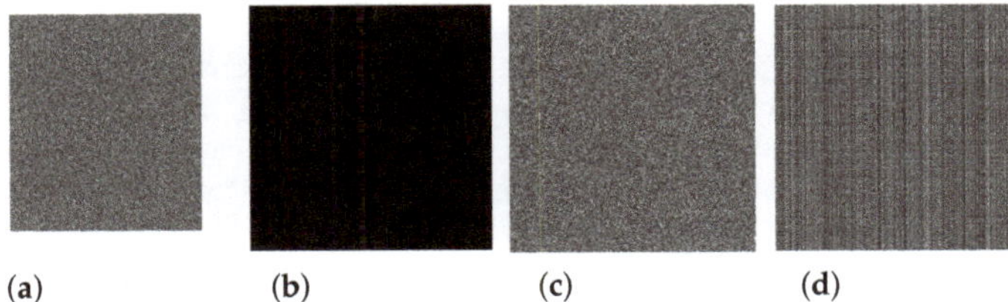

Figure 8. Analysis of chosen plaintext attack: (**a**) Encrypted image E, (**b**) designed image D, (**c**) encrypted image of D, (**d**) Decrypted image D_E.

4.9. Computational Analysis

The average times of encryption and decryption algorithm for one hundred time are 34.11 ms and 30.57 ms (Tested image of size 512×512), respectively. On the other hand, for the tested image of size $M \times N$, the encryption algorithm needs $5\,MN + M + N + 2000$ operations. The complexity time for the decryption algorithm is equal to the complexity time of the encryption algorithm. Table 10 shows that the running time of the **IEGBNDS** algorithm is effective compared to some recent image encryption algorithms such as in Reference [40] by Shakiba and Reference [23] by Cao et al.

Table 10. Running time of the encryption for the **IEGBNDS** algorithm and the recent algorithms.

Algorithm	Image Size	Running Time (ms)
IEGBNDS	512×512	34.11
[40]	512×512	976 ± 24.6
[23]	256×256	32.43

4.10. NIST Statistical Tests

NIST were established to test the randomness of generating cipher images created by encryption algorithms [43]. For the **IEGBNDS** algorithm, it is used to check the randomness of a sequence that consists of 100 cipher images of length $512 \times 512 \times 8 = 2{,}097{,}152$ bits. They were generated by using different random secret keys. Table 11 presents the results for 15 tests and all of them passed these tests.

Table 11. NIST statistical test for 100 cipher images by the **IEGBNDS** algorithm.

Statistical Test	IEGBNDS Algorithm	Result
Frequency monobit test	100/100	PASS
Block frequency test	99/100	PASS
Rank test	99/100	PASS
Runs test	99/100	PASS
Longest runs test	100/100	PASS
Cumulative sums test	99/100	PASS
Discrete Fourier transform	98/100	PASS
Random excursion test	56/58	PASS
Random excursion variant test	57/58	PASS
Universal test	99/100	PASS
Approximate entropy	98/100	PASS
Linear complexity test	100/100	PASS
Serial	100/100	PASS
Non Overlapping templates test	99/100	PASS
Overlapping templates test	100/100	PASS

5. Conclusions

In this article, the **IEGBNDS** algorithm via Newton-Raphson's method and general Bischi-Naimzadah duopoly system (**GBNDS**) has been suggested. Newton-Raphson's method has been used for shuffling the rows/columns of the plain image. **GBNDS** has been used to producing chaotic sequences to diffusion phase of image encryption algorithm. The extracted chaotic sequences from the **GBNDS** is extremely random based on the NIST statistical tests. Many security experiments are applied to evaluate the efficiency of our algorithm. The **IEGBNDS** algorithm has a large key space ($10^{14(k+1)} + 10^{168}(>>2^{100})$, the histograms of the generated cipher images are close to the uniform distributions, all entropy values for the cipher images based on **IEGBNDS** algorithm are close to 8, all correlation coefficient values for the cipher images are close to zero. The **IEGBNDS** algorithm outperforms some recent algorithms at least in one of the two measures, highly sensitive to small changes of the secret key, can be robust against the noise and data loss attacks, and can hold out against the plaintext attacks. In comparison to several recent algorithms, the **IEGBNDS** algorithm has a small running time. NIST statistical tests for 100 cipher images by the **IEGBNDS** algorithm are performed and all tests are passed. Finally, quantum image encryption algorithm based on **GBNDS** will be designed in the future to increase the security of the current algorithm.

Funding: This research received no external funding.

Institutional Review Board Statement: Not applicable.

Informed Consent Statement: Not applicable.

Data Availability Statement: Data sharing not applicable.

Acknowledgments: I deeply thank Shehzad Ahmed for his proof reading the paper.

Conflicts of Interest: The author declares no conflict of interest.

References

1. Arab, A.; Rostami, M.; Ghavami, B. An image encryption method based on chaos system and AES algorithm. *J. Supercomput.* **2019**, *75*, 6663–6682. [CrossRef]
2. Yin, Q.; Wang, C. A new chaotic image encryption scheme using breadth-first search and dynamic diffusion. *Int. J. Bifurc. Chaos* **2018**, *28*, 1850047. [CrossRef]
3. Askar, S.; Karawia, A.; Alammar, F. Cryptographic algorithm based on pixel shuffling and dynamical chaotic economic map. *IET Image Process* **2018**, *12*, 158–167. [CrossRef]
4. Karawia, A. Encryption Algorithm of Multiple-Image Using Mixed Image Elements and Two Dimensional Chaotic Economic Map. *Entropy* **2018**, *20*, 801. [CrossRef]
5. Askar, S.; Karawia, A.; Al-Khedhairi, A.; Alammar, F. An Algorithm of Image Encryption Using Logistic and Two-Dimensional Chaotic Economic Maps. *Entropy* **2019**, *1*, 44. [CrossRef]
6. Karawia, A. Image encryption based on Fisher-Yates shuffling and three dimensional chaotic economic map. *IET Image Process* **2019**, *13*, 2086–2097. [CrossRef]
7. Wu, X.; Wang, K.; Wang, X.; Kan, H. Lossless chaotic color image cryptosystem based on DNA encryption and entropy. *Nonlinear Dynam.* **2017**, *90*, 855–875. [CrossRef]
8. Wu, X.; Wang, D.; Kurths, J.; Kan, H. A novel lossless color image encryption scheme using 2d dwt and 6d hyperchaotic system. *Inf. Sci.* **2016**, *349*, 137–153. [CrossRef]
9. Ivanov, G.; Nikolov, N.; Nikova, S. Cryptographically strong S-boxes generated by modified immune algorithm. In Proceedings of the International Conference on Cryptography and Information Security in the Balkans, Koper, Slovenia, 3–4 September 2015; pp. 31–42.
10. Azam, N.; Hayat, U.; Ikram, U. Efficient construction of a substitution box based on a Mordell elliptic curve over a finite field. *Front. Inform. Technol. El* **2019**, *20*, 1378–1389. [CrossRef]
11. Jia, N.; Liu, S.; Ding, Q.; Wu, S.; Pan, X. A New Method of Encryption Algorithm Based on Chaos and ECC. *J. Inf. Hiding Multimed. Signal Process.* **2016**, *7*, 637–643.
12. Hayat, U.; Azam, N. A novel image encryption scheme based on an elliptic curve. *Signal Process.* **2019**, *155*, 391–402. [CrossRef]
13. Tonga, X.; Zhanga, M.; Wang, Z.; Liu, Y.; Ma, J. An image encryption scheme based on a new hyperchaotic finance system. *Optik* **2015**, *126*, 2445–2452. [CrossRef]
14. Guo, H.; Zhang, X.; Zhao, X.; Yu, H.; Zhang, L. Quadratic function chaotic system and its application on digital image encryption. *IEEE Access* **2020**, *8*, 55540–55549. [CrossRef]

15. Pareschi, F.; Setti, G.; Rovatti, R. Implementation and testing of high-speed cmos true random number generators based on chaotic systems. *IEEE Trans. Circuits-I* **2010**, *57*, 3124–3137. [CrossRef]
16. Seyedzadeh, S.M.; Norouzi, B.; Mosavi, M.R.; Mirzakuchaki, S. A novel color image encryption algorithm based on spatial permutation and quantum chaotic map. *Nonlinear Dynam.* **2015**, *81*, 511–529. [CrossRef]
17. Wu, Y.; Hua, Z.; Zhou, Y. N-dimensional discrete cat map generation using laplace expansions. *IEEE Trans. Cybern.* **2016**, *46*, 2622–2633. [CrossRef]
18. Lian, S.; Sun, J.; Wang, Z. Security analysis of a chaos-based image encryption algorithm. *Phys. A* **2005**, *351*, 645–661. [CrossRef]
19. Skrobek, A. Cryptanalysis of chaotic stream cipher. *Phys. Lett. A* **2007**, *363*, 84–90. [CrossRef]
20. Yang, T.; Yang, L.; Yang, C. Cryptanalyzing chaotic secure communications using return maps. *Phys. Lett. A* **1998**, *245*, 495–510. [CrossRef]
21. Shakiba, A. A randomized CPA-secure asymmetric-key chaotic color image encryption scheme based on the Chebyshev mappings and one-time pad. *J. King Saud Univ. Comput. Inf. Sci.* **2019**. [CrossRef]
22. Xiao, S.; Yu, Z.; Deng, Y. Design and analysis of a novel chaos-based image encryption algorithm via switch control mechanism. *Secur. Commun. Netw.* **2020**, *2020*. [CrossRef]
23. Cao, C.; Sun, K.; Liu, W. A novel bit-level image encryption algorithm based on 2d-LICM hyperchaotic map. *Signal Process.* **2018**, *143*, 122–133. [CrossRef]
24. Shakiba, A. A novel randomized one-dimensional chaotic chebyshev mapping for chosen plaintext attack secure image encryption with a novel chaotic breadth first traversal. *Multimed. Tools Appl.* **2019**, *78*, 34773–34799. [CrossRef]
25. Pak, C.; Huang, L. A new color image encryption using combination of the 1d chaotic map. *Signal Process.* **2017**, *138*, 129–137. [CrossRef]
26. Rajendran, R.; Manivannan, D. A secure image cryptosystem using 2D arnold cat map and logistic map. *Int. J. Pharm. Technol.* **2016**, *8*, 25173–25182.
27. Gu, G.; Ling, J. A fast image encryption method by using chaotic 3D cat maps. *Optik* **2014**, *125*, 4700–4705. [CrossRef]
28. Mohamed, N.; El-Azeim, M.; Zaghloul, A. Improving Image Encryption Using 3D Cat Map and Turing Machine. *Int. J. Adv. Comput. Sci. Appl.* **2016**, *7*, 208–215.
29. Zhao, Y.; Gao, C.; Liu, J.; Dong, S. A Self-perturbed Pseudo-random Sequence Generator Based on Hyperchaos. *Chaos Soliton Fract. X* **2019**, *4*, 100023. [CrossRef]
30. Li, C.; Chen, M.; Lo, K. Breaking an image encryption algorithm based on chaos. *Int. J. Bifurcat. Chaos* **2011**, *21*, 3518–3524. [CrossRef]
31. Wang, X.; Teng, L.; Qin, X. A novel colour image encryption algorithm based on chaos. *Signal Process.* **2012**, *92*, 1101–1108. [CrossRef]
32. Li, Z.; Peng, C.; Li, L.; Zhu, X. A novel plaintext-related image encryption scheme using hyper-chaotic system. *Nonlinear Dynam.* **2018**, *94*, 1319–1333. [CrossRef]
33. Lindell, Y.; Katz, J. *Introduction to Modern Cryptography*; CRC Press: Boca Raton, FL, USA, 2014.
34. Li, C.; Zhang, L.; Ou, R.; Wong, K.; Shu, S. Breaking a novel colour image encryption algorithm based on chaos. *Nonlinear Dynam.* **2012**, *70*, 2383–2388. [CrossRef]
35. Ding, L.; Ding, Q. A Novel Image Encryption Scheme Based on 2D Fractional Chaotic Map, DWT and 4D Hyper-chaos. *Electronics* **2020**, *9*, 1280. [CrossRef]
36. Enzeng, D.; Zengqiang, C.; Zhuzhi, Y.; Zaiping, C. A chaotic images encryption algorithm with the key mixing proportion factor. In Proceedings of the 2008 International Conference on Information Management, Innovation Management and Industrial Engineering, Taipei, Taiwan, 19–21 December 2008; pp. 169–174.
37. Hosny, K. Multimedia Security Using Chaotic Maps: Principles and Methodologies. In *Studies in Computational Intelligence*; Springer: Berlin/Heidelberg, Germany, 2020.
38. Askar, S.; Al-Khedhairi, A. Local and Global Dynamics of a Constraint Profit Maximization for Bischi-Naimzada Competition Duopoly Game. *Mathematics* **2020**, *8*, 1458. [CrossRef]
39. Zhang, L.; Li, C.; Wong, K.; Shu, S.; Chen, G. Cryptanalyzing a chaos-based image encryption algorithm using alternate structure. *J. Syst. Softw.* **2012**, *85*, 2077–2085. [CrossRef]
40. Shakiba, A. A novel randomized bit-level two-dimensional hyperchaotic image encryption algorithm. *Multimed. Tools Appl.* **2020**. [CrossRef]
41. Escobar, M.; Castillon, M.; Gutierrez, R.; Hernandez, C. Suggested Integral Analysis for Chaos-Based Image Cryptosystems. *Entropy* **2019**, *21*, 815. [CrossRef]
42. Ahmad, M.; Shamsi, U.; Khan, I. An enhanced image encryption algorithm using fractional chaotic systems. *Procedia Comput. Sci.* **2015**, *57*, 852–859. [CrossRef]
43. Rukhin, A.; Soto, J.; Nechvatal, J.; Smid, M.; Barker, E. A statistical test suite for random and pseudorandom number generators for cryptographic applications. NIST Special Publication 800-822 2001. Available online: http://www.nist.gov/manuscript-publication-search.cfm?pub_id=151222 (accessed on 15 May 2001).

MDPI

St. Alban-Anlage 66

4052 Basel

Switzerland

Tel. +41 61 683 77 34

Fax +41 61 302 89 18

www.mdpi.com

Entropy Editorial Office

E-mail: entropy@mdpi.com

www.mdpi.com/journal/entropy

9 783036 535159